LUNAR OUTFITTERS

Making the Apollo Space Suit

BILL AYREY

T0076427

University Press of Florida

Gainesville · Tallahassee · Tampa · Boca Raton

Pensacola · Orlando · Miami · Jacksonville · Ft. Myers · Sarasota

Frontis: ILC Industries engineer George Durney (*bottom left*), along with Homer Reihm, studies an Apollo suit during testing on a treadmill. Durney has been referred to as the father of the Apollo suit since many of his early designs during the late 1950s and early 1960s were carried into the later generation of Apollo suits. Homer Reihm would become chief engineer and was among the first generation of college-trained engineers that carried the Apollo suit to the next level while including more rigorous disciplines such as systems engineering, quality control, and control of the configuration. These all lead to a highly reliable product that the astronaut crews praised throughout the Apollo, Skylab, and ASTP programs.

This story is the independent work of the author and was neither sanctioned nor approved by the current management and owners of "New ILC Dover, LP," the successor to ILC Industries. The current management and owners of New ILC Dover, LP neither agree nor disagree with the accounts contained herein, nor are they responsible or liable in any way for the contents of this book.

First cloth printing, 2020
First paperback printing, 2023

28 27 26 25 24 23   6 5 4 3 2 1

Library of Congress Control Number: 2020938192
ISBN 978-0-8130-6657-8 (cloth) | ISBN 978-0-8130-8043-7 (pbk.)

The University Press of Florida is the scholarly publishing agency for the State University System of Florida, comprising Florida A&M University, Florida Atlantic University, Florida Gulf Coast University, Florida International University, Florida State University, New College of Florida, University of Central Florida, University of Florida, University of North Florida, University of South Florida, and University of West Florida.

University Press of Florida
2046 NE Waldo Road
Suite 2100
Gainesville, FL 32609
http://upress.ufl.edu

Lunar Outfitters

UNIVERSITY PRESS OF FLORIDA

Florida A&M University, Tallahassee
Florida Atlantic University, Boca Raton
Florida Gulf Coast University, Ft. Myers
Florida International University, Miami
Florida State University, Tallahassee
New College of Florida, Sarasota
University of Central Florida, Orlando
University of Florida, Gainesville
University of North Florida, Jacksonville
University of South Florida, Tampa
University of West Florida, Pensacola

CONTENTS

PREFACE

This book tells the story of ILC (International Latex Corporation) Industries, one of the many companies that contributed to the success of the Apollo space program, and the unique product this relatively small organization developed and manufactured. The Apollo space suit would appear in the most iconic images used to represent the technological boom of the twentieth century. More important, to the astronauts who wore them, the suit was a miniature spaceship that enabled them to walk on the surface of the moon without fear of failure. Just fifteen layers of thin fabrics and films sewn, taped, and cemented together provided the barrier they needed. If this garment failed, the astronauts would certainly die. The men and women of ILC Industries knew that the lives of these astronauts were in their hands. They never shrank from that responsibility—indeed, they embraced it.

*   *   *

Like many of my generation, I was caught up in the space race, building plastic model rockets and watching TV as astronauts in successive missions climbed down from their lunar modules and walked, hopped, and even fell about the lunar surface in those impressive suits. I felt a special bond to all the missions because, coincidentally, I lived in a neighborhood where several of the ILC space suit engineers and managers also lived. The suits were made within two miles of my home.

Just a few years after the Apollo program ended, in 1977, I came to ILC for what I expected to be a temporary job, intending to save money and return to college for another degree. My first assignment consisted of pulling materials through a sewing machine while the operator stitched together large panels ILC was manufacturing for United States Air Force airships. Shortly after starting the job, I happened to see many seemingly discarded Apollo space suits hanging on a rack in a dark corner of the plant. It was the kind of image an artist or photographer would have relished the chance to capture. In a better-lit and more active section of the building, most of the same engineers who designed the Apollo suits had moved on and were now fully engaged in designing the new space suit for the upcoming space shuttle program. It didn't take long for me to realize that this was a rather unique company.

With jobs expanding quickly at ILC, I soon found other opportunities beyond pulling fabric through the sewing machine. My admittedly luke-warm desire to pursue a degree in psychology was quickly fading away. I was hired into the Test Laboratory by Tom Sylvester, who had spent many of his earlier years as a test subject trying out the latest Apollo space suit designs. Ultimately, I become the manager of the test labs and became re-sponsible for the testing of the materials and the completed space suits as well as all the other company products. Given my love for the manned space program and my interest in everything mechanical, that temp job had evolved into a dream come true for me.

Around the time of the thirtieth anniversary of Apollo 11, I appointed myself as company historian when ILC supported a nine-month display commemorating the event at the University of Delaware museum. Follow-ing the exhibit, all the items came back to ILC, and I found an abandoned room to put them on a more permanent display. The display, which proved to be a hit with customers and visitors, has expanded to include many other company products that are almost as unique as the space suits themselves. The exhibit opened the door to collecting stories from many of the ILC veterans as well as the original documentation from those years that backed up their stories.

My hope is that this book will provide the details that have until now been locked away in obscure files and in the minds of those who worked the program. I was fortunate to have access to thousands of the original Apollo files, and I am grateful I can share them with you and can look back upon the historic chapter in American history that we call Apollo and mar-vel at it all. Whether you are a space nut, are simply curious to learn more about the Apollo program and the suits, or are working in the continuing science and development of space suits, I trust this book will be of interest and provide all you seek to gain from it.

*   *   *

I dedicate this book to all the employees of ILC Industries who focused on the mission of making the space suits that made walking on the moon and returning home safely possible for twelve Apollo astronauts. Their efforts also included nonlunar suits for the command module pilots and the suits for our first space station astronauts on Skylab and the Apollo-Soyuz Test Project where we linked up in space with the Russians, effectively bringing an end to the space race and the Cold War. These ILC employees worked feverishly for long hours. They missed the little league games and school plays of their children. Some neglected spousal relationships, resulting in numerous divorces. That's just the way it was. NASA had launch schedules to maintain. The employees were caught up in the excitement of clothing the men who would fly to the far-off surface of the moon and they were thrilled to be part of it. Thanks to every one of you for making one fantastic space suit.

INTRODUCTION

At the close of the twentieth century, the Pew Research Center surveyed Americans and asked what they felt were this country's greatest achievements since the year 1900. There are a lot of possibilities when you consider all the technological strides we made as a nation during that period. What ended up coming out on top, not surprisingly, was advancements in science and technology and specifically space exploration and humans on the moon.[1]

The culmination of our manned space programs produced results that established this nation as the world leader in science and technology. Arguably, the position we occupy today still puts us ahead of other nations, thanks in large part to these early programs. But this leadership and success was not a foregone conclusion. The Soviets initially dominated the race into space, and NASA and the US industrial and aerospace complex faced staggering challenges in their efforts to meet President Kennedy's 1961 proclamation that we would land a man on the moon and return him safely by the end of the decade. Kennedy delivered that speech only twenty days after our first astronaut, Alan Shepard, made his fifteen-minute flight, barely crossing the boundary into space. Kennedy's pronouncement sent shock waves through those within NASA who knew how difficult it had been just to launch Shepard on his mission. To get men to the moon and back within nine years was considered a joke by everyone who knew better.

The Lunar Suit

The forerunners of the modern space suit were pressure suits developed for early adventurers who flew their airplanes to altitudes higher than the human body could tolerate. Pioneer Wiley Post flew his unpressurized wooden plane to an altitude of 50,000 feet in September 1934 and in the process discovered the jet stream. These early daredevils knew that they had to have pressurized air to breathe or they would likely not survive. Attempting to pressurize a wooden plane such as Post's would have resulted in catastrophic results, so the best option at the time was to offer a suit that could be pressurized. These first, one-off, custom-made suits were awkward and hampered mobility when pressurized, but that was OK because they still afforded the basic protection and motion needed.

With the advent of high-performance jet aircraft in the late 1940s and early 1950s, the military became very interested in pressure suits for pilots who were flying at increasingly higher altitudes. Again, these suits were designed for a sedentary human seated in an aircraft who only truly needed to move their arms and hands. A small handful of companies offered mass-produced suits for such a purpose but their focus remained on the aircrew population.

True space suits didn't receive serious attention until the US manned space program began in earnest. Would the early pressure suits support President Kennedy's challenge to land a man on the moon and return him safely to earth? The pressure suits used in the Mercury and Gemini missions were derivatives of the military pressure suits of the 1950s, but even with various modifications, could they meet the challenges of the Apollo missions that were to come? This is where International Latex Corporation (ILC) stepped up with its design and manufacture of the first suits intended solely for use in space, suits that were not derivatives of military pressure suits.

The journey of the Apollo space suit begins in 1952 when ILC employee Leonard (Len) Shepard set out to build the first true space suit. Four years later, George Durney joined him and further developed Shepard's basic space suit design through the early 1960s. The team worked tirelessly to make this suit perform better than anything else on the market and ultimately win all the marbles in the contest to be the first to safely deliver a man to the moon and bring him back alive.

As the challenges of Apollo grew, ILC Industries hired many people with great talent to reach each successive level. This next generation included college-trained engineers who had great vision. Space suit design was not taught in any of the colleges. Mechanical engineers could transfer what they learned through education and work experience to the world of designing and building rockets and capsules for our space program, but the art of designing and building space suits was very abstract in many respects and only a select few could understand the nuances of it all. There really were no previous space suit models to look at. This was all new territory. These garments were 0 percent fashion and 100 percent function; anything "fashionable" about the suits was arrived at serendipitously.

This business wouldn't have succeeded if not for the women who brought the sewing skills that were necessary to stitch together this life-sustaining garment. Theirs was a women-only trade, learned mostly at home from their mothers and solidified through years of experience in the local commercial garment businesses. Some women simply transferred over from ILC's parent Playtex division, where they made bras and girdles. These suits would have to function in an environment that would take a human life within seconds; the seamstresses could not fail at their jobs of sewing the seams together properly, and everyone involved in this business was aware of that fact, including the astronauts.

Many hurdles faced the men and women who accepted the responsibility of seeing to it that the moon-walking astronauts of Apollo would survive the unknowns of the hostile environment they would face. Some of the more obvious obstacles were described in the following observation I found buried deep within a document written at the time the Apollo suits were being developed. It provides just a slight glimpse into what some of the challenges were. The author, a NASA contracts administrator, was attempting to shed light on the challenges and the unique nature of the work of making the Apollo space suits while at the same time justifying the high costs of the suits.

It has been correctly observed that the suit is the only "spacecraft" to go all the way to the moon and back. Designing it to be so versatile that it is comfortable during a 14-day mission for intravehicular wear and also provide extravehicular capability to let an astronaut on the moon's surface leap, climb, and creep, place his hands on top of his helmet, reach behind his back and touch the opposite shoulder with either hand with 100 percent safety is no easy engineering job. It required the expertise of a strange blend of such diverse disciplines as bio-engineering, metallurgy, fluid control and sewing. After the suit is designed and qualified there are still many recurring problems encountered in custom fitting the suits to the specific astronaut. First, the pattern cannot even be cut until the crews are selected, which is done at the sole discretion of the astronaut office. Second, ILC is required to physically locate the astronaut, make certain plaster casts and tailor measure him. Thirdly, after the suit has been cut and stitched, ILC is required to coax the astronaut to Dover, Delaware, for a fit check at which time special design changes may be generated due to the subject's idiosyncrasies such as weight fluctuations, 99th percentile bicep, extra-long neck, etc. Other problems result from the coordinating, integrating and interfacing with the other Apollo systems such as the Portable Life Support System, Command and Service Module, Lunar Module and the Space Suit Communications Systems.[2]

Of the several varieties of space suits used since the beginning of human space flight, the Apollo suits are among the most iconic objects ever photographed, particularly when seen in the photos from the lunar surface. They represent the human species leaving the gravity of Earth and traveling to another world—an absolute first for humankind. Take for instance that famous photo Neil Armstrong took of Buzz Aldrin posing with the "magnificent desolation" of the moon as a backdrop. As with all the images of the Apollo astronauts taken on the lunar surface, the men who occupied the suits are all but invisible. This is the story of that suit.

Figure 0.1. The iconic photo of Buzz Aldrin taken by Neil Armstrong just minutes after Apollo 11 landed on the moon. Aldrin is wearing the ILC Industries Model A-7L space suit. Courtesy NASA.

I

HUMBLE BEGINNINGS

1

School of Hard Knocks

The story of the ILC space suit has its roots in the early 1950s, when a small group of "Hard-knockers," as they would call themselves, began developing pressure suits to protect humans in space. Their moniker comes from the term "school of hard knocks," which means learning primarily from experience, mistakes, and hands-on work, as opposed to gaining knowledge from college or textbooks. That phrase aptly describes these men and women, as they were not part of the formal aerospace organizations that the air force or NASA usually dealt with. The company was better known for making commercial products such as bras and girdles through a closely tied parent division named Playtex. It probably turned out to be a good thing, because the Hard-knockers frequently thought outside the box when designing space suits and didn't always follow the aerospace community's set of guidelines.

The ILC's work on pressure suits followed the success their small division had had with developing and manufacturing pressure helmets such as the model MA-2 for the U.S. Air Force. Post–World War II jet aircraft were flying at higher altitudes and the demand for crew protection had increased. While the air force gave ILC an opportunity to develop the first pressure suits, those early contracts were just an excuse for ILC to get their "space suit" recognized by industry. It wasn't long before everyone in the business knew the ILC pressure suits as state of the art because they were the best available garments for space travel.

The designers of these new space suits were truly artists who worked in a very abstract world as they visualized how pressurized garments would comfortably fit the wearer and, most important, provide complete mobility for arms and legs (which otherwise stiffened because of the air pressure in the suits). They had to invent and integrate all the mechanics to carry this out while working side by side with very skilled women who painstakingly assembled this life-sustaining, one-piece garment with their Singer sewing machines. Sometimes during the development phases, the women would have to tell the engineers that they could not make the seams the way the engineers envisioned them, but they were always willing and able to offer alternative options. In some cases, the ILC engineers would borrow ideas incorporated in pressure suits others had developed, but they still had to integrate them into their suit, which was easier said than done. The creators of the suit were not seasoned aerospace industry veterans like many others who supported the Apollo program. They were just a close-knit team of very talented engineers, seamstresses, and model makers with some truly great ideas about how space suits should be made.

Abram Nathaniel Spanel: Founder of International Latex Corporation

The success of the Apollo space suit program begins with Abram Nathaniel Spanel, who was born in Russia in 1901 and moved with his family shortly thereafter to Paris. The family then immigrated to the United States in 1911 and soon became American citizens. His father was a tailor who found many opportunities for employment in Rochester, New York.

In his teens, one of Abram's first jobs was delivering newspapers, and one of his customers was George Eastman, the man who founded Eastman Kodak. Eastman took the young Spanel under his wing and became an early mentor to him. Spanel studied engineering in college and tried his hand at studying for the ministry. In 1926, at the young age of 25, he formed a company, the Vacuumizer Corporation, which manufactured moth-proof vinyl garment bags that were used to promote the sales of Electrolux vacuum sweepers. This proved to be profitable to both companies, but Spanel benefited the most from this union; his company grew fast, with annual sales rising to $80,000 ($1.2M in 2019 dollars). This growth occurred at the beginning of the Depression years in America and was all accomplished by three employees in a 10,000-square-foot facility located at 77 St. Paul Street in Rochester. In 1932, Spanel turned his attention to developing products that used latex, a liquid product derived from the sap taken from the caoutchouc tree. It performed much better than the solid, crude rubber others were using at the time. He bought some latex manufacturing equipment in Dayton, Ohio, and had it shipped to Rochester. The purchase included several patents for manufacturing products using latex rubber. Soon after setting the equipment up in Rochester, he began production on latex bathing caps and sold them for 50 cents each while others on the market were selling for 10 to 15 cents each. He called his new company the International

Figure 1.1. Abram Nathaniel Spanel, founder of International Latex Corporation. Courtesy ILC Dover LP 2020.

Latex Corporation. The combination of his natural sales ability and a high-quality product increased his profits. His work force grew to twelve people who not only helped make the product but also acted as the sales staff. The product line grew to include hot pads, bathroom slippers, aprons, and other items he could make using latex.

Integrating himself within American culture at such a young age laid the foundation for Spanel to become a steadfast American patriot who took a strong stand on the side of the anti-isolationism movement just before World War II. Proud of his Jewish heritage and aware of Nazi actions in Europe, he espoused his feelings about anti-isolationism anytime he had the opportunity.

In 1936, Spanel realized that it was again time to grow the business, so he moved his new operation to Dover, Delaware, because of its proximity to Philadelphia, Washington, Baltimore, and New York. Spanel likely moved the operations to the Dover area because the winter weather was much more favorable there than in Rochester, where the vats of latex stored outside were in constant danger of freezing during the winters. If that happened, he would lose the product at great cost.

He found ten acres of relatively inexpensive property in Dover. His research had indicated that he could find a good work force to draw from in the area. He proceeded to erect a 100,000-square-foot building that included many modern amenities.[1] He incorporated the new company in Delaware on September 24, 1937. The ten acres and the new building were known as Play-Tex Park, from the registered trade name he used to market his products nationally. He closed the Rochester plant soon after he opened the Dover facility. He brought about fifty managers and engineers from Rochester and hired another 100 from the Dover area. The Dover plant had air conditioning, showers, large lounge rooms, and even a hospital with a registered nurse on duty at all times.[2] All of this was rare in the late 1930s, as the country was coming out of the Depression.

Spanel's first products consisted of latex-dipped bathing caps that were made by the thousands per hour, baby pants, and lady's hair capes. But International Latex consumer products were becoming well established just at the beginning of World War II. After the war began, the flow of rubber was slowed when Japan invaded and took over areas in and around the Dutch East Indies and Malaysia, from which the United States imported 90 percent of its rubber. As this looming shortage became apparent, Spanel installed large underground labyrinths in his Dover plant for storing all the latex rubber he could amass. The walls of these underground tanks were even coated in beeswax to protect the product. At the start of the war, Spanel had 1,835,085 pounds of latex in those tanks.

As expected, in December 1941, the Office of Production Management in Washington, DC, issued a priorities ban on Spanel and his company, which basically amounted to an order to cease the manufacture of all commercial goods using latex. As with other manufacturers who had stashed a quantity of this valuable commodity, he was forced to sell much of his latex back to the government. His plan at that point was to begin designing and producing rubber goods for the US military. He immediately set out designing and making collapsible canteens, inflatable rafts, and devices such as inflatable stretchers that could be floated across waterways. A news article from that time reported that "Spanel designed an assault boat whose impact has been felt all over the world, floating stretchers which have saved lives of thousands and secret equipment whose story cannot yet be told."[3] The success of the inflatable rafts came about after a young army captain came to Spanel looking for a lightweight collapsible boat. Spanel studied this challenge and after a "feverish month, produced the boat on a daring principal, using plastic bladders covered in fabric."[4]

The move from consumer products laid the groundwork for the eventual founding of ILC's Government and Industrial Products Division, which enabled Spanel to keep the doors open while at the same time supporting

Figure 1.2. The Playtex Experimental Lab where new products were being developed, including high-altitude helmets and eventually pressure suits. Ca. 1947. Courtesy Fred Feldman.

the war effort. As the war was concluding, he turned his attention back to commercial products, but he now firmly believed that he needed to keep the company's product range diversified so that he could weather any financial and economic downturns. Thus, in 1947 he created separate divisions. He named one the Playtex Division, thus establishing the famous brand-named commercial product. His next focus was establishing a division for government and industrial products. Being the tinkerer that he was, Spanel bought an old brick building on Pear Street in Dover, across town from the main plant, and hung a sign across the top that said "Playtex Experimental Engineering Lab." (Past employees also referred to it as the Metals Division because part of its purpose was to make the metal racks to display the Playtex garments in stores.) Spanel used this as his laboratory, where he could play around with new ideas that sprang into his mind. In this building he developed such things as hair cutters and a machine for packaging processed chicken. Somewhere among the odd projects, around 1949, the company landed the contract from the air force to design and build the high-altitude pressure helmets. (Spanel later claimed that he started that division because he was convinced that 10 percent of his business should always be directed to products for defense work.) By 1950, Spanel had organized the Industrial Products Division to provide technical and productive support for the economic and expedient handling of military and other specialized research and development programs.[5]

All this early work laid the foundation for the success of the Apollo space suits. It's one thing to be good as an inventor and develop new products, but one of the greatest qualities that Spanel exhibited was his ability to search out and hire talented employees. This trait was really what made all the difference for the future of International Latex Corporation.

Before he died in 1985, Spanel had 2,000 patents to his name, and throughout his time as president of International Latex, he remained innovative and progressive. An article in *Ebony* magazine in January 1961 praised the company for its practice of hiring, integrating, and promoting black workers at a time when that was not the norm. The company worked closely with Delaware State College (now Delaware State University), a predominantly black college, to develop a program to bring International Latex work to the college to provide students with the training they would need for future employment opportunities.

Over several decades, Spanel spent a great deal of his money publishing editorials and running advertisements in major newspapers that promoted such issues as his staunch views against American isolationism and in support of all programs related to the New Deal. Following the war, he was caught up in controversy because many Republicans linked his liberal-leaning advertisements and the general positions he took to his being awarded large government contracts. However, no connections to any unscrupulous contract awards were ever proven.

Spanel donated significantly to various health research programs focused on childhood cancers and other fatal diseases. In the 1950s, newspapers published articles about the money he donated to fly children suffering from cancer to hospitals across the country for treatment. He is remembered by many ILC employees for the impromptu speeches he made over the company loudspeakers at any time of the day or week. The topic of his speeches varied but tended to relate to world problems and the causes he was concerned about. He could also be found wondering the floor handing out silver dollars to employees in a bid to boost morale. His employees respected and admired him very much.

Leonard Shepard: The Man with a Vision of the Future

Soon after Spanel began operating ILC in Delaware, he established his main residence in Princeton, New Jersey. He would commute to the ILC facility and often spend evenings during the week sleeping in a hideaway room beside his office. (His old Princeton residence is now named Drumthwacket and today serves as New Jersey's governor's mansion.) In 1949, Spanel summoned an electronics consultant from Rhinebeck, New York, named Leonard (Len) Shepard, to construct a special radio and high fidelity record player for his personal entertainment in Princeton. Spanel was immediately taken with Shepard's talents and stayed in touch with him in the hope of eventually getting him to work at his facility. Finally, in 1951 the two came to an agreement, and Len was hired as a project engineer working on the study of latex dipping processes. Years later, Spanel described Shepard as "a man of dimension," an understatement. By 1954, Shepard had become the chief engineer for a division that was working on developing the first pressure garments ILC would make. By 1968, he had become a vice-president of the company because of the success of the space suit program he oversaw.

Len was one of those unique persons who did not have a college degree yet constantly worked at a level that made it appear to others as though he did—and an advanced degree at that. (In several instances, Shepard did claim to have an electronics degree from MIT, but this fact was challenged in a court case, when it was revealed that he had attended MIT for only two weeks in July 1945.[6]) Regardless, Shepard did not need a degree to succeed and do truly great things. One person told me that Len did not graduate from high school. As a young man growing up in the Bronx, he faced many challenges that could have easily set him on a different path. Good fortune intervened, however, and he was exposed to technical training in the electronics field, which he took to rather easily.[7] Shepard absorbed knowledge from everything and everyone around him and put it to good use. Before he died in 1988 at the relatively young age of 61, Len had raised a family, become a private pilot, become a concert pianist, and invented many items outside of ILC such as an electronic fish finder, which won him the immediate friendship of those who fished in the Delaware Bay. In 1980, following his many successful years at ILC, he formed his own company, Time System Technology, where he worked on the development of precision instruments using atomic power to achieve high accuracy. His son-in-law recalls asking Len if it was safe to go into his basement, where his office was set up.

Few who were close to Len during the height of the Apollo years remember his being actively involved in the daily technical aspects of the suit. He did however remain the constant guiding force behind the team at ILC, which needed the wise and broad direction he could provide. But it did not start out that way. Sometime around 1952, Len Shepard conceived the vision of what a space suit might look like and how it should function. Shortly thereafter, he cobbled together his version of the futuristic space suit. Lacking any other way to introduce the suit to the world, he took advantage of an air force contract to provide a high-altitude pressure suit for the X-15 program. It's my supposition that Len was not overly disappointed when the suit failed to win that contract (since it wasn't intended for aircraft use) because his mission was accomplished: now many prominent people in the aerospace world were aware of the unique suit he had developed. As the space race heated up, other companies were still developing hardware for such a suit, but Len's suit was already well ahead of the competition. The others would continue to focus their efforts on modifying existing pressure suits for space use. However, from day one, Len Shepard set out to design a suit that was intended only for space exploration—no modification necessary.

George P. Durney: The Man Responsible for the Development of the Early Apollo Suit

In 1956, a 33-year-old man named George Durney knocked on the door of ILC looking for work. George had mechanical abilities and a knack for getting things done with little regard for annoyances that got in his way. He simply worked his way around them even if it meant stepping on others' toes or upsetting those in higher ranks. Like Len Shepard, he wasn't college educated, but he had the ability to make product ideas work. Little did he or anyone else know at the time that he would take over the space suit work Len Shepard started and receive much of the credit for establishing ILC as the early leader in the extravehicular space suit business.

George was born in Union City, New Jersey, in 1923. His father died when he was just one year old. His mother and he moved to Dover, Delaware, sometime around 1924. At Dover High School, George was a standout student who earned an appointment to the Naval Academy upon graduation in 1941. George, however, was not one to conform to the rigors and discipline a military academy demanded and he didn't call Annapolis home for very long. As World War II gained momentum, George joined the US Army Air Corps and was soon training to fly the P-39 Cobra fighter aircraft. In short order he was transferred to the 15th Air Force and was assigned to a B-17 crew based in Italy. His plane was named the *Lil Joe.*

George soon became the aircraft commander, or flight officer, as the position was known, due in large part to the respect he had earned from his

Left: Figure 1.3. George Durney as a young aviator in WWII. Photo courtesy Edward Biter.

Below: Figure 1.4. Aircraft commander George Durney (*front left*) and his B-17 crew. Photo taken two days before Durney was shot down over Kranz, Austria, taken prisoner, and assigned to Stalag Luft 3. Courtesy Edward Biter.

colleagues. The odds that any B-17 crew would complete the total of twenty-five bombing missions required of them before they could go home to the states or take less-harrowing assignments was slim, but George made it all the way to his twenty-fifth mission. Unfortunately, it was on that mission that *Lil Joe* was shot down by enemy aircraft over Austria, on February 24, 1944. George and his crew bailed out and survived, but the Germans took them prisoner. George was held for fourteen months in Stalag 3. When the war concluded and liberation finally came at the hands of the Soviets,

George and his fellow prisoners were told that they could sit tight for a short time and the United States would make all travel arrangements to get them home. Of course, the other option was that they could leave on their own since the war was now over and the front gates of the camp were wide open. But who in their right mind would want to travel across an uncertain, war-ravaged Europe and attempt to figure out how to get back to the states with questionable support along the way? George was one of the few who did just that.[8]

After he returned home, George bounced around for a few years selling cars. He was an excellent mechanic and could repair anything. He also spent a short time in California working as a crop-dusting pilot before moving back to Dover. In 1956, he had an interview at ILC and was hired by engineer Fred Feldman, whose specialty was helmet development. No doubt the combination of George's determination and self-confidence and his piloting background were the selling points that landed him a job working on the pressure helmets ILC was developing for the air force. Very shortly after he began with ILC, George was helping Len Shepard develop his new pressure suit.

Although Len Shepard blazed the trail for developing the first suits ILC made, he and George Durney worked as a team for many years. One coworker from those earlier years recalled that George did most of the day-to-day development work while Len looked on and provided technical support as needed. The air force continued to look for new pressure suits with advanced mobility and George worked tirelessly through the late 1950s to develop pressure suits for air force contracts and other small projects related to suits. Neither Len nor George ever lost sight of the prize, which was to get their suit into outer space. Between contracts, ILC provided the funding to cover the development work because the company had complete confidence in Len and George and believed there was a good future with a space program that would need pressure suits one day.

As the early Apollo years began to take shape and the technical challenges and pressure to meet tight NASA schedules increased, George had to begin sharing and handing off more of the development work to degreed engineers. There was also an ever-increasing need to interface with higher-ranking NASA administrators, which was not a task George was comfortable with. However, the former aviator did a good job of interfacing directly with the astronauts during their fit checks. For example, George welcomed Neil Armstrong into his house in Dover one day for lunch when Neil was in town for a fit check. Since George was so technically versed on

Figure 1.5. George Durney proudly standing beside one of the earlier pressure suits he developed. Circa 1958. Courtesy ILC Dover LP 2020.

the suit, he would easily win over the astronauts' confidence because they recognized his expertise. When astronauts requested changes that were not possible given the time frame or capability at the time, George had a way of convincing them his way was the best option available; when possible, he would make the changes.

As Apollo matured, other ILC engineers would perform the fit checks on their suits and George was required to mentor many of the new engineers and other technical folks. This was new territory for George and he often found it a tough road. In 1999, former company president, Homer Reihm, who was George's co-worker at one time, summed up what it was like working with George through these years.

> George was a great mentor and teacher to his design team and his other subordinates. He was also unique in his ability to relate to the astronauts, to give them what we could, and get them to buy off on unreasonable requests. On the other hand, George could be hard headed and difficult when customers or associate contractors tried to constrain or challenge what he was doing on "his" suit, also he was

somewhat intolerant of administrative and other non-designated ILC personnel who needed to know certain suit data and issues in order to get their job done.[9]

That analysis confirms input I received from other Apollo employees concerning George. It also sets the stage for a better understanding of the relationship between ILC and Hamilton Standard early in the Apollo program. Regardless of how he was viewed, there can be no denying that George was the driving force who laid the foundation for early suit development and established ILC as the front-runner in early space suit design.

2

Developing the State-of-the-Art Space Suit

Context for the beginnings of the Apollo space suit starts with what was taking place in aerospace technology in the late 1940s and 1950s. The United States was on the verge of flying humans to the edge of outer space using aircraft known as the X-planes, a series of twenty experimental aircraft from the 1947 model X-1 to the 1959 X-15. The air force's experimental aircraft and advancements in rocket technology made space travel seem possible. It also needed some basic equipment to outfit flyers for strategic bombers such as the B-52 and the B-58.

But one might wonder why pressure suits—or space suits—are even needed. Pressure suits were first invented to protect humans who were ascending higher in altitude where the low pressure and lack of oxygen could not sustain life. The purpose of the high-altitude pressure suit and space suit is to keep the wearer adequately pressurized with enough oxygen to breathe while at the same time providing good range of mobility for the arms, the hands, and in the case of the space suits used on the moon, the legs.

Here on earth, at sea level, we are surrounded by an air pressure of 14.7 pounds per square inch or 101.4 kilopascals. Our air consists of roughly 21 percent oxygen and 78 percent nitrogen. Humans—indeed, all breathing creatures on Earth—depend on this air pressure to keep the gas flowing into and out of our lungs as our chest muscles expand and contract. At the extreme high altitudes where jets fly and in outer space, the air pressure is either too low or doesn't exist. Thus, we must take this pressure with us no matter where we go in the universe.

Gas Expansion versus Pressure

I often use the analogy of an unopened soda-pop bottle when I explain what happens to our body when it has no pressure surrounding it. Shake a bottle that has the contents under a higher pressure and then open it. What suddenly appears is an explosion of carbon dioxide gas bubbles from out of the liquid. When you open the bottle, you reduced the pressure inside it and the gas "boils" out of the liquid solution. Our body works in a similar way. We inhale air into our lungs and it is then dissolved in our bloodstream and the heart pumps it throughout our body. The gas is also dissolved in the various cells and joints within the body. If you were to pop a glove off in the vacuum of space, the gas inside your body would want to come out of solution like the CO_2 in the soda bottle since there is no pressure acting against it. This would result in severe pain followed by certain death due to suffocation. Although skin tension would mostly retain the gas within the body, the gas within the tissue would still expand. This is similar to what happens to scuba divers who come up too fast from the watery depths where they were breathing air under high pressure. As the water pressure decreases around a diver's body when it rises too rapidly in the water, they will get what is referred to as the bends, or the expansion of the gas contained within the body.

O_2 Partial Pressure

A whole field is related to what is called the partial pressure of gases. This can get rather complex, but suffice it to say that as the atmosphere thins out at higher altitudes, the gases that consist of mostly oxygen and nitrogen decrease, making it necessary to substitute the nitrogen with oxygen for humans to survive. In Figure 2.1, in the area labeled "Unimpaired performance zone," you can see that as you increase in altitude (right column) the oxygen requirement increases ("Volume percent oxygen in atmosphere" across the bottom). The air pressure of the Apollo space suit on the moon was 3.75 pounds per square inch absolute (psia). Looking at the "total pressure, psia" graph and following it across to the line marked "minimum tolerable total pressure," you can see that the lowest pressure we can survive is just below the 4.0 psia mark. The Apollo suits were used at about the lowest pressures possible; the reduced pressure enabled the astronauts to bend and flex their joints more easily against the suit pressure.

Transitioning from Helmets to Space Suits

In the late 1940s, Fred Andrews and Bill Miller, two design engineers at ILC, connected with an air force equipment development group whose

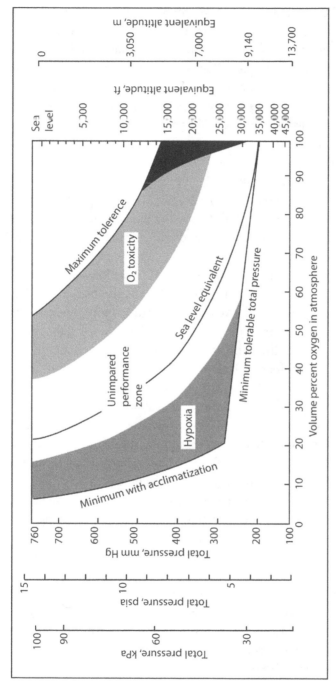

Figure 2.1. This NASA chart shows the unimpaired performance zone at various altitude pressures. At a sea level pressure of 14.7 per square inch absolute, we function ideally somewhere with 18 to 35 percent oxygen. The typical amount found is 20.9 percent oxygen in nitrogen. Source: "Man–Systems Integration Standards," National Aeronautics and Space Administration, accessed January 10, 2020, https://msis.jsc.nasa.gov/sections/section05.htm.

members had concerns about the pressure helmets they were using. Andrews recalled that the air force representative told them that if ILC could come up with a better pressure helmet to solve some problems they were having by the end of that week, they stood a good chance of winning a large contract for helmet production. They worked around the clock through the week and ended up with a better system, which eventually led to the popular K-1 model, which was followed by the MA-2 model helmets. These helmets were attached to the model T-1 partial-pressure suits the air force was using around that time for air force pilots flying high-performance aircraft at higher altitudes. At that time, many of the high-altitude pressure suits, including the popular T-1, were being designed and fabricated by the David Clark Company. Ironically, like ILC Industries and its sister company Playtex, the David Clark Company had started out in the 1930s making materials for undergarments and had evolved into making anti-G suits for military pilots by the mid-1940s.

Len Shepard spent his first few years at ILC working on the K-1 and MA-2 helmets. In 1952, the forerunner of NASA, the National Advisory Committee for Aeronautics (NACA), began studying problems that humans were likely to encounter in space. Shepard's exposure to the pressure-suit business by way of the helmets and the discussions circulating at the time related to human space travel likely played a part in the direction Len Shepard took. He had quickly become familiar with many of ILC's manufacturing capabilities, such as the company's rubber-dipping facilities, sewing capacity, and machining capabilities. Because he was a man of many dimensions who set his sights higher than most, his instinct turned toward imagining what a real space suit might be like and how ILC's facilities and personnel could be used to make such an invention.

As Len was developing his suit, the air force was shifting its focus toward full-pressure suits. Partial-pressure suits consist of a tight-fitting layer of materials with integrated tubes that can be inflated with air, which pulls the fabrics even tighter around the wearer's body. This action puts mechanical pressure on the skin surface that counters the reduced atmospheric pressure at higher altitudes. Experiments in high-altitude flights showed that exposed body parts such as the hands would swell to unacceptable limits because of reduced pressure. Pressure-breathing apparatus could provide protection up to certain altitudes but counterpressure on the skin surface was needed over 40,000 feet or so.[1] Full-pressure suits, in contrast, use compressed air to provide more uniform pressure against the skin. As a

Figure 2.2. Test pilot Joe Walker dressed in the MC-3 model partial pressure suit made by the David Clark Company. Note the lacings and external tubes; when inflated with air, the expansion of the tubes causes the tight-fitting, low-stretch, high-strength fabrics to pull tight and squeeze the surrounding skin surfaces. Walker is also wearing the popular ILC model MA-2 helmet. Courtesy NASA.

result, however, the suit balloons outward due to the large volume of compressed air inside, with the negative result of a more rigid structure.

The suits that were used into the early 1950s were a combination of partial-pressure and full-pressure suits and offered only marginal protection and comfort. The air force and the navy had an agreement to work closely together to develop pressure suits, but the ones the navy offered for evaluation in 1954 were not acceptable to the air force. The best option for the air force was to open the doors to competition that might spawn improvements.[2] In February 1955, the Air Research and Development Command (the research branch of the air force) announced that it was undertaking its own design studies. It asked thirty companies, including ILC, to bid on a contract for a full-pressure suit.[3] The air force liked the K-1 and the MA-2 helmets ILC was producing, so Len Shepard, with the backing of Spanel, felt that this was their chance to get the new space suit ideas moving forward even if it may not have been precisely what the air force was looking for at the time. They rightly assumed that the air force could be the primary government agency that would be responsible for leading human missions into outer space, so if Len could get his suit noticed by those who had influence on selecting space suit contractors for such an eventuality, he would be positioning himself and ILC for a great future in this business.

Neither Len nor anyone else at ILC had experience in the pressure-suit business, so they were starting with a clean slate. This was probably good fortune. Since this was all new to him, Len was not constrained in terms of design or manufacturing methods. He also did not have to take an existing pressure-suit model and modify it. His vision was to create a totally new suit from the ground up—a suit that had full mobility. In a 1972 interview, Len Shepard recalled these early days at ILC:

> In the early 1950s, I was responsible for engineering in this operation here and made a decision at that time that eventually there was going to be a space program and that if we would start now with the space suit technology, as opposed to high-altitude flying technology, we would be in a better position when the United States did develop a space program to be the supplier of the equipment.
>
> The major differences were that our objective was always to enable man to do useful work in outer space or in vacuum while the objective of the pressure suit was to enable a man to survive, bring an aircraft down to a level at which the atmosphere will sustain life and supply only the mobility necessary to operate the aircraft whereas our objective was to enable man to perform tasks, work tasks, even walk around.[4]

Shepard's brilliance was matched by the first-class development facilities of ILC, including the model makers needed to fabricate the tools for making the various rubber-molded parts of the suit. He could also count on the seamstresses who could be borrowed from the Playtex ladies' apparel line. The corporate structure at that time at ILC also benefited the effort. Abram Spanel had sold off the controlling interest to the Stanley Warner Corporation in 1954 for the sum of $15 million, but Spanel had retained his post as chair and continued to steer the direction of the division and, most

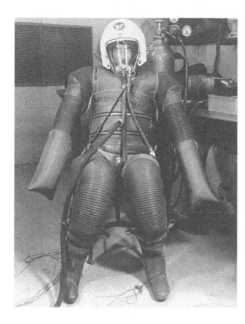

Figure 2.3. The ILC Model XMC2-ILC. The goal of the contract was to demonstrate joint flexibility. Gloves were still in development at the time of this photo. Attached is ILC's MA-2 helmet. Courtesy ILC Dover LP 2020.

important, provided the support Len Shepard needed to pursue the new space-suit technology.

By May 16, 1955, ILC and ten other companies had submitted bids for the air force contract. No options yet existed for submitting a suit designed to meet outer space requirements, so this was as close as Len Shepard could get to showing off his new "space suit." He likely knew from the beginning that his suit would not stand up against the David Clark Company suit, since that was probably the closest to meeting the air force's requirements for aircraft operations. Len was not focused on designing a suit for aircraft operations and he stuck to his vision.

The Air Research and Development Command designated the pressure suit ILC delivered under this contract the XMC-2-ILC suit. A series of model MC partial-pressure suits had previously been developed and were in use, but this would be the next generation; the X stood for experimental. The David Clark Company suit was identified as XMC-2-DC.

Only four organizations supplied the complete suit systems, and ILC was one of them. Not surprisingly, the air force did not choose the ILC suit for the next stage of development for the XMC program, but Shepard and Spanel had accomplished what they set out to do. True to Len's prediction, the XMC-2-ILC suit caught the attention of the people in the air force who counted in terms of the future of outfitting space travelers.

Between the time of the XMC program and the start of the manned space program, the air force provided a few contracts to ILC that enabled it to further develop its state-of-the-art suit. One of those was logged into the ILC contract books as "Air Force Contract, number 33(616)-3192." The contract was technically for the "study and development of various types of controlled surfaces for application to required joint mobility."[5] Translated, the air force was looking for a suit with better joint mobility when pressurized. The overall design objective was a suit made of non-rigid materials except for the helmet assembly.

Key to Making a Successful Space Suit

Len Shepard and others at ILC didn't invent the convoluted joints that made the XMC-2-ILC suit and eventually the Apollo suit so flexible and mobile. That credit should go to Russell Colley of B. F. Goodrich, the maker of the first Mercury space suits. Like many inventions, the design of the ILC Apollo suit was not accomplished in complete isolation. Shepard and the other ILC engineers that followed him looked at all the options, including the designs of others. Throughout the many years that the David Clark Company designed pressure suits, they offered the link-net material that ILC used for the arms of the Apollo suit. Two NASA engineers, Dr. Robert Jones and James (Jim) O'Kane, came up with a new polycarbonate pressure helmet concept around the time the ILC engineers were struggling with the upper torso design. The ILC designers quickly realized that this new helmet could help them out tremendously. They soon found a way to make the helmet work on their new torso and their second submission won the company the prime contractor role in the Apollo contest.[6] The UK's Royal Air Force pioneered the liquid cooling suit design that permitted astronauts to work in the suit without overheating. NASA acquired the rights to this design and eventually handed it over to ILC to produce.

Bringing various ideas together and arriving at a successful outcome required a combination of genius, teamwork, and determination. That's what set ILC apart from the competitors, as two contests would go on to prove. Even when the ideas seemed to come together and appeared to work just fine, rigorous cycle testing that simulated the harsh conditions on the moon's surface often indicated that the design or the materials would not take the abuse. ILC engineers had to go back to their drawing boards on many occasions, often with time running out. The pressure was intense, and to make matters worse, they were working in uncharted territory with no previous examples to use as guidelines.

While making his early pressure suits at B. F. Goodrich, Colley learned through trial and error that to get a joint to properly flex while under relatively high internal pressures, the air inside had to have some place to go rather than simply getting compressed as the joint muscles flexed. Many companies and engineers did a great deal of work to better understand the joints that would allow a space suit to flex where it needed to. One of the best examples is the RX-2 hard suit made by Litton Industries that is now in the collection of the Smithsonian Air and Space Museum. Much like a suit of armor, a suit made of metal joints does not compress like a soft suit does. Thus, the person in the suit does not have to work as hard to bend the pressurized joints. The obvious problem with the hard suit is that it weighs a lot and does not fold and store easily aboard small spacecraft. A hard suit also does not provide a great deal of comfort to the wearer unless it is heavily padded. More about that later.

Russell Colley learned the challenges he faced with poor flexibility early on when he made the first pressure suits for Wiley Post in 1933. Post was one of the Golden Age aviators who sought to set new speed records with his airplane, the *Winnie Mae*. Post realized that he needed to fly at higher altitudes where the air is much thinner and has less resistance. He could not pressurize his Lockheed Vega airplane because the frame was constructed of wood and could not take the internal pressures of high-altitude flight without blowing apart. Post realized that he would need to find someone to design and fabricate a pressurized suit for him. This search landed him on the doorstep of B. F. Goodrich, since they were a large, well-known company with a great deal of experience with rubber, which seemed like a natural material for making a flexible suit. Russell Colley made three suits for Post. The first one blew apart as it was being pressurized during an unmanned test. The second one had to be cut off Post; the day he evaluated it was tremendously hot and he sweated so much that he became stuck inside the suit. The third one finally did the job, and Post wore the suit when he set altitude and speed records and when he discovered the jet stream.

These early suits, however, had almost no mobility when pressurized and were extremely unpleasant to work in. Because of his early experimentation, Colley discovered the idea of the convolute. Some stories say that he hatched the idea of an accordion-looking convolute as he watched a tomato worm in his garden. This was a design that he continued to use in some of

Figure 2.4. Wiley Post in the last of three pressure suits made for him by Russell Colley of B. F. Goodrich. Courtesy NASA.

his advanced suits. Colley figured that if accordion-looking sections could collapse on one side while expanding on the other when being flexed back and forth, the volume inside would remain somewhat constant. With this design, the muscle would not need to do as much work to compress the air in the joint section. This convolute structure is also called a constant volume joint since the volume inside the convolute changed very little when it was flexed. Because these joints had cables that ran down each side of the convolute (on the inside and outside of the knee and elbow, for example), they were considered unidirectional, which meant that they would flex in only one direction. This was fine for the elbow or the knee, neither of which are normally flexed sideways. The shoulder and the wrist were more complicated since they had to flex in all directions. These were referred to as omnidirectional.

Len Shepard had studied the earlier B. F. Goodrich suits to see what did not work so he could try to improve upon them. He noted obvious deficiencies in the shoulder design. The shoulder is one the most dynamic parts of body in terms of mobility. It required a lot of effort to develop a shoulder assembly that was considered omnidirectional. In a 1972 interview, Shepard talked about this issue:

They [B. F. Goodrich] accomplished a lot of the objectives that we had set for ourselves in the form of incorporation of bearings; however, there was one basic problem which they did not solve and that was the problem of bi-directional mobility in the shoulder—allowing the arm to elevate and to rotate forward and backward at the same time. We accomplished that by modifying the joint to allow the cable to slide back and forth within a restraint tube. We got one plane of motion by bending the cable, the other plane of motion by allowing the cable to slide in a tube.[7]

The task of turning Len Shepard's conceptual ideas into convolutes fell to Nate Brown, an expert in the latex-dipping process. Brown formed the bellows by bonding a lightweight tricot fabric and nylon tape in the valleys of the convolute for radial restraint and steel cables for longitudinal restraint between dipped layers of neoprene.[8] Documentation shows the many failures ILC experienced throughout the development process of the early convolutes.[9]

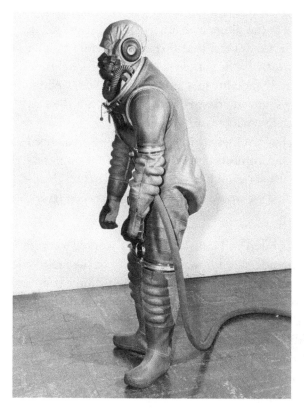

Figure 2.5. Russell Colley of B. F. Goodrich developed the model XH-5 pressure suit by for the Army Air Force in 1943 at a cost of $3,000 ($44,000 in 2019 dollars). Note the flexible convolutes at the elbows and the knees. Courtesy NASA.

Figure 2.6. Dip molds used to make the convolutes in the Apollo space suit. Courtesy ILC Dover LP 2020.

Len Shepard set out in 1954 to make his first suits unencumbered by the many layers in the existing pressure-suit designs. In his 1972 interview he recalled:

> We weren't considering at the moment a lunar landing. This was in the very dim future. But we did feel that there was a future of man working in space. Very early in the game we came to the conclusion that the only way to achieve mobility in a joint structure was to eliminate volumetric changes and interlayer friction. These are the two major deterrents of mobility in a pressurized garment.
>
> So we accomplished both of these things by designing a balanced bellows for the joints and combined all the layers of fabric necessary—the gas bladder, the abrasion bladder and the structural bladder—into one layer so that we would have no sliding between layers, thus eliminating friction.[10]

After George Durney was hired in 1956, he and Len worked very hard through the late 1950s to understand how the mobility of the suit could be improved when it was pressurized. Under one of the air force contracts that ILC delivered in November 1960, George demonstrated the mobility he could achieve using double-exposed film. He would take a time-lapsed photo as he held a flashlight and rotated it in the widest arc that he could to demonstrate his full range of motion. He used a grid-board backdrop to show this range of motion while he was wearing a shirt. He would then get into the suit and perform the same motion with the flashlight while the

Figure 2.7. George Durney demonstrates range of motion while using a flashlight in this extended and double-exposure photo. The inside light image demonstrates the range of motion when inside the suit while the outside image was made while wearing a short-sleeved shirt. This tool helped the ILC engineers determine the optimal design for pressure suits. Courtesy ILC Dover LP 2020.

Figure 2.8. George Durney in an early pressure suit. These rubber convolutes were in the areas of the suit where bending motion was needed, including the wrist and neck area. Courtesy ILC Dover LP 2020.

suit was pressurized. The photo lab personnel would then superimpose the suited photo over the shirt-sleeve photo to show the difference between the two conditions. Of course, the suited, pressurized condition provided less mobility, but this photographic method provided a clear example of the suited range versus what was called the nude-body range. George was able to demonstrate and provide clear evidence that his suit had very good range of motion that was not greatly inhibited by the pressurized suit.

During the 1950s, George and Len laid the foundation for what would ultimately be the nation's first lunar space suits. They obtained patents for helmets and hardware connectors that were ultimately not used as filed. But each step helped establish these underdogs as serious contenders.

Essential Material Layers Used in Pressure Suits and Space Suits

What are the most essential layers that make up a pressure suit and a space suit? In both cases, the suit needs to be leak free, so it has a pressure bladder that is made of a coated cloth that prevents the oxygen inside the suit from leaking to the reduced pressure or vacuum of space. For Apollo, the bladder layer was made of a neoprene-coated nylon fabric. The bladder layer is surrounded by what is known as the restraint layer, which is responsible for holding the shape of the suit and keeping it from expanding when it is pressurized. This layer takes a good deal of the loads brought about by the internal pressure, so it needs to be lightweight but very strong and it also needs to have very low stretch. If the restraint material stretches when pressurized, the suit will get bigger, thus making it very difficult if not impossible for the wearer to work in. A good analogy for the bladder and the restraint layers is an inner tube and a tire. The inner tube holds the air in and the tire provides the shape. For Apollo, the restraint was made of a blue nylon fabric. The convolutes acted as both the pressure and restraint layers of the suit. Dipping the convolutes to form a single layer eliminated friction between multiple layers.

The pressure bladder and the restraint layers are technically all that is needed for a simple pressure suit that would be used aboard a high-flying aircraft, for example, or even a launch within a spacecraft where no spacewalks are needed. If, however, the mission requires a spacewalk, a third layer is needed to reflect solar and cosmic radiation and to guard against the resultant solar heat loads. This layer, called the thermal micrometeoroid garment, or TMG, typically consists of the outer cloth layer that provides abrasion and fire resistance followed by several layers of thin aluminized

Mylar film. Each layer is separated by a light material that reduces heat conduction through the layers. The Mylar acts as a mirror that both reflects heat away from the suit during solar exposure and keeps some heat inside the suit when it is not exposed to solar loads. This is followed by an inner liner. Certain sections of the very outer layer may also contain additional material to protect against abrasion in susceptible areas. During Apollo, this outer layer had to be designed to protect against the possibility of flames should a fire break out in the command module (as happened in the Apollo 1 fire that took the lives of the three astronauts). Owens Corning Corporation developed a material that was eventually selected as the outer fabric. It consisted of woven fiberglass strands that were coated with Teflon. These outer layers also absorb small micrometeoroid impacts—small dust particles that travel though space or could perhaps be thrown up from the lunar surface should a meteor strike the surface of the moon and spew up the particles at very high speeds. These small micrometeoroids travel at speeds of 10 kilometers per second, or 22,000 miles per hour. On impact with the surface layer of the space suit, the particle essentially explodes. The subparticles that are created as a result may pass through the next few Mylar film layers and are stopped when they reach a neoprene "bumper" layer that serves as the inside liner of the TMG.[11]

ILC Presents Its First True Space Suit

In the mid- to late 1950s, everywhere one turned, people were talking about space travel. Cars sported fins that resembled rockets on their tail sections. Buick had a model called the Rocket 88. On TV, Walt Disney included space travel in much of his programming. Helping him as a spokesman was none other than Wernher von Braun, the developer of the V-2 rockets Germany used in World War II. (Von Braun later spearheaded the development of the Apollo Saturn rocket for NASA.) The space race was under way. Given the many Russian firsts the world had witnessed, it would not be much longer before the American government-industrial machine would catch up, and both George and Len knew they could play a big part in that adventure. Other engineers had been added at ILC by that time, including John Malinowski, Rodney Hill, Harry Gottwails, and a staff that specialized in the chemistry associated with making the rubber convolutes under the direction of Nate Brown in the Playtex Laboratory. Although ILC's engineers stayed focused on space suit development, the Government and Industrial Products Division was almost forced to close its doors

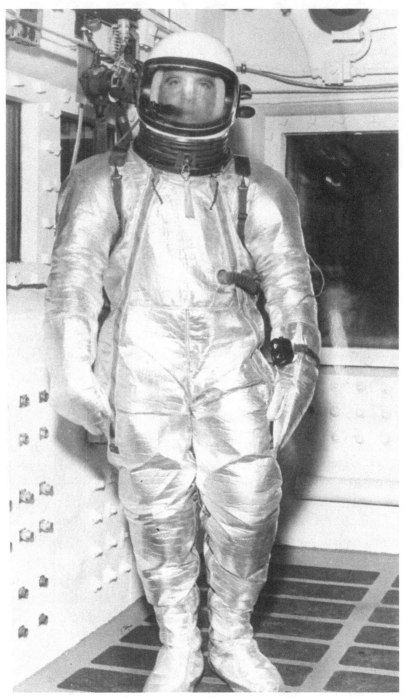

Figure 2.9. The ILC model SPD-117 suit with its thermal cover. Courtesy ILC Dover LP 2020.

in 1959. To keep government projects flowing as Spanel had envisioned, the company had been taking on too many stitched-goods products for the war department, competing against other companies that charged pennies per piece. In the period 1957 to 1959, the division lost $2 million. Corporate owner Stanley Warner suggested liquefying the division altogether, but Spanel fought vigorously against such a move and installed a new treasurer who quickly turned things around. He immediately dropped stitched-goods products and focused all the resources of the Government and Industrial Products Division on the research and development group that included Len, George, and the other engineers who were working on space suit development.[12]

That R&D group continued to get contracts the air force funded. The air force was immersed in the development of a program known as DYNA-SOAR that would provide a combination spacecraft and winged airplane. In relation to this program, the air force was looking for space suits to support activities such as satellite maintenance, aerial reconnaissance, and space rescue. In a 1961 article, ILC was recognized as the developer and provider of what was called a "quick change" space suit. The article referred to a six-year effort under three different contracts from the air force's Wright Air Development Division that totaled $250,000. The article did not mention that ILC funded half of the costs, so the total was more like $500,000 ($4.3 million in 2019 dollars). Photos of the suit that ILC provided clearly show the model SPD-117 suit with a silver TMG (Figure 2.9). This suit appears to be identical to the suit ILC had proposed to NASA for the Mercury mission just months earlier. One way or the other, ILC was intent on getting its suit into space. In the article, an ILC spokesperson is quoted as saying that the suit was "adaptable to prolonged usage outside the space vehicle, on the moon, or for interplanetary travel."[13]

The Mercury Suit Contract

Six days after NASA was created on October 1, 1958, project Mercury was approved; NASA officially announced it on December 17, 1958. Just months later, in 1959, ILC entered the contest to develop a space suit contest for the Mercury program. What a big moment that had to have been for the small ILC staff that had worked diligently on the suits over the past several years. It was clear that this new civilian agency would present many opportunities for ILC. Len and George were supplemented by a small handful of engineers from time to time and they also relied on a few talented seamstresses, but that was it. Several of the seamstresses were recruited from the Playtex Division, where they were sewing commercial products. Others came from small clothing and drapery manufacturing businesses in the Dover and Milford, Delaware, area. The seamstresses would take the ideas from the engineers and assemble suits using Singer industrial sewing machines. This was different from sewing drapery or men's shirts. Poorly made seams that failed to hold a shirt sleeve together could mean death to an astronaut on the moon.

The ILC contracts office referred to the company's first space suit effort as Contract AF33(600)-39314. The contract was for the development and fabrication of full-pressure garments to be evaluated for the Mercury program. Other contenders were B. F. Goodrich and the David Clark Company. The contest had a very tight six-week delivery schedule, which ILC met. ILC called the suit Model SPD-117 (the 117th product ILC listed on its books). Because this model did not vary much from the suits the company had developed for air force mobility studies, fabricating this suit was relatively simple since much of the design work was already completed. ILC included the aluminized cover layer to reflect the heat that was anticipated during reentry based on the early engineering studies NASA provided. This silver outer layer looked fashionably cool for a space suit. It is believed that ILC made only one suit for this contract primarily because of the tight delivery schedule.

NASA asked two organizations to make recommendations about the best suit; it chose the Aero Medical Laboratory at Wright Air Development Center in Dayton, Ohio, and the Navy Air Crew Equipment Laboratory (NACEL) in Philadelphia.[14] As expected, the Wright Air Development Center recommended the David Clark Company's X-15–style high-altitude suit and NACEL recommended the B. F. Goodrich Mark IV suit. The ILC suit was still a curiosity to those in the business; it was much bulkier but had great mobility when pressurized. Most of the Mercury astronauts were navy veterans and were familiar with B. F. Goodrich's Mark IV pressure suit the navy used. This familiarity and proven track record likely factored into NASA's decision to use the Mark IV suit in July 1959. The Mercury program did not include a spacewalk and the capsules were designed to maintain pressure throughout the mission, so the suits were to be used only for emergency pressurization should the craft lose pressure at altitude. If that happened, the suit would be inflated to provide the life-sustaining pressure. The Mark IV suit was the best choice, given the selection. The ILC's SPD-117 suit probably would not have fit into the tiny Mercury capsule anyway.

Although the basic design was a sound one, much more work was required to reduce the bulk of ILC's suit.

Len Shepard was now advertising his suit to anyone who would listen. He called it the only "operational piece of equipment rather than an emergency device," meaning that it was a functional space suit rather than a pressure suit that would stiffen and essentially make the wearer a "prisoner."[15] Many who worked in the pressure-suit business were calling the ILC suit the state-of-the-art suit because it was the only true example at that time that demonstrated a vision of what total suit mobility would look like.[16] Although Len and George did not get a shot at making a space suit for the Mercury astronauts, they needed to wait only another twenty-eight months to fulfil their vision.

Kennedy's Challenge: The Countdown Begins

On May 25, 1961, in a speech to Congress, President John F. Kennedy declared that before the end of the decade, the United States would land a man on the moon and return him safely to the Earth.[17] Although Kennedy and Vice President Johnson had been in discussions with NASA director James Webb and others about this plan, it still took the vast majority within NASA by surprise. It wouldn't be a stretch to say that this proclamation set off a wave of near-panic throughout the engineering hallways of NASA, since Alan Shepard had made America's first fifteen-minute suborbital flight into space only twenty days earlier. Certainly there would be job security for many in the business but how was this huge industrial machine going to design and manufacture all the things necessary to get a human to the surface of the moon and back in just nine years, when all the United States had accomplished to that point was to launch a man 90 miles up into the air with a splash-down fifteen minutes later in the Atlantic?

By November 1961, NASA had asked eight companies to provide the space suit for Apollo. Two of the contenders were no surprise. ILC would compete once again, as would B. F. Goodrich. Hamilton Standard, which manufactured the primary life-support system, or backpack, as it is sometimes called, was going into the competition as a primary contractor with the David Clark Company providing the suit for them as a subcontractor. Others included Arrowhead Rubber Products, which had a brief history making pressure suits when they competed against B. F. Goodrich for the Mark IV Navy suit in 1955 (they would never win any contests with their suit), National Textile Research, Rand Development Corporation, and

Rubbex Fabricators, all of which had done pressure-suit research at one time or another but did not have much to offer in terms of an actual pressure suit. Another Apollo contender was Mauch Laboratories, which had done some work in the late 1950s on a mechanical counterpressure suit. Instead of using air pressure contained within a suit, the idea of a counterpressure suit was to use a tight-fitting elastic suit to squeeze around the body's skin, thus decreasing suit bulkiness and perhaps increasing joint mobility. That idea still gets kicked around today but continues to be a long shot in terms of leading to anything meaningful. As technology and advanced materials move forward, it is not out of the question that this technology will work someday, but the challenges are significant.

3

The Turbulent Years, 1962–1965

The project Apollo suit is no doubt the most complicated suit and supporting equipment development program ever attempted. The lunar suit will require as much or more engineering effort than the combined pressure-suit-development efforts by the Department of Defense, Industry, and NASA from Wiley Post days to the present. Most important, is that the suit must be truly a space suit and removed from the status of an emergency garment to that of the primary "life" sustaining ensemble.[1]

ILC's Proposal to NASA for the Apollo Suit

On March 28, 1962, Len Shepard, ILC's director of research and engineering, sent a cover letter with the company's proposal to NASA for the Apollo space suit that summarized the work he and others had done over the previous decade.

International Latex Corporation has been actively engaged in the supporting research, design and fabrication of the pressure suits unassociated with any aircraft-oriented needs. This direction was established prior to 1953 as the result of recognition that requirements for a true space suit would eventually exist. No deviation from this direction was ever made despite funding limitations, lack of immediate need, and the temptation to concentrate on development of a pressure flying suit with large production potential. It is gratifying to see this need finally develop.[2]

Len's letter made it clear that at ILC, all suit development since the early 1950s had been dedicated to the goal of developing a space suit for extravehicular use—a rather significant statement. What is also significant was his statement that the company had not deviated from the goal of designing a future space suit, even when the temptation arose to focus more on pressure flying suits such as the earlier XMC-2-ILC model. That's what organizations like the David Clark Company did. They went on to be very successful with their high-altitude suits but would not be known for highly mobile extravehicular activity space suits.

Other Suit Concepts from the Aerospace Community

Examining other extravehicular space suit concepts from the period provide a sense of how aerospace designers thought about protecting astronauts. Grumman Aircraft was the future designer and builder of the Apollo lunar modules that successfully carried US crews to the surface of the moon and back. Their idea of what a space suit would look like, however, is perhaps what you might expect from an aerospace company better skilled at designing and fabricating rigid structures. Republic Aviation, another aerospace company, proposed a suit that was similar to Grumman's design. ILC had teamed up with a division of Republic earlier to help with the more technical aspects of the ILC suits related to thermal and altitude testing—areas ILC was not familiar with at the time. I'm sure that one look at these tin-can contraptions from Republic and Grumman gave both Len and George the reassuring feeling that they were on the right track. NASA did not totally ignore hard suits such as the one in Figure 3.1, as you will learn later in the story. It seriously considered more refined hard suits that two other contractors made during the early stages of Apollo.

NASA's Initial Perceptions of ILC

No doubt when Len and George heard Kennedy's speech that ensured the need for the moon suit, they knew that their long-awaited dream had come true. They knew they had the basics of what it would take to win the space suit contest, but they probably did not realize the challenges that lay ahead. They would now be playing ball in the big leagues. But did they truly have the rigor NASA required to produce such a life-critical device? NASA doctrine required all contractors to employ business models and quality standards typically used by all in the aerospace world. It was a world where mistakes would cost lives if the proper rigor was not employed and strictly followed. Few, if any, at ILC at that time had experience working in this culture.

Configuration Management

If you had asked Len or George in early 1961 what configuration management or systems engineering was, I'm not sure they would have had

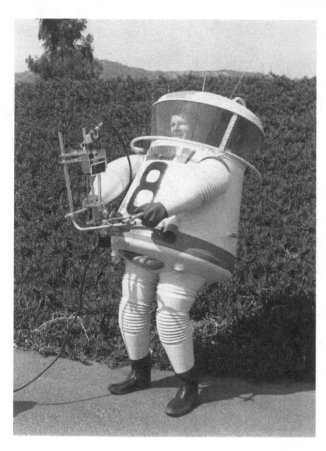

Figure 3.1. Early con-
cept of what a moon
suit might look like,
circa 1961. Manufac-
tured by Grumman.
Courtesy NASA-JPL.

a meaningful answer. These disciplines emerged in the 1940s and 1950s within the aerospace industry. Although the Department of Defense was using them, both were still considered relatively new as business principles. Both Len and George came from the Playtex culture that was immersed in producing commercial goods. In that business culture, these ideas only added unnecessary costs. The earlier air force suit contracts ILC had received were for relatively small quantity builds, so the company did not feel pressure to conform to the rigors of the aerospace world at the time. Helmet production was an exception, but these programs were fairly clear cut and relatively simple to control.

Unfortunately, for Len and George and their division, configuration management and systems engineering were extremely important to NASA administrators and were routinely followed by all within the aerospace community at this time. Early in their development process, neither George nor Len cared about documenting the changes they made to the suit,

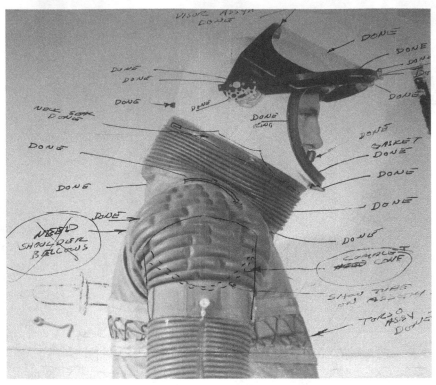

Figure 3.2. A photo of an early air force development suit showing the lack of rigor of the small ILC Industries team in the early years as they developed the first suits. Courtesy ILC Dover LP 2020.

as configuration management demanded. Len might have asked George about the details of how they had made an arm assembly the week before and George might still have had the patterns that made the fabric sections of the arm. One of the challenges with the design and fabrication of a space suit is that the parts are made up of individual pieces of fabrics that are cut to size based on patterns. Thus, in effect, the pattern acts as the "drawing" that documents the process and is used to repeat it when needed.

The companies that made the hardware for the rockets that would take humans into space made mechanical drawings for every single part so that they could be fabricated and inspected in comparison to the related drawing. NASA liked this process: it gave them the ability to control the configuration and to order parts as needed. One of the problems NASA had early on with space suits was the lack of mechanical drawings. A new group of relatively young NASA engineers was academically trained to read mechanical drawings, but working in the world of soft, flexible, fabric suits was something quite new for them. In the end, NASA settled on using the patterns as the controlling process for managing the configuration of the suit. The patterns were each given unique part numbers and any changes to the patterns had to be documented and approved by higher authorities so that nothing arbitrary could take place that would lead to problems such as poor fit or worse. A seam could fail because of improper fit when pieces were sewn together and pressurized. If that happened, the result could be fatal for an astronaut.

The Apollo space suit was custom made based on almost seventy different dimensions gathered from each astronaut. These measurements determined the dimensions of the torso, the arms, and the legs of a suit. Because hands were critical to carrying out a mission, plaster hand casts were made of the right and left hand so that the ILC model makers could have exact dimensions to use when making the dip molds for the rubber glove bladders. Even though the Apollo suits were custom made, the suits were made up of a torso that was chosen from small, medium, and large sizes; the same was true for the arms, legs, and boots. Modifications could be made to any of the components based on the body dimensions or based on feedback from the astronaut after fit checks. The individual sections were selected based on an individual astronaut's dimensions and were put together to make one custom suit.

Systems Engineering

The other major discipline ILC lacked at the time was systems engineering. The NASA Systems Engineering Handbook defines it as a methodical, disciplined approach to the design, realization, technical management, operations, and retirement of a system. It focuses on how the various parts are interconnected, the role each part plays in the performance of the final product, and how the product relates to and meets the customer requirements.[3] It takes a good deal of engineering management skills plus experience to look at the big picture of what you are designing in such a way that it functions as intended when all the parts are working together in the environment it is intended for. The best that George and Len could do during the early days of suit development was to focus on the small details of the various parts and hope they would all come together at the end to work as one. It wasn't a bad approach given where they were at the time, but NASA would expect much more in the coming years, and ILC was not capable of meeting NASA standards quite yet.

Good News, Bad News

When ILC delivered Technical Proposal 160 to NASA in March 1962, many young and newly hired engineers at NASA were about to be engaged in selecting the space machinery necessary to carry out President Kennedy's challenge. One of the people NASA hired to oversee the development of the space suit, Joe Kosmo, admitted years later that at the time neither he nor those who hired him knew much about space suits.[4] That is understandable, since the basic mission requirements were still being formulated and there was no real history of extravehicular activity space suits to draw from.

NASA's evaluation and selection of the Apollo space suit systems were conducted during the summer of 1962. ILC submitted their AX-1L suit, which used a life-support system manufactured by Westinghouse. AX-1L was the designation NASA gave it: the "A" referred to Apollo, the "X" referred to the experimental nature of the suit, the "1" indicated that the suit was the first model ILC delivered, and the "L" stood for ILC.

ILC also used its own internal method of identifying suits using another number that was easier for employees to track for accounting purposes. They used the model number SPD-143 for the first Apollo suits: "SPD" stood for "Specialty Products Division" and the number referred to the fact that the suit was the 143rd product ILC entered into its registers. This number appeared on the labels attached in the suits.

The outcome of the selection process was quite clear to the folks at NASA. The ILC suit had several advantages over other competitors, including its mobility while pressurized. However, the Westinghouse backpack was not what NASA was looking for, particularly given that Hamilton Standard had designed a more advanced backpack. The technical proposal ILC submitted contained details of a basic system design by Westinghouse and addressed how they would make two prototype systems that would focus on thermal control.[5] As the NASA selection team dug deeper into its evaluation of the proposals, it was increasingly clear what the outcome had to be. As smart as Len Shepard and George Durney were and as clear as it was that they were more than capable of developing a space suit, the depth of their ILC division was seriously lacking. It did not have the quantity of personnel necessary to turn out space suits and it did not use configuration or systems management. It was also probably obvious that Westinghouse had less to offer in terms of life-support systems compared to aerospace contractor Hamilton Standard. Hamilton Standard was a large aerospace company filled with very educated aerospace engineers who applied all the textbook solutions to aerospace problems and designs and who understood systems engineering and configuration well. Based on these facts, there was only one outcome to this contract award.

In 1972, Len Shepard recalled the moment when he learned about NASA's decision:

> We submitted this proposal and I guess one of the high points in my career was a telephone call which I received in my kitchen about 6:30 one evening—I remember it very clearly—it was from Dick Johnson, at that time Chief, Crew Systems Division, in which he said "Len?" I said "Yes? Hi, Dick." He said, "We like your proposal; we want you to come down and talk." That was a very exciting moment, obviously.
>
> So we got our team together and went down to Houston and it was at that time that we found out that NASA desired to award the prime contract to Hamilton Standard Division of United Aircraft and wanted us to negotiate with Hamilton Standard as a subcontractor to supply the suit.[6]

It was a true roller-coaster ride of emotions for the ILC team. A memorandum released by the NASA Public Affairs Office on October 12, 1962 announced the award to the public.

> The White House formally announced this afternoon, the award of a $1,555,970 contract to Hamilton Standard, Division of United Aircraft, of Windsor Locks, Connecticut, for the development of a "Moon suit"—a space suit to be used by crew members during trans lunar flight and lunar landing in the Apollo spacecraft.[7]

In 2019 dollars, that award was worth $12.8 million. The press release did not even mention ILC. It's not hard to imagine the pain Len, George, and many others at ILC felt after learning of the news. Another company would henceforth get top billing, even though it had never made a single stitch in a space suit. Moreover, the suits would now be designated the A-H series to identify them as Hamilton Standard suits.

As the prime contractor, Hamilton Standard would be responsible for the overall management of the suit program and the life-support backpack it would design and manufacture. Efforts were immediately made to get the upper management of the two organizations together so that the details of the pricing and structure could be worked out. As part of the negotiations, Republic Aviation was kept as a subcontractor to ILC to help with human factors and with the environmental testing of the suits.[8]

In short order, Hamilton Standard engineers began knocking at the front doors of ILC to get the "rogue artists" from the school of hard knocks in line and teach them a thing or two about real aerospace engineering. At first, George and Len and the others probably shrugged it off and continued with their work, even after spending stressful hours in meetings with Hamilton engineers who were driving their efforts in directions they did not care to pursue. No one at ILC would have disagreed with the assessment that much more needed to be done to make the AX-1L a more mobile lunar space suit that could fit into the spacecraft NASA had on the drawing boards. But this was an environment that reeked of strong egos; the credit bestowed upon those who developed the space suit to allow man to walk on the moon would be significant and everyone was aware of that.

ILC and the air force had shared the costs of developing the suit over the previous eight years and the work was mostly done on a shoestring budget. Now, with more proper funding provided by NASA through Hamilton Standard, the ILC suit was expected to progress much faster. This included work ILC was putting into the helmet system. Fred Feldman and a few other ILC engineers were proud of their helmet work. They had had a long run with the popular MA-2 helmet for the air force and were looking forward to topping the Apollo suit with their fine craftsmanship.

Team Partnering NASA's Way

ILC work on the Apollo suit initially started out with the intention of designing and manufacturing three models. The first prototype model yielded two individual suits. One suit would be delivered to NASA for mobility studies and the second would be delivered to Hamilton Standard for interface trials with the primary life-support system. These two suits would meet the agreement to deliver this phase of the contract within ten months. The results of these tests were supposed to drive improvements to the second models. Concurrent with the building of the second models, the third model was to be built and used for environmental and physiological testing.[9] By the time the second prototype was delivered, however, it was obvious that little improvement was being made. Unfortunately, many issues conspired to doom this program almost from the start. ILC was not used to turning out more than one or two suits, and that was over the period of a few months. ILC also found that the challenges of ramping up for a higher level of production while maintaining a quality product was daunting. Trying to design and manufacture gloves and other support gear just added to the problems.

While ILC engineers were battling their own demons, the aerospace engineers at Hamilton Standard wanted to see many improvements that included redundant pressure bladders and aircraft-grade hardware that was fit for high-end aircraft but not necessary for the space suit. They were not in favor of using the cables and pulleys that ILC engineers were using on the suit. In 1972, ILC senior engineer Mel Case summarized the relationship:

> They gave us some requirements to meet that were in our opinion, a lot of them were not related to the space suit field. They were related to the aircraft industry. And we just can't live with those requirements all of the time when you are doing something that's not really related. And they had requirements like the smallest screw you could use anywhere was a No. 6-32 screw, say. Well, we just couldn't live with that—or we couldn't use any roll-pins or all kinds of requirements which probably stem from specifications that they had been using for years in the aircraft industry.[10]

Work progressed through 1963 as ILC developed the prototype models that suffered from poor mobility and bulky shoulders. The ever-changing requirements from NASA dictated that the maximum width between the shoulder and arms should be no greater than 24 inches and the shoulders of the early Hamilton Standard series suits were much wider than that.

There have been various observations among space suit historians as to what or who was responsible for the misdirection of the suit development at this time. Some put the blame on ILC, some on Hamilton Standard, and some perhaps on the fact that NASA forced ILC to team with Hamilton, thus causing a loss of focus. In December 1964, a NASA memorandum noted that "it is unfair to blame ILC for all our ills; it is a combo of blame, ILC, HSD [Hamilton Standard Division], CSD [NASA's Crew Systems Division]."[11] I take no sides in this matter but can see where all the participants played a part in the problems. As with many of the challenges we face in life, it sometimes takes a major turn of events to drive significant change, and that was about to come.

Late in 1963, things took a serious turn that would drive the two organizations even farther apart. In a December 18, 1963, meeting at Hamilton Standard, ILC was advised that Hamilton would be taking the space-suit helmet business away from them. Their reasoning was that it would free up

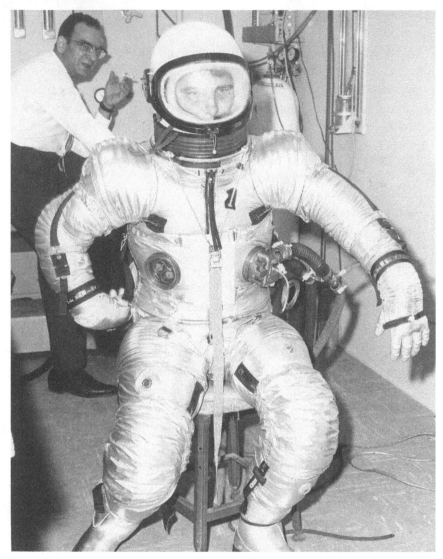

Figure 3.3. George Durney pressurized in an early ILC suit made under the Hamilton Standard contract. Len Shepard is in the background. Courtesy ILC Dover LP 2020.

some personnel at ILC to focus on the suit. However, ILC's staff included men like Fred Feldman who had extensive background in the helmet business and were not involved in the day-to-day work on the suit. The idea that the prime contractor would dare take work away from the subcontractor infuriated ILC's management. The company's upper management immediately called for a meeting between the two organizations to protest.[12] ILC president Wally Heinze and Len Shepard met with the management of

Hamilton Standard in New York on January 8, 1964, to let them know that they considered this action a serious matter that violated an agreement ILC had made with the president of Hamilton Standard, William E. Diefenderfer.[13] This meeting did not sway Hamilton's managers, who made it clear to Heinze and Shepard that Hamilton would press on with the development of the helmet. This was a significant blow to ILC, since it was the helmet business that had played a big part in getting ILC involved in space suit work. In addition to the fact that ILC people felt that Hamilton was stealing business from ILC, there was the matter of pride and respect—or now the lack of it—between the two organizations. Heinze and Diefenderfer exchanged several letters over the next few months as tensions grew. The ILC president was asked on several occasions to come to the Hamilton facility to meet with Diefenderfer, but those plans never materialized, likely because Heinze knew that no good would come out of it. A Hamilton Standard internal memo of February 7, 1964, outlined four approaches that Hamilton could take, including the outright takeover of the helmet work. The conclusion of the memo backed up that approach. The only concern Hamilton seemed to express was that NASA might see that action as a disruption to its space program.[14]

At roughly the same time all of this was taking place, NASA asked Hamilton Standard to submit a proposal for thirty-eight training suits. The cost that Hamilton Standard proposed to NASA was well beyond what it expected and NASA pulled the plug on the proposal until more serious negotiations could be worked out. NASA eventually reduced the number of suits to fourteen. During these negotiations, NASA agreed that Hamilton Standard could take over the helmet business. However, it wasn't long before the Crew Systems Division of NASA stated that NASA should give some money to ILC so it could continue to do some helmet development work. This probably annoyed the Hamilton Standard folks who thought NASA was doing too much to appease ILC, adding more tension to an already bad relationship.[15]

The Gemini space missions began in April 1964 with the David Clark Company intravehicular pressure suits. Also in April 1964, NASA made the decision to continue to use a version of the David Clark Company's Gemini suit for the initial Apollo missions to the moon because of poor reviews of the ILC/Hamilton Standard Apollo suits. This gave NASA engineers the extra time they felt was needed to fully develop the more technical lunar suits. Because of this decision, NASA called the first Apollo missions the Block I missions. The Block II missions would be the lunar missions (Apollo 11 and up) that would require the extravehicular suits—from the company that could supply the best suits when the time came.

In their haste to get suits for training, NASA administrators went around Hamilton Standard midway through this teaming agreement and provided a contract directly to ILC for three suits. ILC management saw this as an opportunity to win favor with NASA. One way to do that was to lower their labor rates.[16] They hoped that this would make an impression on the customer and open the door for discussions on making ILC a prime suit provider. ILC called these suits the model A-2L. This was perhaps a small shot in the arm for ILC but certainly not enough. There was not a real opportunity to turn this A-2L suit into something radically different given NASA's time frame for producing the suits.

One Crew Systems Division schedule dated September 23, 1964, shows that three model A-2L suits existed, serial numbers 001, 002, and 003. Serial numbers 001 and 003 were listed as supporting zero-gravity flights while suit 002 was listed as being at Hamilton Standard. The schedule shows that all three suits supported testing or training at Grumman Aircraft Engineering Company (the designers and manufacturers of the lunar modules) for two weeks and would then be shipped to Holloman Air Force Base in New Mexico.[17]

The Hiring of Dr. Nisson Finkelstein

ILC president Wally Heinze made the fortuitous decision in January 1964 to hire Dr. Nisson Finkelstein as a new vice-president. Finkelstein became a buffer between the front-line ILC team and people at Hamilton Standard and NASA. Heinze likely felt the need for backup from a top-level manager who was closer to the action after his recent encounters with Hamilton Standard management. Finkelstein came from General Dynamics, where he had served as the vice-president of their electronics division. He had received a BA in physics from Harvard and a PhD from MIT, so he had credentials that would help open many doors for ILC. He may have come a bit too late to help patch together the relationship between Hamilton Standard and ILC, but he provided support for opportunities to forge a new path with NASA. His help would be much needed in the coming months and years.

Hamilton Standard issued an internal memorandum dated March 31, 1964, that described a new task force the company was setting up whose purpose was "to assure the operational success of our first space product—

the Apollo suits." Hamilton's Robert Breeding, who headed this task force, was instructed to operate out of the ILC plant along with the other four members until ILC could provide an acceptable suit. This action was accelerated by the delivery of a model AX-3H suit, serial number 024, that several astronauts used on March 26–27, 1964, to evaluate a model of the lunar module. The astronauts all protested that this suit could not be used on a space mission because of its poor mobility and the many other quality issues it suffered from. Astronaut Gordon Cooper proclaimed outright in front of many high-level NASA administrators that he would not go to the moon in that suit.[18]

In his book *Carrying the Fire*, astronaut Mike Collins recalls a memo written by astronaut Pete Conrad around the time that refers to the early ILC suits. Conrad said, "Based on the zero-G work last week and some work . . . this week, I've concluded that the ILC suit is useless and should be abandoned." Conrad added "I would like to take a Gemini Extravehicular suit up there (to Grumman) for direct comparison."[19] The input from astronauts carried a lot of weight with the NASA management team that made decisions that affected the direction of suit contracts.

With all the turmoil taking place between ILC and Hamilton Standard, it was becoming more apparent to many of the top space suit managers in NASA that the David Clark Company's Gemini suits was a good foundation to build on and that perhaps greater effort should be put into refining that suit for future Apollo missions. NASA's astronaut office was making it clear that the proven Gemini suits should be considered based on the ILC/Hamilton results to date. On December 21, 1964, Matthew Radnofsky, NASA Apollo Support Office assistant chief, initiated a memorandum that said that "negotiations [will] be initiated directly with David Clark for suits as described."[20] On January 25, 1965, Richard Johnson, chief of the Crew Systems Division, issued a memo that made it clear that NASA was beginning the negotiation process with the David Clark Company to undertake a backup plan to get David Clark started on the development of a prototype suit to be used for Block II (lunar) missions. The cost was estimated to be $150,000 ($1.2 million in 2019 dollars) to get the work started and another $176,000 ($1.4 million in 2019 dollars) to obtain a development suit. The memo stated that "this backup development has been revealed to Hamilton Standard Division and they are aware that they are in competition. This information will be made available to International Latex Corporation by Hamilton Standard."[21] The mere thought of this no doubt sent chills down the spines of the Hamilton Standard management.

These events are likely what led Hamilton Standard to hire Dr. Edwin Vail, a noted air force veteran. Vail had gained much of his experience while assigned to the X-15 program, where he worked closely on the pressure suits the David Clark Company made. Vail set out to design a suit with a new arm configuration that would be put into competition with an ILC suit. Vail's suit was known as the "Tiger suit." Hamilton Management saw this as another option should ILC continue to falter. This further infuriated the ILC folks, since it demonstrated the inability of the two organizations to work together. Hamilton's idea for a better mousetrap did not proceed very far. In a memo written soon after NASA evaluated the Tiger suit, Matthew Radnofsky of NASA wrote "Tiger suit a bust, forget it."[22]

"Play Suits" in the ILC Back Room

While ILC was trying to make the best of its shotgun marriage to Hamilton Standard, things were taking place in a corner of ILC to cobble together a suit that incorporated various concept ideas. One version that was told to me by ILC engineer Bob Wise, and subsequently corroborated by a few others, was that ILC was secretly putting together a suit in the back room to hide new breakthroughs from the Hamilton Standard engineers. Apollo suit engineer and future ILC Dover president Homer Reihm does not recall that specific story, but he does not discount the fact that it happened. Reihm does recall a "play suit," as they called it, for trying out different ideas.[23]

One of the first examples of the ILC play suit is displayed in Figure 3.4. It shows a new rear-entry–style zipper closure that ILC designers had added. Although this was not a new concept—this approach had been used earlier by other manufacturers—this style of closure system ended up becoming the mainstay for the early Apollo suits.

In November 1964, ILC's Finkelstein sent a letter to Richard Johnson, the chief of NASA's Crew Systems Division, expressing his disappointment that NASA was not interested in pursuing a "V.I.P" helmet design that ILC had been working on.[24] Finkelstein asked if he could send John Hughes, a newly hired engineer at ILC, to Houston to discuss other helmet work that ILC could possibly be involved with. Finkelstein assured Johnson that the helmet project was not impacting the production of model A-5H and training suits that Hamilton Standard was working on.[25] At this time, two NASA engineers, Jim O'Kane and Dr. Robert Jones, were developing and testing a new polycarbonate bubble-style helmet made with the help of Texstars, a

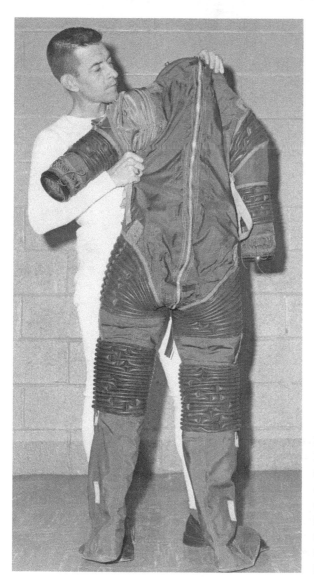

Figure 3.4. ILC Employee Earl Williams demonstrates a zipper closure on an ILC play suit. This closer system was used in the model A7L suits for Apollo 7 through 14. Courtesy ILC Dover LP 2020.

company in the Houston area. O'Kane and Jones would ultimately deliver a helmet that would prove to be a savior for NASA and ultimately for ILC.

Photo evidence shows that in early 1965, George and the others had put together a complete play suit that involved NASA; we know this because the suit included the new bubble helmet. It is likely that the December 1964 meeting Finkelstein asked for between Hughes and the NASA engineers resulted in an opportunity for ILC to incorporate the new NASA bubble

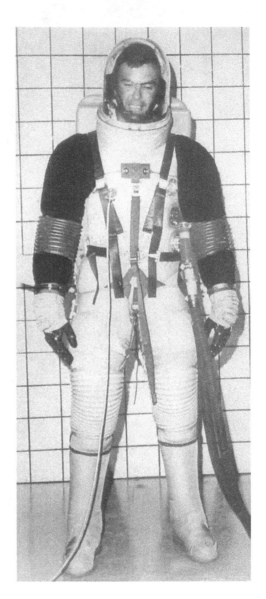

Figure 3.5. The new NASA bubble helmet shown on the ILC development suit for the first time. Note that the details of the new arm design are blacked out. ILC would use this suit for the next competition and did not want to give away the design details. Courtesy ILC Dover LP 2020.

helmet in the ILC play suit. The modern manufacturing techniques and materials available to form the polycarbonate bubble helmets were quite an improvement over the older style. These new helmets gave crew members the ability to move their heads around inside and not feel the claustrophobic effects of the earlier, tight-fitting design that offered a poor range of visibility and were very heavy. This new, simplified helmet also solved many problems ILC was having with the shoulder and torso designs, since it eliminated the need for large neck/helmet convolutes and the restraint

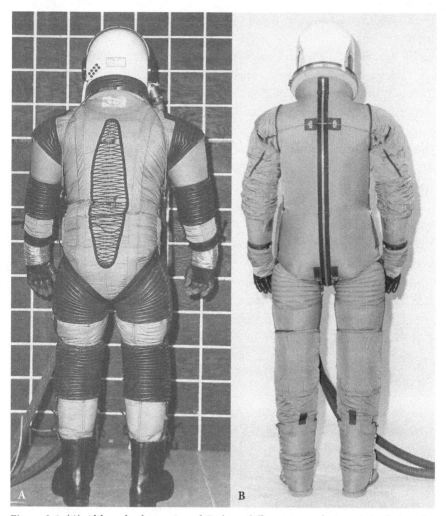

Figure 3.6. (A) Although photos A and B show different torso designs, it is obvious by looking across the shoulders of both suits that the suit on the right (B) uses a more tapered cone section at the base of the neck than the suit in photo A. This was possible once the lower-profile polycarbonate pressure helmet took the place of the older-style helmet, thus reducing the distance across the shoulders. Courtesy ILC Dover LP 2020.

hold-down cables that directly impacted the shoulder design. The timing was excellent because ILC now had an opportunity to radically change the torso structure of the suit and present a suit that was much more in line with what NASA was looking for.[26] The new helmet was shown on a pressurized ILC suit that was made during the period when ILC was under contract to Hamilton Standard; in the photo in Figure 3.5, ILC blacked out the shoulders and arms so as not to show the details of the team's development work.

By January 1965, Hamilton Standard was sharing details with NASA about how it would make its second state-of-the-art model A-5H suit. Through the first quarter of 1965, Hamilton set out to make five new suits that incorporated many of the new features they desired, such as improved joints, decreased leakage, and increased comfort.[27] B. F. Goodrich provided the soft-goods part of the suit work while Hamilton Standard developed many hardware items in house. Hamilton was pressing ILC to provide a molded thigh convolute for evaluation, so it was evident that Hamilton wanted to use multiple suppliers if necessary to arrive at the best possible suit configuration.[28] But Hamilton also realized that they could survive without ILC's help, so they weighed the consequences and set their sights on dropping ILC altogether.

In March 1965, the schedule that NASA's Crew Systems Division circulated outlining the suits it needed for training through 1966 listed Hamilton model suits: seven model A5H suits, six model A6H-1 suits, fourteen model A6H-2 suits.[29]

Parting of the Ways

In February 1965, Hamilton Standard executives R. E. Breeding and E. V. Marshall traveled to Houston and met with Richard Johnson NASA's crew systems chief. Also in attendance were NASA's Robert Smylie and Matthew Radnofsky. Breeding and Marshall brought along a book filled with evidence supporting their decision to drop ILC as their suit subcontractor and move development work over to B. F. Goodrich, including issues such as cost overruns, delinquencies related to qualifying the suit to meet various mission requirements and reliability reports, delinquencies in hardware delivery, and details of the personnel clashes (of which there were countless numbers to choose from). Radnofsky wrote a memo outlining the meeting that made it clear that NASA's managers neither agreed nor disagreed with that decision but simply that they understood. Radnofsky told the Hamilton Standard group that NASA was still willing to give ILC another chance if they wanted one.[30] It provides a hint that the NASA group felt some sense of responsibility for the fallout that developed after it teamed Hamilton Standard and ILC together following the 1962 contest.[31] It also indicated that NASA still regarded ILC as a potential contender.

Shortly after this meeting, Hamilton Standard notified ILC by telephone that it had decided to drop it.

NASA looked at its options for lunar (Block II) suit providers and realized there were two possible outcomes. The first was that Hamilton and B. F. Goodrich might provide the best lunar suit; the second was that the David Clark Company could upgrade its already proven Gemini suit and excel over Hamilton. The best path forward for NASA was to establish a contest between the two organizations that would once and for all determine the best suit. The suits would be sized for astronaut Mike Collins and undergo a comprehensive evaluation that included twenty-four major tests.

Immediately after learning of NASA's decision to hold a contest between Hamilton Standard and David Clark, the ILC team started moving forward with plans to participate in the competition. ILC's Finkelstein visited NASA's Manned Spacecraft Center in Houston after Hamilton Standard had dropped ILC and made it clear to Richard Johnson that ILC was still very interested in joining in the competition once again. NASA managers were aware of the development work ILC had been doing outside the Hamilton Standard relationship prior to the breakup, so it was in their interest to see what ILC might deliver. Johnson told Finkelstein that ILC could join the contest but that it would have only six weeks to provide something radically different from the A-H series of suits made to date and that ILC would have to bear the full cost of developing and producing this suit.[32] That second part was fine with Finkelstein because he knew that ILC was close to providing a radically different suit that could win and that doing so with ILC money meant that there would be no constraints about how the suit had to be made. There would no longer be any Hamilton Standard folks poking around and directing how the suit should be built.

On September 17, 1965, Robert Gilruth, the director of NASA's Manned Space Center in Houston, sent a memo to the attention of Dr. George Muller at NASA Headquarters. This memo best summarized the events that had taken place over the previous years from the perspective of NASA with details about the later part of 1964 and early 1965, when things came to a head between ILC and Hamilton. It reads as follows:

In May 1965 a procurement plan was submitted to NASA Headquarters outlining the previous development efforts that had been accomplished in connection with the Apollo Extravehicular Mobility Unit (EMU). This plan noted in some detail the prior competitive actions which had led to selection in early 1962 of Hamilton Standard Division for development of the portable life support system with the International Latex Corporation as a subcontractor to Hamilton Standard for the development of the space suit. Hamilton Standard was designated as the system manager with responsibility for system development, integration, and qualification of the EMU.

These two contractors had not initially proposed together but were selected on the basis that Hamilton Standard had submitted the best program management plan and technical approach to the development of the portable life support system and International Latex had submitted the best technical proposal for the development of the space suit.

The combination of these two contractors into a prime-subcontractor relationship was accomplished by negotiating an agreement with both parties. However, since its inception there have been problems in fostering a team approach to the EMU development effort. Deliveries of the initial prototype hardware in early 1964 indicated that the development program was not progressing satisfactorily. As a result of a command module mockup review in April 1964 it was determined that the Hamilton Standard/International Latex Corporation space suit was not compatible with the spacecraft and a decision was made at that time that the Gemini space suit should be used in Block I flights. This decision gave additional development time for the final lunar space suit. The Hamilton Standard/International Latex Corporation effort was restructured to the Apollo Block II flight program. Initial suit deliveries under this plan were called for in December 1964. These deliveries were slipped to March 1965 due to deficiencies in the suits.

Hamilton Standard consulted with MSC [Manned Space Center] and some of the apparent relationship problems with their subcontractor were emphasized to the Manned Space Center at that time. Hamilton Standard, in its role as system manager, had taken a number of actions designed to improve the progress of the development work. These actions included efforts by Hamilton Standard to undertake the development of the space suit helmet in-house rather than through the subcontractor as originally contemplated. During this period, Hamilton Standard also had increased their inhouse efforts on development of space suit joints and other components to provide backup to their suit subcontractor. *The increased emphasis by*

Hamilton Standard on backup efforts finally reached a point where the subcontractor, International Latex Corporation, was literally in competition with its prime.[33]

In February 1965 Hamilton Standard informed MSC of its intention to cancel the International Latex Corporation subcontract on the grounds of poor performance, cost overruns and delinquent deliveries. MSC concurred in this action on the basis that the contractual and program responsibilities for the EMU development rested with Hamilton Standard and the grounds for the cancellation were factual based on our own observations of the program.

MSC had also become increasingly concerned about the success of the Hamilton Standard/International Latex Corporation effort and in January 1965 had instituted a backup effort with David Clark Company, developers of the Gemini suit, to assure meeting Block II requirements. Delivery of the David Clark suit was effective in June 1965.

At the time this backup effort was established, Hamilton Standard was advised that a competitive technical evaluation was contemplated in June 1965 to select the Block II space suit contractor. Hamilton Standard had already cancelled the suit efforts by International Latex Corporation and was proceeding with their own efforts. International Latex Corporation was aware that there was to be a continuing development effort. They were also aware of both Hamilton Standard's and MSC's dissatisfaction with the prior performance which had resulted in cancellation of their subcontract by Hamilton Standard. However, they felt they still had something to offer and requested an opportunity to submit a suit for evaluation. The procurement plan previously referenced specifically had excluded International Latex Corporation from consideration as a source. However, in order to foster competition and to assure fair treatment of International Latex Corporation, their suit was accepted for evaluation in June 1965.[34]

4

Second Chances

The Model AX-5L and A-5L Suits

The new suit ILC Industries made for the contest was identified as the AX-5L suit. The "5" meant that it was the 5th model ILC made (the company had made the AX-1L, A-2L, A-3H, and A-4H models in the prior three years). The "X" was used in the aerospace and NASA world to denote an experimental version. The six-week time frame was a bit of a challenge, but Dr. Finkelstein and the team figured it was certainly worth trying. They had the crew in Dover that could pull it off. ILC president Wally Heinze approved the funding, hoping that ILC could pull this one out and redeem the company in the face of all the adversity the company and he had suffered. Based on the budget that was allocated and the deadlines ILC had to meet, the company manufactured only one model AX-5L suit.

The competition would be scored based on a host of criteria NASA had generated to meet its latest suit requirements. Hamilton Standard proposed a new suit that was essentially the A-5H model suit. Hamilton was likely concerned about the David Clark Company and the David Clark Company was likely concerned about Hamilton, but neither of them probably gave much thought to ILC, since they assumed that it would not be allowed to compete or, if it was, the product would not meet the challenges of the contest. By April 1965, Hamilton Standard was partnering with B. F. Goodrich and was building five new suits at the B. F. Goodrich facility. This new model had improved joints and redesigned neck openings and other hardware pieces to reduce the leakage that had plagued Hamilton's earlier models.[1]

The David Clark Company entered their extravehicular, or "E-suit," as they called it, into the competition. This was the suit it was developing for the advanced Gemini missions, which would include the first extravehicular activity (EVA). NASA's experience with the mobility of the joints of these suits while pressurized was not stellar. That became apparent when Eugene Cernan (Gemini 9A), Richard Gordon (Gemini 11), and Eugene "Buzz" Aldrin (Gemini 12) did a total of six EVAs between the three of them. The Gemini suits proved very difficult to work in, but the point of Gemini was to learn what needed to be improved to do EVAs on the moon and other things. The clear lesson learned in Gemini EVAs was that a lot needed to be gained in terms of suit mobility and physiological performance.

During this period, ILC had perhaps 100–120 employees in the entire organization with about ten or fifteen working on the space suit development.[2] Once the word got back to the team at ILC that they still had a chance, slim as it was, it was full speed ahead. Len Shepard selected a team consisting of Mel Case, Homer Reihm, Bob Robbins, and George Durney. In a later interview, Case gave high marks to Bill Ditolla, whom he credited with being the project lead engineer. Five full-time research and development seamstresses did the bulk of the manufacturing on the suit. The engineering team set goals about what they wanted to achieve in terms of mobility and weight and other positive characteristics they had developed with their research for the play suit. Case later referred to that play suit as a "test-bed" suit they made to try out the designs.[3]

With six weeks and a limited staff meant that there would be little time off. If anyone walked into ILC at any hour day or night during this period, it would have looked like a typical workday. There was an occasional break at 2 or 3 A.M. to go home to clean up and get a few hours of sleep, but otherwise people were at work. Len and George focused much of their time on the shoulders of the suit to narrow the width and to provide the best mobility they could squeeze out of them. The supplies they needed were often unavailable or were locked away in the storeroom after hours. Mel and the others often climbed over a tall chain-link fence that "secured" the storage area in the wee hours to collect the parts needed. Bob Penny, who was an engineer at ILC during this time, recalls having to break apart a motor that he had in his garage at home to obtain some ball bearings needed for some piece of hardware used in the suit.

The six weeks ticked along rather quickly. There was no stopping to look at the clock. As the model started to look more like the suit the team had wanted all along, their spirits lifted and they thought they might have a shot at winning, particularly because they felt that anything the David Clark Company or Hamilton Standard/B. F. Goodrich produced would be less mobile and suffer from other issues. The significant features the team incorporated into this new suit included a neck and shoulder cone that

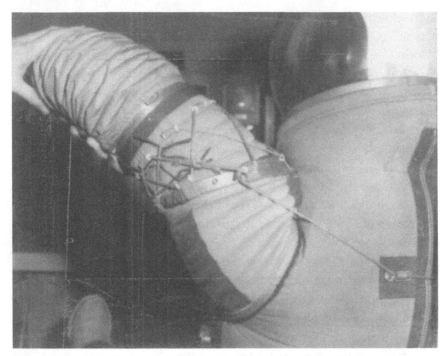

Figure 4.1. The AX-5L shoulder design. This feature outperformed the Hamilton Standard/B. F. Goodrich and David Clark Company shoulder in terms of range and ease of motion. The wire cable you see on the back slides through a tunnel on the upper arm that served as a guide and connected to hardware on the chest. This allowed for forward and backward flex and extension. The rubber shoulder bellows enabled the arm to raise and lower, so motion was achieved in all directions with minimal effort (omnidirectional was the term used to describe it). Note the tapered neck cone where it intersects with the top of the torso. Courtesy ILC Dover LP 2020.

was narrower than earlier designs, eliminating the earlier issue of a shoulder cone that rested on the wearer's shoulder, causing obvious problems. It had tapered knee and elbow convolutes, but Mel Case later said that that was a mistake; the team eliminated them on future suits. The wire hold-down cable for the helmet was eliminated. The patterning of the torso and neck cone was designed to carry loads without the need for the cable. This tapered neck cone with the new bubble-style pressure helmet made for a much more refined and contoured-looking suit; it also created a better transition to the shoulders that reduced the width and started to drive continuous improvements to the arm mobility. The tapered look also extended down the torso to the point where the legs were joined. The team spent a lot of time working on the leg restraint cables that proved to be beneficial in the competition.

There was a feeling among the ILC troops that many key players at NASA believed in their talents and their ability to deliver a good suit, despite some of the shortcomings of previous suits they had worked on. In a 1972 interview Len Shepard commented that he had always felt that Richard Johnson, chief of the Crew Systems Division in Houston, was very fair in his dealings with ILC and that perhaps Johnson had a feeling all along that ILC could pull it off.[4]

Brand-X Declared the Winner: The Model AX-5L Suit

When the contest deadline arrived, ILC's suit was still not ready. ILC engineers and seamstresses needed another week to put the finishing touches on it. NASA approved the brief delay. A photograph of the AX-5L suit at the ILC Dover plant shows model maker and part-time suit subject Richard Ellis in the suit as Len Shepard and Marty Schwartz measure the width of the shoulders (Figure 4.2). They had to be pleased with the outcome of this suit since it was by far the best they had come up with to date.

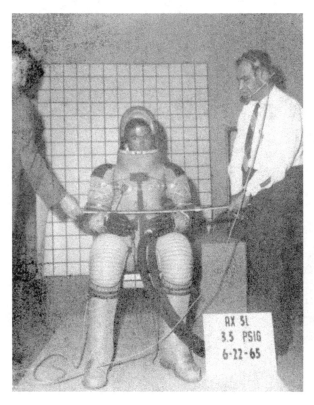

Figure 4.2. Len Shepard (*right*) and Marty Schwartz (*left*) hold the measuring stick to evaluate the final width across the arms of the suit ILC designed for NASA's contest; test subject Richard Ellis is in the suit. The date in the photo was one week past the due date and one day before the absolute start date for the contest, but NASA allowed the delay because its engineers wanted to see the resulting ILC suit. Courtesy ILC Dover LP 2020.

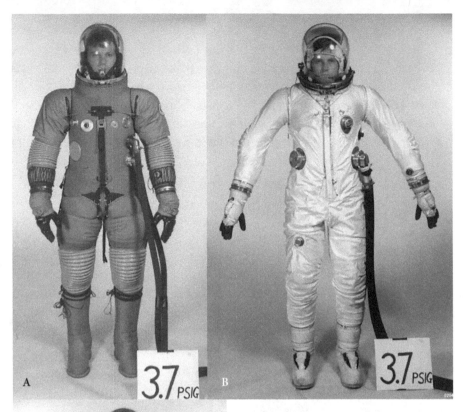

A

B

C

Figure 4.3. The three suits entered in NASA's 1965 competition: (A) ILC's model A-X5L suit; (B) The David Clark Company suit; (C) The B. F. Goodrich/Hamilton Standard suit. Courtesy NASA.

The suit was carefully packed up and delivered on a chartered jet directly to NASA. It was taken to a small office ILC had at the Space Center, where ILC's lead suit technician, Bill McClain, and a few other ILC technicians worked. Bill was responsible for taking care of the suit throughout the evaluation process and making sure NASA was happy, including changing sizing adjustments as necessary to make the evaluators comfortable in the pressurized suit. Astronaut Mike Collins was originally selected to evaluate the suits, but plans changed and a NASA technician, Jack Mays, stepped in to fill the void. This was not a problem for ILC because Bill had become good friends with Mays and although the merits of the ILC suit would drive the outcome, that relationship was not going to hurt ILC in the least.

Homer Reihm, who within a few years would be the chief project engineer for the Apollo suit program at ILC, tells the story of how just before getting into the suit to demonstrate the mobility to the NASA judges, Mays directed McClain to "tighten this suit up because I'm going to beat these Hamilton bastards."[5] Perhaps this attitude summed up the innermost feeling of many of the personnel at NASA, the ones in the trenches, who truly understood the makings of a space suit.

As confident as the ILC team may have been, ILC's suit was by no means a shoo-in. There were sixty-six separate evaluations for each of the three suits that included functional and engineering tests, comfort evaluations, and field maintenance tests. The suits were tested under controlled conditions to make sure that each suit was always exposed to the same conditions. NASA anticipated that problems with the suits would develop during the evaluation process, so they set rules related to how the teams could do repairs without jeopardizing the schedule. This contest had to be completed in a reasonable time frame because of the Apollo program schedule and the costs that were involved in running it. The memo that outlined what a team could do when a suit failed a particular test read as follows:

a. If a failure occurs during test, no more than one hour will be allowed for repair. If the repair cannot be accomplished within that time period, the contractor receives a zero rating for that test.

b. If a failure occurs during test, a maximum total elapsed time of 48 hours will be allowed to effect repair, with the penalty cited in a, above. If repair cannot be affected within the allotted 48 hours, zeros will be given for all subsequent tests missed during down time.[6]

The point was that the contractor had one hour to repair their suit or receive a zero for the missed test. They had another 48 hours to complete the repair while further tests were placed on hold. If they did not make the repair in those 48 hours, all testing resumed with the remaining contractor suits and the suit under repair would be given a zero for further missed tests.

The helmet neck ring of the David Clark Company suit detached from the suit while the wearer was pressurized inside it. The Clark team completed repairs within the 48-hour time frame. The zipper closure of the ILC suit became separated from the torso while Jack Mays was donning the suit.[7] The ILC team resolved the issue within 48 hours, but it created quite a concern. Since these suits were essentially development suits that had undergone very limited testing to identify any design and manufacturing flaws, it was not unusual to see these and other failures. Once the contest winner was selected based on the best performance and design, NASA would pour greater sums of money into the further development of the suit to eliminate problems and assure that the final design would sustain human life on the moon.

Mel Case later recalled that he and the other folks at the Dover plant had very little feedback from Houston during the six-week contest, so they had no idea where they stood. In his 1972 interview, Mel said that as good as they thought their suit was, everyone at ILC realized that the odds were stacked against them because they didn't have anything in Dover to offer, meaning no configuration management or systems engineering, let alone the staff to further develop the suit and then manufacture it. A mood of despair began setting in among the ILC team because they started to realize that NASA would basically be starting over from scratch if they selected the ILC suit, whereas if it selected the Hamilton Standard/B. F. Goodrich suit or the David Clark Company suit, it would only need focus on some basic upgrades to existing suits that had been proven on earlier NASA missions.

To everyone's surprise, at the end of the competition, the ILC suit earned twelve first-place finishes in the various categories, compared to eight for the David Clark Company suit and two for the Hamilton Standard/B. F. Goodrich suit. NASA gave the ILC suit high marks for showing outstanding characteristics in operational mobility, vision capability, hand dexterity, suit adjustability to crewmen, overall dimensions, and physiological tests.[8] NASA gave points for second- and third-place finishes, but in the end none of that mattered. ILC took the only prize up for grabs, and that was the opportunity to make the space suits for Apollo. Soon after the contest, on July

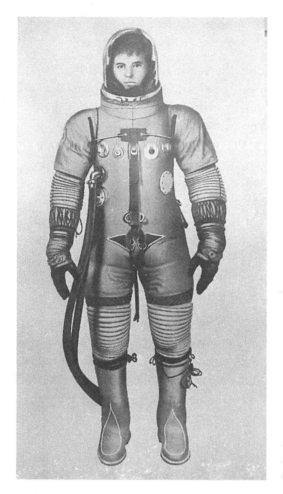

Figure 4.4. Artists' rendering of the model AX-5L ILC suit entered in the competition. Courtesy NASA.

1, 1965, Vice President Hubert Humphrey was touring the space center in Houston, Texas and was told about the competition and how the underdog ILC Industries suit took the top prize. His comment was that "it looks like Brand X won the contest."[9]

ILC engineers were aware of the few shortcomings of the AX-5L suit, but they knew they were on the right track. The scary question now was, what would NASA do about the contract? Would ILC be forced once again to be a subcontractor, and if so, to whom? No doubt NASA managers were left scratching their heads too. Certainly they had learned their lesson about forcing two contractors with such diverse ideas to join forces. It was now time to take a different approach. ILC had come out on top in two out of two competitions, proving its ability to fashion the space suit NASA needed for future lunar explorers. The conclusion of a 1965 NASA memo between Robert Gilruth and Dr. George Low summarized the outcome:

> On the basis of this evaluation, the space suit submitted by International Latex Corporation was the clear, first place choice, David Clark's suit was second, and the suit manufactured by Hamilton Standard was last. Based on this evaluation, MSC recommends the following actions.
>
> Amend the existing Hamilton Standard contract to provide for the development, qualification, and fabrication of the portable life support system and associated equipment for the Apollo flight program.
>
> Award a separate contract to International Latex Corporation for the development and fabrication of test and flight space suits and associated equipment
>
> MSC would assume responsibility for total program management, systems integration, and space suit qualification.[10]

In July 1965, Crew Systems Chief Richard Johnson outlined the justification for noncompetitive procurement of two A-5L suits to be delivered by September 6, 1965, so NASA could use them for the astronauts participating in the design review of the command module at North American Aviation. This would provide for the latest required Apollo suit/vehicle interface simulation. This was a critical milestone for NASA, and they really wanted the suits.[11] This would give NASA time to work out the details of the Apollo suit contract to be provided to ILC.

In 1972, Len Shepard made the following comments about his views of Johnson and how he made the decision to award the contract to ILC:

> I think it required a tremendous amount of courage on the part of Dick Johnson to give us the contract in 1965. He was committing that $40 billion program.
>
> How easy it would be for him to stay with Hamilton Standard and say "Gentlemen, we gave this contract to a division of United Aircraft, one of the largest companies in the United States. What better judgment could I have shown in the choice of the contract?"

Len went on to say that as Johnson made decisions, he visualized himself testifying before a committee to justify them.[12]

By September 1965, Dr. Robert Gilruth, director of the Manned Spacecraft Center, issued a memorandum to the chief of the NASA Procurement and Contracts Division stating his case for selecting ILC Industries.

Evidence to date and particularly that arising since June 1965 supports the conclusion that International Latex Corporation is clearly a more desirable source for the specific effort contemplated than any other firm in the same general field. The evidence includes the first competitive procurement in October 1962 and the competitive technical evaluation of space suits in June 1965. Technically this procurement must be handled on a noncompetitive basis but in fact competition established the basis for the selection of International Latex Corporation[13]

NASA selected General Electric to integrate Hamilton Standard's primary life-support system and the ILC space suit and generate a single specification that both companies could work to.[14]

Apollo 11 astronaut Mike Collins summed up the competition and ensuing years this way:

The suit completion was a good idea because it spurred ILC on to a higher plateau of performance, especially in the area of shoulder mobility and comfort. The ILC suit was clearly superior to the other two, and it got even better in the four years between the 1965 competition and the first moon walk. By the end of Apollo, astronauts were spending long hours in lunar EVA with no apparent discomfort, a fact beyond our wildest expectations during 1965, when we got our first look at the lunar EVA hardware.[15]

Hiring the Staff to Design and Manufacture the Suits

In the period 1965 to 1968, ILC hired many new employees to support the suit business. At the peak of Apollo in 1968, the workforce numbered around 900 people. Most of the employees worked on the Apollo program, while a small handful supported the division that focused on various protective helmets and visors for police departments and motorcyclists. In addition, a production line made inflatable structures that could be erected over tennis courts or large swimming pools. Several other small contracts would ebb and flow through the production areas during the Apollo years, but the space suit is clearly what it was all about at ILC.

Many within the labor force—including the model makers, the machine operators, and the seamstresses—came from the small towns surrounding Dover. They included Milford, Felton, and Harrington in Delaware and towns in nearby Maryland such as Denton and Greensboro. The professional staff were typically recruited from all over the nation. Many had college degrees and some had experience working on suits at B. F. Goodrich and the David Clark Company.

One former employee told me that ILC Industries did not pay the best wages compared to some of the other aerospace companies that were hiring during the Apollo years. While it was true that the cost of living was low in the Dover area, ILC could draw the talent it needed because it offered the exciting challenge of working on the space suit program. On top of that, it offered employees the chance to participate in a company profit-sharing program. (This sounded great at the time, but when the company closed its doors at the end of the Apollo era, much of that profit sharing was nowhere to be found). No one I know who worked at ILC during the Apollo years ever complained about the wages. They only remembered how great it was to be involved in making the space suits that enabled our astronauts to walk on the moon.

Once ILC became the prime contractor for the suit, the pressure was on to make deliveries on time and meet all the quality requirements. The budget NASA provided was not an issue. ILC ran ads for employees in major aerospace publications such as *Aviation Week* next to ads from companies such as Pratt & Whitney, McDonnell, and Boeing. Many ads displayed some sort of rocket or space-age contraption. One ILC employment ad that ran in 1968 did not even mention the space suit but did specify that applicants needed to have experience with soft-goods and hard-goods materials and know how to test them. The ads conformed to the gender roles of the time with statements such as "to be specific, we need men with the above qualifications to fill the career openings listed below."[16] And as was the case at many of the other aerospace organizations at the time, few if any women filled the more technical roles at ILC. Arguably, however, the technical roles at ILC included the skilled seamstresses who had significant input in the design and assembly of the Apollo suits. These women came from a variety of backgrounds and had the skills and patience needed to make these life-critical products. This was the first kind of clothing used to permit humans to walk on another celestial body other than Earth and provide the thin barrier between the vacuum of space and the life-sustaining environment sealed within the suit. These women made it all happen.

Company founder Abram Spanel was an advocate of hiring African Americans at a time when other industries would not. Many photos from the Apollo years show black employees who played a prominent role in the making of the Apollo suits, such as Walter Lee, who was one of the Apollo

Mission managers, and Mrs. Iona Allen, who could take on the toughest sewing challenges with her Singer industrial sewing machine. In my earlier years at ILC, I always looked forward to talking with Iona, who was sewing the space shuttle suits at the time. She was very petite and very soft spoken. She would light up when she told you that she made Neil Armstrong's lunar boots. She passed away many years ago, but those boots still sit on the moon not far from where Neil Armstrong took his first historic steps. Unfortunately, I never knew Walter Lee. He passed away at a relatively young age while working on the Apollo suit program. Lee figured prominently in the management role for Apollo and was very highly respected by everyone who worked with him. He was one of the first in a series of professionals hired after the 1965 suit contest and helped elevate the standing of ILC in NASA's eyes. He was fondly remembered for the fishing boat that he had called the *Leaking Lena*. Many ILC employees were invited out for a day of fishing on that boat—that is, when they had a day off.

Engineering and Support Staff

By 1965, ILC had begun hiring some of the best college-trained engineers. They included the aforementioned Melvin Case and Bob Wise, who both had experience working at B. F. Goodrich, manufacturer of the Mercury space suits. Their job title was "softgoods engineer." Others included John Schieble and John McMullen. Homer (Sonny) Reihm started with ILC in 1960 after graduating from the University of Delaware with a degree in electrical engineering. He initially worked on other ILC products, but during the first contract win in 1962, he worked on the thermal properties of the suit with the help of a subcontractor, Republic Aviation. He quickly moved up the ranks to become chief project engineer and then Apollo engineering manager. He was also well respected by the NASA customers. A former employee recalled that Reihm could distill all the technical details and provide a clear and concise report to the NASA customer in a very well-organized and well-thought-out presentation that won the respect of his audience. This was tricky business given the pressure to meet schedules and tackle the serious technical problems that emerged during the development and manufacturing processes and were continuous. Reihm had the ability to project confidence and professionalism at every turn.[17]

Quality assurance was always a top priority particularly as the program gained strength. As ILC was ramping up, they sought to hire seasoned professionals with experience. They found Richard McGahey, who was hired in January 1966 as reliability coordinator for the Apollo program. He rose to the position of quality engineering supervisor and then, in August 1969, he became the Apollo product assurance manager and directed quality assurance, inspection, and reliability activities. Before coming to ILC, Richard had worked as the reliability control manager for the David Clark Company on their Gemini suit program, so space suits were not something new to him. I'll always consider myself very fortunate to have worked for McGahey for many years during the space shuttle days. All of us in the Quality Department learned a great deal from him and will always remember his firm but fair handling of issues when they arose. He loved to read books, as was apparent once you got into conversation with him. He would always impress me with his wide range of knowledge and his good sense of humor.

Along with helping where they could, NASA managers made it clear to the ILC folks that it was now time for them to step up and fix many of the management and organizational problems that plagued them before and during the union with Hamilton Standard. It was encouraging that professionals like Nisson Finkelstein and others were on board, but much more needed to be done. It was not long before ILC started a conversation with aerospace giant LTV, located in Grande Prairie, TX. LTV stood for Ling-Temco-Vought, which manufactured aircraft and other aerospace items. It was ranked as one of the top forty industrial corporations by the late 1960s. As I understand it, Len Shepard had a contact at LTV that enabled him to inquire about a contract agreement to provide the training ILC needed in configuration management and systems engineering. As big as LTV was in the aerospace business, they had to know a thing or two about these disciplines. Not long after Len made that call, approximately fifty LTV contractors descended on ILC to help get things in order. Some of the LTV employees were loaned out to ILC for several months while one employee, Gary Raper, resided at ILC for several years.

As was the case at many of the companies that supported the growing space program, ILC needed new engineers to fill new roles. Colleges could not turn out graduates fast enough to meet the demands of the space industry. ILC hired Tom Pribanic, who learned from the LTV contractors and became the manager of configuration management and technical services, the group that would oversee all the paperwork associated with space suit production. And there was a lot of paperwork. Tom's department was responsible for controlling all documentation including patterns, drawings, specifications, and tables of operations, or TOs, as they were referred to. These tables provided the written details for people on the floor about how to build the suit. If anything changed for whatever reason, all paperwork associated with the change had to be updated or else a deficient, nonconforming product would be built.

ILC hired John McMullen in 1966. He served in various engineering roles during Apollo. He ultimately worked his way up to manager of systems engineering in 1969. John came from General Electric, where he had worked on various electronic projects for both the Mercury and Gemini programs. As was the case with many of the new engineers and others ILC hired, space suits were something entirely new to him. John recalls that when he was promoted to systems engineer, George Durney's first response was "What the (expletive) is a systems engineer?" George had little time or patience for this new breed of employee.

John soon became involved in and responsible for the cycle qualification testing of the Apollo suits in the ILC labs. This started out with John and some NASA personnel looking at films taken of suited individuals as they simulated expected lunar activities. They would literally count the numbers of flex cycles of all the suit joints, such as the elbows and knees, during these activities and arrive at an expected number of flexes per joint. They would then multiply these cycle counts to meet what would amount to a total of twenty missions' worth of cycles. Quite often, particularly in the beginning, suit failures meant that John had to go back to the engineers to have them resolve the issues.

McMullen enjoys telling the following story about his initiation as a new ILC employee and his early dealings with George Durney. As John was settling into the program, he was instructed to go to North American Aviation in Downey, California, the manufacturer of the command module capsule, to witness a test in which the three astronauts would perform an interface test between the latest version of the space suits and the command module couches. John and others were looking through the top hatch of the capsule as the suits were being pressurized. Although the shoulder width was borderline acceptable, the arm bearings were rubbing together between the three suits. This caused the crew member in the center couch to pop out of his seat like a slice of bread in a toaster because of the limited space between the men. The test clearly showed that changes needed to be made to the suits if three guys were going to be sitting side by side in such tight quarters. Not long after witnessing the problem, John heard an announcement over the loudspeaker at the North American facility that said, "Would John McMullen from ILC Industries please report to the NASA Office?" Immediately he knew that he was in for some trouble. The NASA representative who was there told John that he should plan on going back to ILC in Dover and telling his co-workers that there was more work to be done to get this problem fixed. John delivered the bad news in a meeting at ILC, and although it was obvious to everyone else in the room what had to be done, George slammed his fist on the table and told John that word had to be delivered to NASA and North American Aviation that they had to start work to make the capsule wider. Making a capsule even an inch wider would have increased the width of the Saturn rocket, which would have added more weight, which would have required more fuel, which would have added more weight, and so forth. George had a way of putting fear into anyone who questioned his word, so John was not sure how to answer him. Later, outside the conference room, Len Shepard pulled John aside and told him not to worry because this was a cost-plus program and ILC would be paid all costs for making the changes needed to the suits. The result was that the arm bearings were removed from the crew member's suits for Apollo missions 7–10. By the Apollo 11 mission, ILC had designed and manufactured a new low-profile arm bearing that was used on all future EVA suits, from Apollo 11 through Skylab. The command module pilot was in the center couch and did not have an arm bearing until the Apollo 15–17 missions.

Model Makers and Other Craftsmen

Several model makers were true craftspeople when it came to making the molds for rubber parts and fabricating structural brackets, pulleys, and zipper closure hardware for the suit. Because of Richard Ellis's size and dexterity, NASA chose him to demonstrate suit mobility on several occasions during the space suit development process. The original AX-5L suit was sized for astronaut Mike Collins. Richard was similar in build and stature and had the ability to bend and contort his body so that anything he did in the suits looked easy. This was important when demonstrating it to the folks at NASA.

The best craftsmen were assigned the task of making the tools needed for glove fabrication. Model makers Kenny Dennis, Harry Saxton, Julius Herrera, and Tommy Townsend were tasked with making the dip molds for rubber gloves based on the hand casting of each astronaut. This was a challenging job and they excelled at it.

Draftsmen were brought in from M&T Contracting, located in Philadelphia. The suit would have to be documented with many sketches and mechanical drawings of the hardware found throughout the suit. Included in this group of contractors was Sid Williams, a 19-year-old draftsman. The drafting department included many talented individuals, who all had a sense of humor that could drive any supervisor crazy at times. This was partially attributable to the many hours they had to work. Although everyone in the organization worked long hours, the draftsmen were all housed in one small area together; if not for the diversion of humor, they would have all gone nuts (many in the company thought they were already there). They were the best at making the technical drawings that accompanied the Apollo suit program. There were no CAD programs to make drawings. To this day, I find it interesting to look at the many detailed mechanical drawings and sketches this fine group of "artists" made and see the care and attention to details they put into them.

Ultimately, the Apollo program at ILC employed more than 900 people, but the increase was not immediate. NASA turned on the money gradually as the budget people there tried to get a handle on how the contract would be set up now that ILC was selected as the prime contractor for the space suit. Hamilton Standard would be the prime contractor for the primary life-support system and NASA would oversee the integration of the suits and the primary life-support system with the aid of contractor, Westinghouse. Many details had to be worked out between NASA and ILC such as the quantity of suits needed and the sizing schemes based on the selection of astronauts and test subjects the various contractors needed.

As the A-6L suit was being developed through 1966, ILC continued to hire staff that was very diverse. This growing team would soon come together and form a bond and forge memories that would last the rest of their lives. Fifty years after the height of Apollo, you can ask any ex-employee what the best job of their life was and they would all proudly say that it was during the few years working at ILC when they made the Apollo space suits. Especially in Delaware obituaries, you can occasionally find the names of past ILC employees. Although many went on to run their own businesses or retired as schoolteachers after working at ILC for only five years during the Apollo program, you can bet there will be mention of the ILC-Apollo connection in their obituaries. You'll find this to be true for many of the aerospace employees who supported Apollo across America. The sense of pride that these folks had is seldom topped by any other form of industry.*

And what of the fate of George Durney and Len Shepard, who laid the groundwork for the winning suit? Both men were obviously the brightest when it came to the initial space suit develop. George Durney was like an automotive designer who made a sleek-looking automobile from a chunk of raw clay. The challenge was to take it to the next step—the next generation would transform it into a fully functional vehicle that could be operated reliably for many miles in conditions that were as yet unknown and unheard of.

As the trained engineers settled in at ILC and contributed their vast talents, George would slowly fade into the background, but his skills and knowledge were always being tapped one way or another. I recall that even after he retired, George would come back into ILC and help the younger engineers as they worked to advance the space suit in the 1990s. Len used his brilliant management skills to oversee the Apollo program throughout its maturation process. Homer Reihm has little recollection of Len Shepard working directly on the Apollo suits at that point,[18] but Shepard had already left his mark and was becoming more involved in program management and in interacting with people at NASA.

II

BUILDING THE MOON SUITS

5

The Model A-5L Space Suit Contract

Work Hard, Play Hard

The early period after ILC won the contract as the prime supplier of the suit came with a lot of pressure from NASA, and it was not undeserved. There were ongoing problems with critical issues such as the paperwork not matching the suits and lack of discipline such as a banana being found in the clean room where the suits were being tested. NASA's Richard Johnson told Len Shepard that one of his memorandums written in 1966, "Reliability Engineering Status Report for Apollo Block II," was "poorly written in that it conveys little meaningful information to NASA MSC [manned spacecraft center]."[1] At one point, NASA sent a team from Houston to put Len Shepard in the hot seat, threatening to take the contract away from ILC if they did not shape up. Len came out of it OK, but the organization was under constant pressure to conform.

Various milestones during the Apollo years—such as delivering suits critical to the Apollo launch timelines or perhaps completing some rigorous testing schedules to certify new upgrades—were accompanied by celebrations that were known as "splashdown" parties. For months on end, employees worked fifty, sixty, or more hours a week, involved in such chores as measuring astronauts for a mission, making the patterns, cutting the parts, sewing the parts together, installing the hardware, and inspecting and testing the suits. A corps of support personnel was responsible for tracking the paperwork that accompanied each suit. Thus, when the initial suits were delivered, hopefully on time, or the missions ended with the safe splashdown of the crew in the Pacific Ocean, many people had ample reason to celebrate, release built-up stress, and let loose. Splashdown parties were not unique to the ILC workforce; all the aerospace workers in America who supported the Apollo program worked extreme hours. In today's more socially responsible business world, people might go to the local gym to relieve stress, but in the 1960s, neighborhood bars were the place, and bars across the country did well after a splashdown. Of course, these folks would start their work routines all over the next day in support of the next mission, picking up where they had left off. This routine led to a great deal of alcoholism, divorce, and other maladies that accompany high-stress work and long work hours, and the culture in ILC was no different than that at any of the other aerospace companies that were constantly under pressure to deliver.

In 1965, following the big contest win for the Apollo suit, Sambo's Tavern saw the first of the big celebrations. Sambo's was a simple restaurant that specialized in steaming-hot crabs and cold beer. It was perched along a creek in Leipsic, Delaware, just a few miles outside Dover. Leipsic had about 150 residents who lived along seven main streets in older, well-kept modest homes. In addition to Sambo's, the town had a volunteer fire department and a post office. Many Leipsic residents were watermen who made their living harvesting crabs in the Delaware Bay just a few miles downstream and docked their well-used crabbing boats behind Sambo's. Many ILC parties were held at Sambo's over the years, as there were many milestones to celebrate. John McMullen recalled using the company charge card to pay the bills. On at least one occasion, he had to run back to the bar when it opened at 11 the next day because he had forgotten to pay the rather hefty bill he had left the previous night. That was no problem for the proprietor because he knew the bill would eventually be paid if the ILC group wanted back in—and they certainly would.

The Model A-5L Suit: NASA Contract NAS9-5332

On July 31, 1965, NASA sent ILC a request for proposal for the A-5L pressure-garment assembly. Just a day before that, Richard Johnson of the Crew System Division at the Manned Space Center issued an internal memo titled "Justification for Noncompetitive Procurement" for two model AX-5L suits to support North American Aviation's final critical design review of the command module. The memo stated that NASA needed the suits by September 6 of that year.[2] This was extremely important if NASA wanted to stay on track with advancing the Apollo program.

ILC responded to the request on August 24th with a ten-page purchase description. It outlined the first two suits that ILC would build and size to astronauts Richard Gordon and Dave Scott. (Astronauts Joe Kerwin and

Michael Collins were added on later.) The proposed price was $89,981 for the first two suits ($711,000 in 2019 dollars), or roughly $45,000 per suit. That included all materials and labor. ILC assumed that many hardware items such as the Gemini pressure-sealing zippers and gas connectors would be provided to ILC as government property.[3] On October 9, NASA issued the contract to ILC as the prime provider of the Block II Apollo suits. This would be known as contract number NASA 9-5332. NASA proclaimed that the A-5L suit would be advanced to become the A-6L suit, which would be used on the Block II missions that would land the first men on the lunar surface. The David Clark Company would continue to provide the Block I suit.[4] This would give ILC and NASA time to work out the remaining technical issues with the A-5L suit. NASA was comfortable with using the David Clark suit for intravehicular use. Early comments from the astronaut corps was that the ILC suits were great when pressurized but when they wore them unpressurized as they laid down in the command module seats, they felt all the restraint cables, miscellaneous hardware, and zippers under them and it was very uncomfortable. In contrast, the David Clark suits were very comfortable in the unpressurized mode, the mode for much of the early part of the mission prior to and throughout launch, when astronauts were pressed into their seats.

Contract number NAS 9-5332 called for ten model A-5L suits at a total cost of $1,950,149 ($15.4 million in 2019 dollars). This included provisions to pay ILC for any new technology that it developed as part of the process, since NASA wanted to claim ownership to all new development. A NASA memo dated September 1967 stated that NASA was withholding a total of $54,965 ($412,000 in 2019 dollars) from ILC that was owed at the time because it believed that ILC had withheld some of the technology it had developed during that earlier NAS 9-5332 contract. The only items ILC was reporting were the shoulder and arm assembly. Later, after ILC agreed that the hip and thigh joint should be included, NASA paid the balance on the contract.[5]

Like the AX-5L suits, the A-5L suit had the rear-entry zipper-closure assembly that made donning and doffing the suit easier and freed up the front chest area. The pressure-sealing zipper for the A-5L was like the one the David Clark Company used on the G-4C Gemini suits and had reinforcing patches added to each end. The suit also contained an inner nylon liner to make it easier to put on. Because the lining was held in place with Velcro tabs, it could be removed for cleaning. After hearing feedback from astronauts, ILC engineers added a lacing system to the arms and legs that was similar to how shoelaces function. This added sizing adjustment. The helmet disconnect ring was angled slightly forward to provide what NASA called the "correct eye-to-heart angle" that was required during launch acceleration. Having learned after the June contest where many of the wear points were in the suit, ILC added abrasion patches to the convolutes and the pressure bladder where they were needed.[6]

ILC delivered two A-5L suits, serial numbers 002 and 003, to the U.S. Army Natick Laboratory outside Boston, Massachusetts, in November 1965. Natick Labs is a government facility that is well known for its ability to support the army and other government agencies by testing new materials and systems. It was too early in the program for ILC to have any kind of handle on how to test the suits to the level that NASA expected, and even NASA itself was not prepared yet with suit-testing facilities. ILC eventually gained strength in this area because system-level testing was extremely important in the development process of the Apollo suit.

NASA coordinated with Natick to see that the initial A-5L suits could take the abuse that would come with basic wear. A NASA status report dated January 6 to 13, 1966, indicated that ILC had to make repairs and design updates to correct problems uncovered during endurance testing. The restraint cables were routed through Teflon cable guide tunnels, and according to the status report, these guides were failing when the suits were cycled under pressure.[7] This should not have been a surprise, since this immature design had seen relatively little endurance testing up to this period. A May 5, 1966, status report indicated that Natick Labs had passed the A-5L suit and neck dam in water floatation testing.[8]

As ILC started delivering more of the A-5L suits to NASA, it began to receive a lot of feedback about problems with the suits. The quality suffered in many areas. An unidentified employee, likely at NASA, made a list that detailed the problems with several A5L suits. They ranged from improperly installed neck rings to loose screws on connectors to excess cement on glove disconnects. It appeared that there was more wrong with the suits than right. The ILC quality and reliability team had their work cut out for them in the months and years ahead as they transitioned from building one development suit at a time to producing multiple suits on tight schedules while meeting rigorous quality standards. Suit A5L-007 was made for Gus Grissom to evaluate, and handwritten ILC notes from that time show the many problems with the suit suffered after pressurized evaluations. Many were general workmanship problems that indicated that ILC was overwhelmed with the workload and understaffed at that time. Production

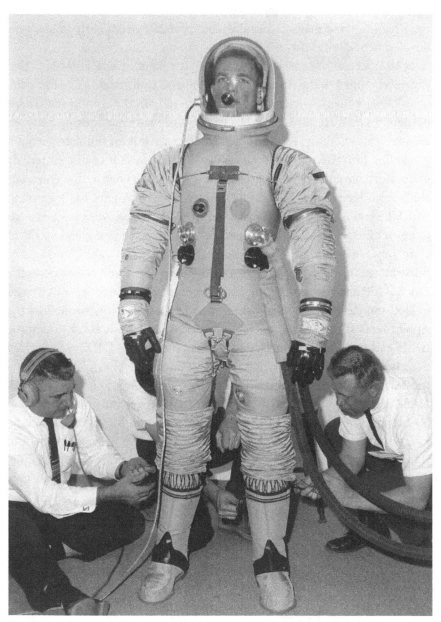

Figure 5.1. The model A-5L suit with astronaut Dave Scott inside. George Durney (*left*) and softgoods engineer Bob Wise (*right*) are shown adjusting the cords (like lacing a shoe) that lengthened or shortened the legs for a better fit. Courtesy ILC Dover LP 2020.

staff and the company's quality inspectors were relatively new to the process and although many of them went on to excel in their assignments, they were learning and making mistakes that typically occur early in any new development effort.

NASA's Crew Office assigned Apollo 11 astronaut Mike Collins to follow the development of the Apollo space suit at ILC Industries. As he has written, "Developments were not all rosy, especially in the Apollo suit development. In the first place there is a kind of love hate relationship between an astronaut and his pressure suit. Love because it is an intimate garment protecting him twenty-four hours a day, hate because it can be extremely uncomfortable and cumbersome. Generally, as time goes by, the emphasis shifts from hate to love, so that by flight day the astronaut has a garment which he regards as an old friend. One which has been worn long enough to be comfortable, but not long enough to suffer undue wear and tear."[9]

In July 1965, personnel were shuffled within NASA ranks, and one of those moves included the appointment of James McBarron II to oversee the Block II suit program as lead suit engineer at the Manned Spacecraft Center in Houston. All of the ILC staff who worked with McBarron throughout the Apollo years spoke of how fair he was as he helped guide ILC through the twists and turns of the contract as it constantly evolved. ILC people tried their best to stay on top of things, but it was a constant struggle and McBarron was always there to help point ILC in the right direction when needed. NASA needed their suits, so there was mutual respect and great interest on NASA's part in seeing ILC succeed. The pressure on McBarron had to be intense at times, however, since he was stuck between the management of ILC and his managers at NASA, who were following a very tight schedule.

A NASA trip report generated on August 31, 1965, details observations about a visit to ILC. (This would have been one of the first visits NASA management made to ILC after it won the suit contest.) In the report, it was clear that the visit was intended to address critical areas that NASA had concerns about. "Generally, ILC management was cooperative and appeared to be open to suggested changes and constructive criticisms. An inspection of the facilities indicates that a high degree of effort must be exerted early in the program to build up capabilities for the production phase. In addition, work and test areas were observed to be in poor condition from lighting, ventilation and cleanliness standpoints." It went on to address the organization and personnel buildup that would be required to get ILC up to speed and concluded by saying that extensive efforts would be needed to meet the requirements of the Apollo Block II program.[10]

The government inspectors who worked out of a regional office in Phila-delphia but resided at ILC to oversee the contract expressed understand-able concern to NASA. These government personnel were present every day to inspect and stamp off on the critical paperwork involving everything from inspecting sewn seams to the shipment of finished suits. They raised the issue to NASA that ILC was not adequately staffed to handle the con-tract. In response, NASA's Charles Lutz, chief of the Apollo Support Office, outlined how his office had taken steps to relax the ILC paperwork and configuration management efforts of the PERT (Program Evaluation Re-view Technique) reporting program, which was time-consuming yet popu-lar for laying out the flow of work in the 1960s.[11] NASA had also agreed to loan ILC government-furnished computer programs for the Apollo Re-quirements Document Reports. However, NASA could not find a computer program that was compatible with the system ILC had, so they gave ILC the option of going to an outside facility to run their weekly reports to NASA.[12] Ultimately, ILC Industries relied on using the Playtex Company computers to solve this problem. This was all in conjunction with ILC hir-ing Ling-Temco-Vaught (LTV) staff as temporary contractors to help set up the configuration management and systems engineering departments until ILC could learn from them and take it over.[13] Since ILC was going to be responsible for all the major suit components minus the backpack, the company needed to find subcontractors for equipment such as the helmets. Along with all the engineering management training LTV provided to ILC, LTV also ended up providing the extravehicular visor assembly and lunar extravehicular visor assembly that would cover the bubble helmet during extravehicular activity. This was a deal that ILC worked out with LTV as the two organizations grew closer.

In late February 1966, a NASA status report indicated that delivery of the first ten A-5L suits was approximately two months behind schedule, but ap-parently the delay did not impact NASA at that time. There was also a note that the first model A-6L suit for qualification testing would be delivered sometime around late June 1966.[14] ILC had already done a lot of engineer-ing work by spring 1966 on the A-6L suit development that began with a model A-5L suit that ILC had made for in-house evaluations (sized for ILC employee Richard Ellis) that was modified with the features NASA needed for the A-6L suit. That was the way it was typically done as ILC evaluated new upgrades. ILC made another suit later in the program for employee Tom Sylvester, who had his own personal, custom-made suit complete with his name stitched on the thermal micrometeoroid garment (TMG). That suit was used exclusively for evaluation purposes.[15] These suits were modified repeatedly to evaluate the upgrades in the ILC Laboratory.

The Smithsonian Air and Space Museum has many A-5L suits in its collection, up to serial number 019. These early A-5L suits played a critical role in getting NASA the first suits it needed to train astronauts and provide the understanding that North American Aviation and Grumman Aircraft needed about how the space suits interfaced with their command modules and lunar modules.

Throughout the Apollo contract, it was not uncommon for ILC to make a suit or two that was used to support some in-house mobility testing and was never assigned a serial number. Instead, the suit labels contained an entry such as "DMU-1" (design mockup unit). These suits are difficult to find in the contract information and are otherwise lost to history.

6

The Model A-6L Space Suit

Unveiling the First Moon Suit

NASA Contract NAS9-6100

ILC reengineered the contest-winning A-5L suit to fix several issues that were uncovered during its relatively short life-span and limited endurance testing. It morphed into the model A-6L suit, which included an outer cover layer of fire-resistant Nomex that functioned as the thermal micrometeoroid garment (TMG). This cover layer protected the astronauts during their lunar extravehicular activities. ILC also designed and added rubber convolutes to the ankles. The "Apollo Experience Report" by four high-level NASA engineers suggested that this ankle joint was changed to make it easier to put the suit on and take it off, and although perhaps that is true, my belief is that this joint was even more necessary to improve the ability of astronauts to walk on the lunar surface. I have several photos and videos of subjects walking on treadmills in the ILC Laboratory in efforts to understand how to improve the comfort and mobility of the suits. Without the ankle convolute, the natural flex of the ankle would have been compromised. The pressure-sealing zipper was also lengthened to make it easier to don and doff the suit. The locking neck ring and suit-side glove disconnects were improved to assure better locking capability.[1]

The model A-6L suit fell under a new NASA contract that remained in force during several contract change authorizations throughout the length of the program. Contract NAS9-6100 was carried out based on incentive awards that were tied to schedule, cost, and technical performance. This all sounded good, but as ILC and NASA would find out over time, the making of space suits did not lend itself to meeting schedules and costs in a world that was constantly changing. NASA later reviewed this system and made adjustments that were fairer and more equitable for ILC.

NASA's demand for the improved A-6L pressure suit was great because it needed it to support various training and interface testing. This model suit was the first to be qualified for the lunar missions and NASA needed to start testing as soon as possible. Testing was to be conducted using suit serial number A-6L-001. One NASA memo addressed the slippage of the qualification testing date because delivery of the extravehicular visor assembly from LTV Corporation was estimated to slip to November 11, 1966. NASA responded to ILC and made it clear that that date was not acceptable and that it required delivery by September 15, 1966, so that qualification testing could be carried out. It is unclear if that schedule was met, but these details highlight the dynamic nature of the contract and the pressure everyone was under to meet the deadlines.[2]

Continuous Improvement: Building the A-5L Suit

Both the A-5L and A-6L suits were vast improvements over the initial air force versions and the early space suits ILC made. These suits were much more tailored. This translated into improved comfort, mobility, and fit inside the command and lunar modules. The zipper placement made a big difference, as did the new bubble-style pressure helmet, which resulted in significant improvement of the neck joint. Other obvious changes included the metal restraint cables across the back and around the shoulders. These cables reduced the shoulder width and provided the means for omnidirectional flexing. They also restrained the loads imposed on the suit due to the internal pressures and loads caused by using tools or climbing up or down the ladder on the lunar module. ILC engineers and NASA staff continuously scrutinized the steel restraint cables and guide system. The cables were frequently mentioned in documentation that captured the feedback of the astronauts about the comfort of the suits. The cables were repeatedly listed as a constant annoyance because they dug into the shoulders of the wearers. This discomfort tended to go away once the suits were pressurized, but for a good portion of the mission, they would be unpressurized while the astronauts were lying in the couches of the command module for launch operations. This feedback drove NASA to conduct a study in 1966 to evaluate the pros and cons of using two separate suits on each mission; the David Clark model G-5C for intravehicular use and the ILC model A-6L for extravehicular use. An undated and unsigned NASA report noted that evaluations clearly indicated that the ILC suit was much more mobile than the David Clark Company suit when pressurized but that the David Clark Company suit was more comfortable at vent pressure.[3] In the end, NASA decided to go with the ILC suit and concentrate on reducing these pressure points. The factors of cost and storage space likely quashed any thoughts about procuring two separate suits for Apollo.

The Design Mock-Up Unit Suits

The first tests of the new A-6L model suit used what was called DMU (design mockup unit) Mark 1 and DMU Mark-2. These two suits were built to establish a baseline for the A-6L configuration. They also made possible a comparative evaluation of torso assemblies fabricated during the performance of the following development tasks: hip, thigh, and waist mobility and comfort improvement and optimization of the AX-6L helmet-to-head position.

Manned testing started on October 21, 1966. ILC test subject Tom Sylvester was used to evaluate the different designs.

Two years later, notes in the September 1968 technical paper that accompanied ILC's demonstration to NASA of what would become the model A-7LB suit show that a sustained engineering task had begun in 1966 to evaluate an improved version of the frontal entry method. ILC engineers George Durney and Mel Case were attempting at that time to eliminate the zipper from the waist area, as they later would in the model A-7LB suit. The technical paper offered comments on two suits that had been built with successively improved frontal entry hardware and waist mobility. Suit DMU Mark-1 was one of them. This 1966 evaluation concluded that the frontal entry concept was unreliable and difficult to close. It also commented on the poor neck mobility of the suit.[4]

A set of photographs show the DMU Mark-1 being testing throughout its range of motion while pressurized with Tom Sylvester inside. The DMU Mark-1 was likely a standard model A-6L suit modified to contain a front zipper closure, while the DMU Mark-2 suit was built from scratch with the rear torso closure.

A closeup photo shows the front entry system. In Figure 6.2, it's possible to see the zipper-pull tab that terminates just below the front of the neck ring. The fact that the neck ring is a compression band makes it clear that this suit was designed to open by releasing this band clamp around the neck. Once released, the hardware is removed and the zipper, which

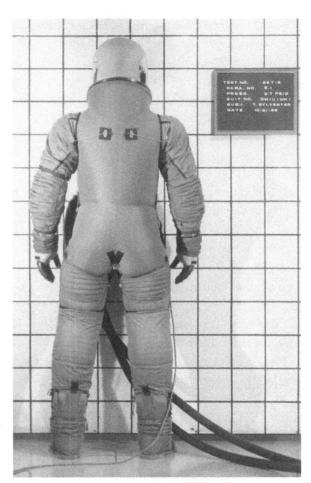

Figure 6.1. This design mockup unit (DMU) Mark-1 for the model A-6L suit shows the absence of a zipper closure in the rear torso. This was one of George Durney and Mel Case's first attempts to eliminate the zipper in the waist area to provide improved waist mobility. Courtesy ILC Dover LP 2020.

Figure 6.2. Front view of the early model A-6L DMU Mark-1 suit shows the zipper pull tab just below the neck ring and a unique clamping-style helmet hardware connection. The zipper is hidden behind the torso pull-down webbing. ILC employee Tom Sylvester is in the suit. Courtesy ILC Dover LP 2020.

Figure 6.3. Photo taken two weeks after the photo was taking of the DMU Mark-1 suit, which used a front zipper closure. This is model A-6L suit serial number 006. It shows the tapered and flared neck section that was used only up to the model A-6L suit and included the rear-entry zipper that would be used in the A-6L and the A-7L suits. ILC test subject Tom Sylvester is in the suit. Courtesy ILC Dover LP 2020.

terminates at the top of the torso under the compression band, can be pulled down, fully opening the front of the suit to make it possible for the astronaut to put on or take off the suit through the neck opening. Extra restraint-material fabric in the neck area made it possible to open the neck wide enough to allow entry from the neck opening. The zipper closure is hidden behind the pull-down webbing on the torso.

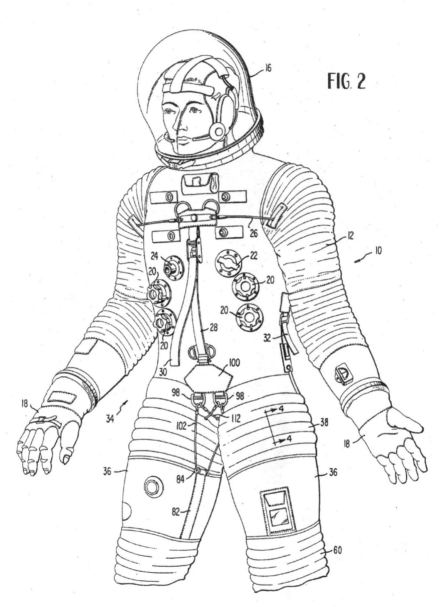

FIG. 2

Figure 6.4. Sketch included in George Durney's patent for the thigh-restraint cable system. Note how the cables from the thighs are all connected to D rings attached to the large patch located over the crotch. This patch also carried the loads of the torso strap from above. Source: U.S. Patent Office.

Improving Leg Mobility

One of the most challenging issues related to suit design was how to provide the best mobility for the legs while still providing the restraint needed to hold the loads the suit would encounter. The main purpose of the steel cables located throughout the suit was to carry the primary loads that were spread throughout the suit due to internal oxygen pressure. In addition, the person inside the suit was creating a lot of physical loads as they kicked their legs fore and aft or as they bent down and in the process spread their legs apart at the crotch, for example. It was this crotch area that took a lot of punishment and could hinder the ability of the astronaut to perform basic walking steps.

George Durney focused a lot of his attention on solving this problem. He did so by developing a unique crotch cable system that he eventually patented.[5] The early design had nylon-coated stainless steel cables that passed through Teflon guides on the thighs, but endurance testing showed that the cable could cut through the Teflon guides. The fix for that problem was to replace the Teflon guides with pulleys.

Additional Changes

Other prominent issues surfaced during the changeover from the A-5L to A-6L suit:

Nylon coating on the shoulder restraint cable was abraded when it rubbed against the sharp edge of the guide tube for the arm cable. This was remedied by adding flare to the end of the guide tubes.

The vent system ducts collapsed (kinked), shutting off the gas flow. The fix was to add nylon coil springs inside the vent ducts that prevented further kinking. This change resulted in Patent US3,717,530 titled "Method for Forming Crush Resistant Conduit," which Mel Case and George Durney filed in 1970.

The bladder sustained several leaks because of abrasion in high-wear areas. This was remedied by adding reinforcement patches in those areas.

The Gemini-style zippers showed excessive leakage. The fix was to switch over to a B. F. Goodrich pressure-sealing zipper, which had tighter tolerances.

Improvements were made to the hardware, including using retract-able latches that were safer and more reliable on the neck ring. Other hardware was updated to provide better ease of operations.

The A-6L Thermal Cover Layer

Throughout the design and development process, ILC worked closely with and depended on NASA engineers and managers to make sure they were on the right track when they selected and worked with the materials in the space suits. NASA eventually had—and still does have—test laboratories that are ahead of all others for testing materials in harsh, space-like environments. During Apollo, they concentrated on building their laboratory resources to simulate the lunar environment and the other space-like conditions astronauts would experience. This is not a cost that private industry would want to bear.

The TMG is exposed to the greatest abuse suits encounter in outer space and on the moon, so these exterior suit materials underwent a lot of testing at the NASA labs. The TMG is typically the last part of the space suit to be designed since the mission requirements dictate the layup of the various thermal and protective layers. I would like to say that it is the easiest part of the suit to design and manufacture, but I'd be lying. The outermost layer of the model A-6L TMG was made of Nomex because it was the best off-the-shelf material for protecting against any high temperatures that would be expected—or so the thinking was at the time. The suits that the David Clark Company were supplying for the Gemini missions had Nomex and it was working just fine. Below the Nomex were seven layers of very thin perforated Mylar, each layer separated by a layer of polyester nonwoven scrim and two layers of ripstop nylon coated with eight ounces of neoprene to act as a micrometeoroid-stopping layer.

Since it was reasoned at the time that the layers of the TMG were only needed during extravehicular activity, there was no reason to add bulk to the suits astronauts wore as they sat in the rather cramped seats during all the phases of the mission when suits were required. In addition, crew members needed all the side-by-side arm room they could get. A NASA memo dated January 17, 1967, just ten days before the fatal Apollo 1 fire, suggested that ILC consider using a single Nomex cover layer on the Apollo suits to prevent snagging and tears on any sharp objects or edges inside the command module during intravehicular wear.[6] This was probably the last

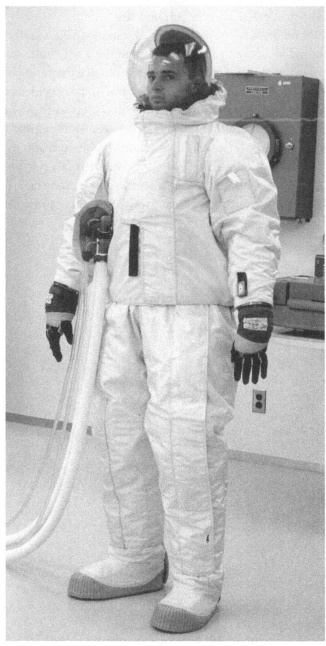

Figure 6.5. The model A6L slip-on jacket, pants, and lunar boots, all
of which were covered with Nomex. Courtesy ILC Dover LP 2020.

idea that was pitched concerning the cover layers before a more radical approach was investigated after the Apollo 1 fire.

To reduce the bulk of the suit, the A-6L model TMG was designed as a slip-on, slip-off jacket and pants arrangement that astronauts would put on only before extravehicular activity and remove once they were back inside the space craft. Otherwise they would be stowed away. Videos of tests in the ILC laboratory that demonstrated how the TMGs were put on over the pressurized suits are comical to watch as the wearer attempted to slip the pants over their lower torso. Had this scheme come to fruition, we might have been watching Neil and Buzz tripping over themselves in the cramped quarters of the lunar module as they attempted to put on their TMG coats and pants before their extravehicular activity in their pressurized space suits.

7

The Model A-7L Space Suit, 1967–1971

Adversity has the effect of eliciting talents, which in prosperous circumstances would have lain dormant.

ROMAN POET HORACE

Apollo 1: Learning from Mistakes

After making several trips to North American Aviation in California to check out the capsule as it was being manufactured, Gus Grissom and the Apollo 1 crew knew that things were being hurried along too fast and the quality was suffering. To make his point to NASA officials, Grissom picked a lemon from his Houston backyard before leaving for the Kennedy Space Center on January 22nd and hung it over the capsule simulator after he arrived. Sadly, that point was made all too clear by the end of the day. During the evening hours, as the Apollo 1 crew was working through many technical problems that kept popping up, an intense fire erupted on the spacecraft that was fueled by the 100 percent oxygen it was pressurized with, killing the crew within minutes.

On that day, all three astronauts were wearing the David Clark Company Gemini suits since that was the plan for the missions before the moon landings.

That event had a significant influence on the development of the Apollo space suit program. Various committees made up of the best NASA and industry experts available reviewed the entire system the astronauts interfaced with. A primary focus was the issue of the 100 percent pure oxygen environment inside the capsule. To make matters worse, the oxygen was pressurized to 16 pounds per square inch (psi) while on the launchpad, thus adding more oxygen molecules and raising the combustibility of anything even slightly flammable. The initial response was to reduce the amount of flammable materials on board the spacecraft. After working diligently to meet this reduction, NASA ended up failing in thirteen of twenty-eight simulated fires intentionally set on board a command module test mockup. Following these tests and further research, NASA finally announced in March 1968 that the command module would be pressurized with 60 percent oxygen and 40 percent nitrogen, which would greatly reduce the chances of a spreading fire in the command module while it was on the launchpad. The capsule was redesigned so that as the Saturn rocket launched and soared into space, the 60-40 mix of gas was gradually vented and replaced by 100 percent pure oxygen, but at only 5.6 pounds per square inch absolute pressure.

In addition to eliminating the 100 percent pure oxygen environment, NASA engineers began looking at obvious things that needed to be changed, including reengineering the capsule door so it opened outward so the crew had a chance of escaping the pressurized craft on the launchpad if another fire erupted. Trials of the outward-opening hatch indicated that when the crew was strapped onto the couches, it was possible for all three of them to egress from the capsule within fifty seconds if such an incident were to happen again. Fortunately, it never did.

While much of the investigation and testing was focused on the command module, the suit program was put on hold with a stop-work order from late January 1967 through March 1967. Once that order was lifted, all the focus was placed on the redesign of the suit based on the findings and suggestions of the review boards and NASA engineers. ILC worked with NASA to investigate new materials for a fireproof space suit. During this time, NASA decided that the idea of separate Block I and Block II suit programs was basically dead. There was no point to redesigning two separate suit systems. The Gemini program had ended and Apollo was still considered as being in the development stages. NASA would now focus on wrapping up the contract with the David Clark Company and beginning Apollo with the ILC Industries suits. This was no doubt a blow to the David Clark Company, since it had played a significant role in the early Apollo program. On March 11, 1966, John Flagg, president of the David Clark Company, proposed that his company take over all the NASA efforts to oversee the ILC Space suit contract and be the integrator of the suit related systems. You couldn't blame him for asking. NASA's chief of the Apollo Support Office, Charles Lutz, quickly shot down that idea in a very brief response.[1]

During the Gemini program and before the Apollo 1 fire, NASA had begun looking into a new fabric Owens Corning, a company well known for its fiberglass-related products, was making for the commercial market. This new fabric was called Beta cloth, and Owens Corning was touting it as a new fireproof material for use in bedspreads, draperies, and so forth in places such as hotels where the chance of fire could be high. Because the glass fibers are a product of the oxidation process, they would not burn and could take temperatures up to 945°F before the strength of the fiber would even begin to yield. It was listed as having a "service temperature range" of 1,250°F before off-gassing, meaning it would start to change properties. This was considered adequate to get an astronaut out of a burning spacecraft within 45 to 50 seconds. You might think that a glass fiber would break when it was flexed but not in this case. The individual Beta fibers were only 3.8 microns in diameter, which translates to a diameter of 0.00015 inches. A human hair by contrast might have a diameter in the range of 0.002 inches. This very thin cross-section gave Beta fibers the ability to bend easily without breaking, as a thicker piece of glass fiber would.[2]

As NASA worked on eliminating the risks associated with the 100 percent oxygen pressurization on board the capsule, they also looked at replacing more than 150 different materials throughout the spacecraft with this new Beta cloth. NASA began reaching out to the various contractors, including ILC, to get them on board with switching over to this new material. However, the suit would be a challenge because of all the precision sewing that had to be done with it and there was little history of working with this material.[3] Unlike many of the materials used aboard the spacecraft, the space suit materials would get a lot of flexing and abrasion that had to be evaluated if the Beta cloth was to be used.

Fireproofing Space Suits

New Beta-Cloth Thermal Micrometeoroid Garments

The transition to a new thermal cover layer began early in March 1967, when Jim O'Kane and William Gill, both from NASA, came to ILC Industries. O'Kane and Gill were accompanied by a man named Dillon who was a representative from Owens Corning. The purpose of the meeting was to advise ILC that NASA was prepared to offer all the help it could with sewing the new Beta cloth fabric to form a new protective cover layer for the suit. Dillon returned to ILC on the week of March 20, 1967, with many yards of Beta cloth to train the ILC seamstresses.[4]

Initial results indicated that it would prove tricky to sew the Beta cloth because the glass fibers were somewhat fragile and difficult to work with. To weave the fiberglass strands into this cloth, Owens Corning had to use light silicone oil that made it feel sleazy to the touch. The oil also meant that the seams would pull out after they were sewn.[5] At first, Owens Corning coated the surfaces of the Beta cloth with Teflon, since the Beta fibers did not hold up well to abrasion under dynamic conditions. However, coating the surfaces caused the material to have a significant decrease in tear strength, since the Teflon coating locked the fibers together and made the fabric easier to tear. The Apollo 7 mission flew with this Teflon-coated cloth, and the results proved that the material did not hold up well during a mission. Ultimately, that was resolved by applying the Teflon to the individual yarns then weaving it into the cloth. That fixed the tear-strength problem. This new fabric became known as Super Beta cloth and was used on all the subsequent missions. To stop the fibers from pulling out of cut edges, a cement made by blending together 20 grams of a material known as KEL-F-800 and 100 milliliters of methyl-ethyl-ketone (MEK) was brushed along the edges to bond the fibers in place.

The challenges of making and integrating this new cover layer were numerous. The underlying Kapton film was very brittle and even more fragile than the mylar films that were also used as one of the thermal layers. Another thermal layer known as Beta marquisette could not even stand up to having lines drawn on it when tracing around patterns so they could be cut. This was resolved by using a chalk bag to dab blue chalk along the edge of the pattern that was laid out over the marquisette. The operator would then cut along the blue chalk line after the pattern was lifted. Initially, ILC was supplied with Beta threads to join the Beta cloth for the thermal micrometeoroid garment (TMG) together, but the tensile strength of this fiber was too low. To resolve this issue, ILC seamstresses used a quarter-inch double seam using Nomex thread for strength along with a third seam of Beta thread. To inspect it so it could be verified that both threads were used, quality inspectors placed the TMG under an ultraviolet light. The Nomex thread would glow whereas the Beta thread would not.

The final step of interfacing the finished TMG to the pressure garment presented its own set of challenges. Among them was making sure all the pass-through holes for the gas and water connecters would line up properly to the torso of the pressure garment. ILC overcame this challenge by making an adjustable jig that could match the different torso sizes and the variable spacing between the connector hardware. The holes in the TMG were slipped over the jig set to match the torso size being manufactured and the

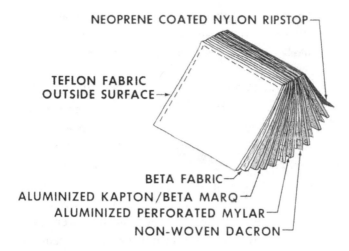

NEOPRENE COATED NYLON RIPSTOP

TEFLON FABRIC
OUTSIDE SURFACE

BETA FABRIC
ALUMINIZED KAPTON/BETA MARQ
ALUMINIZED PERFORATED MYLAR
NON-WOVEN DACRON

Figure 7.1. This diagram shows the various materials that made up the layers of the thermal micrometeoroid garment. Courtesy of NASA.

layers were bonded at the access holes before closing the edges of the Beta cloth.[6]

By late summer of 1967, work to design and fabricate the new Beta-cloth cover layers was in high gear. The Dover plant was running out of production space and ILC needed more room. The pressure-garment assembly was arguably the more complex part of the suit design and fabrication, while the new Beta-cloth TMG had its own set of challenges that needed attention. Thus, it was correctly reasoned that both could be manufactured at separate locations and brought together for integration when both were completed. In September 1967, Homer Reihm and Len Shepard from ILC, NASA's Crew Systems chief Dick Johnson, and NASA's lead suit engineer Jim McBarron visited a vacant building that was for lease in the town of Frederica, Delaware, about 15 miles south of Dover. It had previously been used to manufacture thread. ILC leased the facility and by October 1967, production equipment was being moved into the Frederica building, which ultimately became the plant where the Beta-cloth TMGs were fabricated for all of the Apollo, Skylab, and Apollo-Soyuz Test Program missions. At the final stage of assembly, the TMG was integrated with the pressure-garment assembly at the Dover plant. The fit checks with the astronauts were also conducted in the Dover plant in a special room set up for that procedure.

The first Beta cloth thermal micrometeorite garments for the Apollo A-7L suit were made by the end of November 1967. This is confirmed by a photo of the Apollo 7 crew taken at North American Aviation on or around December 8, 1968, in Apollo space suits with TMGs. They were missing the boot TMGs; likely the design and production process for these items was still under way.

An internal memorandum was released to the folks at the Frederica facility in early January congratulating them on the rapid production of the first TMGs.[7] Completing all the work necessary to create this new, almost exotic, outer cover layer of Beta cloth marked a major milestone. Each TMG consisted of approximately 1,600 pieces of material assembled to form the outer cover layers.[8] Once the process of producing TMGs leveled out, the Frederica facility could turn out one set every twenty days.

Initial testing of the insulation showed that there were heat leaks; the insulation was not providing the thermal insulation needed when solar loading was introduced. The solution was a new TMG that was made with increasingly larger-sized cut parts within each layer from the inside out so that the outermost layer was larger than the innermost one. This increased

Figure 7.2. Apollo 7 crewmembers Wally Schirra, Walt Cunningham, and Don Eisele pose before entering the command module at North American Aviation for training on or around December 8, 1967. Courtesy Tom Shustack.

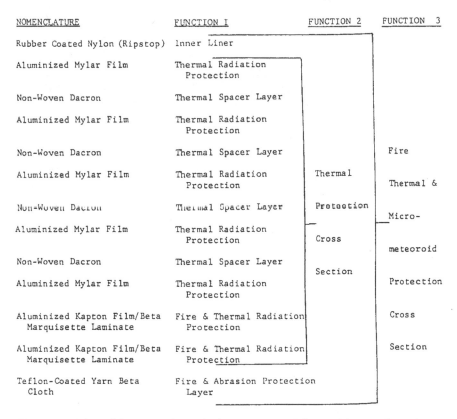

NOMENCLATURE	FUNCTION 1	FUNCTION 2	FUNCTION 3
Rubber Coated Nylon (Ripstop)	Inner Liner		
Aluminized Mylar Film	Thermal Radiation Protection		
Non-Woven Dacron	Thermal Spacer Layer		
Aluminized Mylar Film	Thermal Radiation Protection		
Non-Woven Dacron	Thermal Spacer Layer		Fire
Aluminized Mylar Film	Thermal Radiation Protection	Thermal	Thermal &
Non-Woven Dacron	Thermal Spacer Layer	Protection	Micro-
Aluminized Mylar Film	Thermal Radiation Protection	Cross	meteoroid
Non-Woven Dacron	Thermal Spacer Layer	Section	
Aluminized Mylar Film	Thermal Radiation Protection		Protection
Aluminized Kapton Film/Beta Marquisette Laminate	Fire & Thermal Radiation Protection		Cross
Aluminized Kapton Film/Beta Marquisette Laminate	Fire & Thermal Radiation Protection		Section
Teflon-Coated Yarn Beta Cloth	Fire & Abrasion Protection Layer		

Figure 7.3. A list of the materials in the lunar version of the model A-7L thermal micrometeoroid garment from the inside outward (i.e., from top to bottom) that includes the function they served. Source: "Familiarization & Operations Manual," Change 5, ILC document no. 8812700149B, October 1969.

the bulk but eliminated the compression or squeezing together of the layers. The downside was that it added a bit more bulk, although that was not reported anywhere as an issue.

One of the earliest versions of the extravehicular model A-7L TMG is shown in Figure 7.4. This model was flown on Apollo 9. However, that mission did not use the connector cover; in fact, that cover was never flown on a mission. This TMG was identified as part number A7L-200000. The design had a few features that would disappear a short time later. The Model A-6L and the early model A-7L suits had the ability to disconnect the shoulder restraint cables that came across the upper arms and terminated at an attachment point on a piece of hardware located on the upper chest. The TMG provided a shoulder disconnect access cover for this hardware.

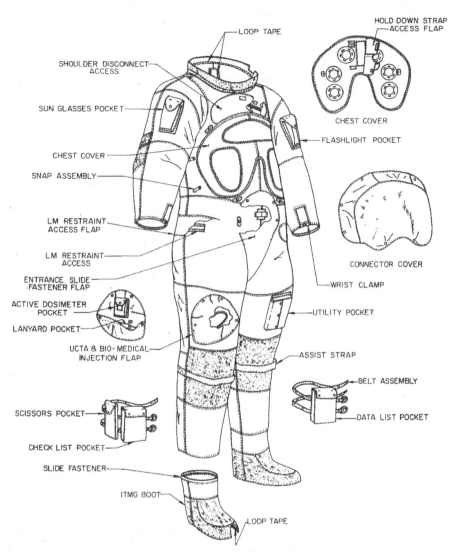

Figure 7.4. Details of the early model A-7L thermal micrometeoroid garment cover for extravehicular use. The connector cover and shoulder-disconnect access flap were eventually eliminated.

Once accessed, the hardware could be pulled forward and rotated, which enabled the astronaut to unhook the cables from pins beneath. When the cables were disconnected, there was no longer any restriction to shoulder mobility, which was thought to be a positive feature in certain cases. However, before the first missions, this feature was eliminated because the cables provided necessary restraint against the pressure loads inside the suit. This was critical for the safety of the wearer. Serious issues could result if the cables were not connected when the suit was pressurized.

This early TMG also had a laced-on chest cover that would permit the attachment of a connector cover to shield the hardware connectors from the sun. However, engineers found that these covers were unnecessary; testing determined that the heat loads were acceptable without them. They were difficult to work with and got in the way when mating and de-mating the connectors.

The next generation of TMG for the model A-7L suit was identified as part number A7L-201100. That included Velcro strips and snaps for the connector around the perimeter of the TMG. However, as previously mentioned, the cover would not be used on any missions. The Velcro and snaps for the connector cover were installed on the TMG through Apollo 12. They were deleted on later suits since there was never a need for the cover.

Thermal Micrometeoroid Garment Pockets

ILC engineer Bob Wise told the story of how he had a lot of difficulty attempting to make the astronauts happy with the placement and numbers of pockets on the TMG. Astronauts liked their pockets. There was a pocket for pens, one for sunglasses, one for checklists, and even one for rocks. For one of the scheduled reviews with NASA, Bob and the development team made up a TMG that had so many pockets on it there was little room to squeeze another one in. His objective for the meeting was to start with more pockets then was needed and then have the NASA evaluators tell him which ones to remove. It just seemed easier that way based on the difficulty he previously had with trying to get them to commit to the locations and numbers. At the review, the evaluators looked it over and all agreed that we were getting closer to the number of pockets they wanted.[9]

An ILC internal memo generated in November 1968 provides evidence that the specific pocket locations were always highly personal and were based on crew input during these early stages of TMG development. Systems engineer Al Gross, the author of the memo, suggested that pockets should be a "crew preference" item, meaning that the crew member could choose which pockets they wanted. However, it turned out that everyone had to agree on specific locations for all pockets because reinforcements had to be added to the TMG at the factory in Dover during manufacture.[10] The pockets were designed into the suits and kept standard through the program.

The Model A-7L Space Suit

Patenting the Apollo Suit

The Model A-7L suit became the foundation for the space suits used throughout the Apollo, Skylab, and Apollo-Soyuz Test Program missions (Figure 7.6). The main core of ILC designers patented it, but many others played a significant role in its development and production. Since NASA eventually paid ILC for all work accomplished under ILC funding during the initial suit competition of 1965, NASA became the assignee the patent was granted to. The ILC individuals listed as the inventors included Leonard Shepard, George Durney, Melvin Case, A. J. Kenneway, Robert Wise, Dixie Rinehart, Ronald Bessette, and Richard Pulling. Many others whose names were not on the patent were involved in the early designs of the suit.

United States Patent [19]

Shepard et al.

[11] **3,751,727**

[45] **Aug. 14, 1973**

[54] SPACE SUIT

[75] Inventors: Leonard F. Shepard; George P. Durney; Melvin C. Case; A. J. Kenneway, III; Robert C. Wise; Dixie Rinehart, all of Dover; Ronald J. Bessette, Wyoming; Richard C. Pulling, Dover, all of Del.

[73] Assignee: Granted to The United States National Aeronautics and Space Administration Under The Provisions of 42 U.S.C. 2457, Washington, D.C.

[22] Filed: Aug. 5, 1968

[21] Appl. No.: 750,031

[52] U.S. Cl. .... 2/2.1 A, 2/81, 128/1 A
[51] Int. Cl. A62b 17/00
[58] Field of Search .... 2/2, 2.1, 2.1 A, 2/6, 3, 81; 128/2.06, 2.05, 2.1, 283, 1.01, 142, 2.95, 285, 1 A

[56] References Cited

UNITED STATES PATENTS

1,490,470	4/1924	Laubach	2/227
2,954,562	10/1960	Krupp	2/2.1 R
3,432,860	3/1969	Durney	2/2
2,404,020	7/1946	Akerman	2/2.1 X
2,749,558	6/1956	Lent et al.	128/283 X
2,842,771	7/1958	Foti	2/2.1 UX
2,939,148	6/1960	Hart et al.	2/2.1
2,966,155	12/1960	Krupp	2/2.1 X
3,000,014	9/1961	White	2/2.1 X
3,067,425	12/1962	Colley	2/2.1 X
3,221,339	12/1965	Correale	2/2.1

3,286,274	11/1966	O'Kane	2/2.1
3,315,272	4/1967	Olt et al.	2/6 X
3,362,403	1/1968	Fleming et al.	2/6 X
3,409,007	11/1968	Fuller	128/2.06
3,463,150	8/1969	Penfold	2/2.1 X

FOREIGN PATENTS OR APPLICATIONS

957,085	5/1964	Great Britain	2/2.1 R
957,688	5/1964	Great Britain	2/2.1
666,671	9/1964	Italy	2/2.1

OTHER PUBLICATIONS

International Science and Technology Publication, February 1967 (page 33 relied on), by M. I. Radnofsky

Primary Examiner—Jordan Franklin
Assistant Examiner—George H. Krizmanich
Attorney—Leonard Rawicz, Neil B. Siegel and Marvin F. Matthews

[57] ABSTRACT

Disclosed is a pressure suit for high altitude flights and particularly space missions. The suit is designed for astronauts in the Apollo Space Program and may be worn both inside and outside a space vehicle, as well as on the lunar surface. It comprises an integrated assembly of inner comfort liner, intermediate pressure garment, and outer thermal protective garment with removable helmet and gloves. The pressure garment comprises an inner convoluted sealing bladder and outer fabric restraint to which are attached a plurality of cable restraint assemblies. It provides versitility in combination with improved sealing and increased mobility for internal pressures suitable for life support in the near vacuum of outer space.

11 Claims, 25 Drawing Figures

Figure 7.5. A copy of the patent for the Apollo space suit and the names of ILC inventors. Source: U.S. Patent Office.

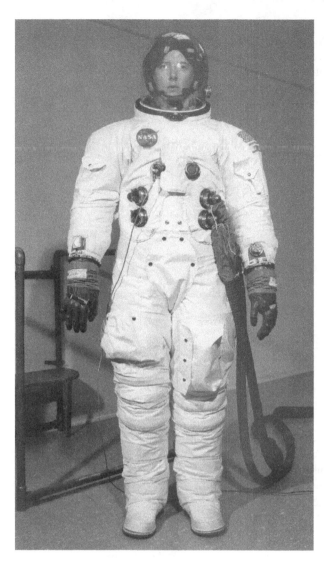

Figure 7.6. The model A-7L extravehicular suit as photographed on April 1, 1968. ILC test subject Ken Shane is in the suit. Courtesy ILC Dover LP 2020.

Those whose names were on the list proudly displayed a copy of the patent on their living room wall.

Design Review of the New Model Suit

The critical design review for the new model A-7L suit was held at ILC on May 22–24, 1967. The Apollo 1 fire had occurred only four months earlier, on January 27, 1967. It was also only less than seventeen months before Apollo 7 would be launched and the first space suits would be needed. Things were on a fast track.

Figure 7.7. This sketch George Durney provided at the model A-7L suit design review
shows the old pattern outline of the upper
torso and neck ring versus the new outline
(the solid lines). Initially, George favored
the tapered look of the neck area, but after
crew members complained about comfort
issues it was opened. Sketches attached to
"Minutes of Delta Critical Design Review
for A-7L Pressure Garment Assembly for
Apollo EMU Garment CEI Program," ILC
Document 8812700409, May 22–23, 1967.

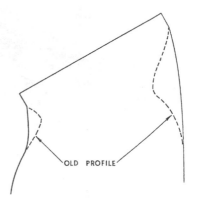

OLD PROFILE

Figure 7.8. This George Durney sketch
shows the new A-7L pattern layout for
the upper torso arm-scye openings, both
pressurized and unpressurized. A more
vertical opening provided greater comfort
and mobility, particularly when the suit was
pressurized. Sketches attached to "Minutes
of Delta Critical Design Review for A-7L
Pressure Garment Assembly for Apollo
EMU Garment CEI Program," ILC Document 8812700409, May 22–23, 1967.

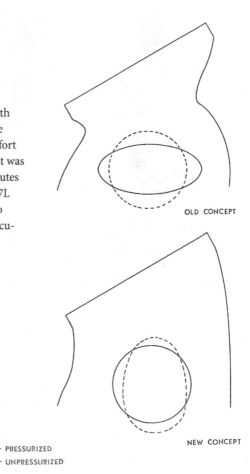

OLD CONCEPT

NEW CONCEPT

— PRESSURIZED
-- UNPRESSURIZED

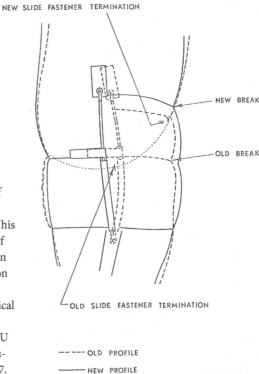

NEW SLIDE FASTENER TERMINATION

NEW BREAK

OLD BREAK

OLD SLIDE FASTENER TERMINATION

– – – – – OLD PROFILE

——— NEW PROFILE

Figure 7.9. The proposed details of the new A-7L lower torso and hip area, including the slide fastener. This sketch shows the new patterning of the hip area that George focused on to provide an increase in hip flexion and improved comfort. Sketches attached to "Minutes of Delta Critical Design Review for A-7L Pressure Garment Assembly for Apollo EMU Garment CEI Program," ILC Document 8812700409, May 22–23, 1967.

Prior to the critical design review in May, ILC upgraded the A-6L suit serial number 009 to include many of the proposed changes for the A-7L model. Many of these advances were not directly related to the fire but were instituted after much evaluation and testing of the A-6L suit. Included were several new patterning changes that addressed feedback from crew members who complained of fit and comfort issues. Figures 7.7, 7.8, and 7.9 show sketches of the changes that were presented as part of the critical design review.

One of the sketches George Durney offered illustrated a small wedge-shaped piece of restraint material added under the upper arm that provided a slight increase in range of motion when the arm was lifted over the head.

The shoulder-cable tube guide was redesigned to include a machined radius in the tube ends. Testing results for the A-6L suit showed significant wear to the coating on the cables as they slid into and out of the tubes. This flair eliminated that problem. The tube profile was also changed to eliminate some curves that resulted in increased friction.

Quality Assurance

Testing Space Suits in the Laboratories (and Elsewhere)

Homer Reihm once explained to me that all the testing done on Earth could not duplicate the actual use of the suits on the lunar surface. The first time Neil Armstrong ventured outside the lunar module was, indeed, the first true test of the Apollo suit, which explains why Homer and George Durney were so nervous as they watched both Neil and Buzz on the lunar surface from mission control on July 20, 1969. You could put a suit inside a vacuum chamber or you could see how it performed in 1/6th G on the KC-135 aircraft that NASA flew to simulate zero or reduced gravity (for all of 45 seconds), or you could subject the materials to solar loads, but you could not do all these things at the same time on Earth. All of this testing still had to be performed on Earth and then the results had to assembled so engineers and designers could somehow arrive at the conclusion that the suit would not fail on the moon.

The subject of space suit testing is vast. I'm not going to dwell too much on all the development testing. Instead, I address individual challenges ILC faced in the process and touch on related testing throughout the book. Testing started on day one when a concept of a suit came together and each of the parts and assemblies had to be tested for structural strength or leakage. As the suit was assembled during the design process, many tests were conducted along the way to make sure that it would hold up to the rigorous demands. Would the sewn seams take the structural loads? Would the hardware perform as intended? Sometimes the ILC engineers had to develop the test requirements, but all the top-level requirements were developed at NASA and passed down to ILC. Testing of space suits determined how a suit would hold up over time and use. It also generated comments by the astronauts that drove changes to the suit based on subjective criticism and general observations. The astronauts had review power and input over the development of all systems used on Apollo, from the Saturn rockets to the lunar module, but the space suit was the one item where their comments flowed like water falling over Niagara Falls. The suit was one item that they intimately bonded with and it became their home while on the moon. They obviously wanted it to work properly and provide as much comfort as possible. Testing provided the foundation to make sure it all came together as a system.

During the height of the Apollo missions, three primary manned tests were taking place. First was the design verification testing (DVT), followed closely by certification testing of a new suit design. That could be the model A-7L or the A-7LB. DVT challenged the design and verified that it was adequate. If problems were discovered during testing, changes could be made. Once DVT concluded, certification testing took place. During that phase of testing, no changes could be made to the design without a lot of explaining about why the problem had not been uncovered during DVT.

The first A-5L model suit did not require any DVT, certification, or qualification testing because everyone understood that this suit was a development model only and the next generation, the A-6L suit, would be the first flight-qualified model. That designation changed when the Apollo 1 fire happened, prompting the design of the new model A-7L suit.

The third primary type of testing was qualification testing, which consisted of manned testing of a suit following detailed cycles of routines that were predicted for specific missions, such as so many elbow flexes or so many walking steps. The numbers predicted for a mission were then multiplied by ten to determine how many total cycles were required for a test. The suit had to stand up to this factor of ten to be qualified for a mission. Apollo suit serial number 039 was used to perform qualification testing to demonstrate that the Model A-7L suit would perform adequately for the Apollo 11 mission.

A lot of documentation suggests that astronaut John Young and his Apollo 16 mission generated unique qualification tests because he liked to push the limits of the suit's performance. He devised a way in the laboratory of bending the legs while wearing the suit to simulate attempts to get down to the lunar surface to gather samples. This action ended up failing parts of the lower torso cable restraint that held the suit together. This outcome really concerned NASA, ILC, and of course Young himself. Some changes were made to the suit, but in the end, engineers realized that in the 1/6th gravity of the moon, Captain Young would not be capable of applying the loads into the suit, so he could not cause the damages that were encountered in the 1-G on Earth.

After the suits for the Apollo mission were assembled, they were pressurized and tested for leakage in the ILC clean-room facility. Suits were not allowed to leak more than 180 standard cubic centimeters of air per minute. If the leakage was excessive, the flaw had to be located and repaired. The pressure-sealing zippers were often the culprit, but leaks could be found anywhere in a suit, although it was not common to fail leakage on a brand-new suit. Leakage in the zipper could be reduced by applying a DuPont fluorinated grease known as Krytox that was compatible with oxygen.

One of the challenges ILC faced in the design of the suit was meeting the requirements of what was called the pressure-drop test. The suits were attached to a test panel and subjected to a flow of 6.0 cubic feet per minute of oxygen into and out of the suit. This is the flow that the primary life-support system provided. The vent system located inside the suit directed the air to flow back out of the suit.[11] Because this vent system had to be flexible, ILC used a series of flexible coils bundled together to act as small tunnels that allowed the air to flow (see Appendix A for details of the vent system). This vent system did not always meet NASA's requirements for the proper amount of air flow within the tubes without causing some pressure buildup. This was due to restrictions within the tubing system and was attributed to design issues. These NASA requirements were developed based on the amount of work the primary life-support system could do to create the airflow. High pressure drops resulted in insufficient airflow. I have many folders containing reports that detail the problems ILC faced concerning the ventilation system. It appeared to be more significant in the model A-7LB suits.

Once the basic acceptance testing was completed on the suits, an ILC suit subject would wear the suit while it was pressurized. Engineers would then evaluate it to make sure many basic characteristics passed their inspection. This was done before fit checks with actual astronauts, which took place once the missions had been lined up and the selection process had begun.

Many other manned pressurized tests were done to test minor design changes or to evaluate specific issues that the customer in Houston might raise. Throughout the program, several male test subjects hired on a temporary basis would test a suit that represented the next design or contained the next changes in the ILC laboratory. They were sometimes referred to as the "in-house astronauts" but they were also called suit subjects. They were either airmen from the nearby Dover Air Force Base or local college students. ILC would put out calls in the hope of locating men who met specific body dimensions and average size and weight. For the airmen who were selected, their participation on the suit testing was not to interfere with their assignments at the base. Many of these subjects ended up spending many hours supporting the testing, often under very tight schedules. I heard one story that ILC rented hotel rooms at the local Holiday Inn during a series of qualification testing so the subjects could get a quick nap and shower between test cycles. That way, ILC knew where they were and could deliver and pick them up on a schedule that worked for everyone. Sometimes, ILC needed the subjects on site while rotating between them, so a sleeping cot was set up in the room across the hall so they could be as well rested as possible. Sometimes the testing continued through the entire night and into the next workday. No one ever complained about the schedules.

ILC Apollo systems engineer John McMullen told me that sometime around 1967, he and a few other ILC engineers traveled to the Houston space center and spent over a week looking at film footage of crew members and other test subjects inside the Apollo suits practicing the tasks they would perform on the lunar surface. NASA had a large play area on its land that simulated the lunar surface in many respects so the crew and the engineers could evaluate the suits and the tools while walking through the mission timelines. John and his team sat through each movie reel and counted every cycle of every joint so that they understood how much wear and tear the suits might see. Even if an elbow was slightly flexed, they would count that as one full bending cycle. In the end, they came up with the numbers of cycles that each joint of the suit would be subjected to. ILC engineers reached an agreement with NASA that to certify the suit model[12], they would multiply the basic numbers of cycles expected by a factor of ten to verify that the design worked. All of the testing performed in the laboratory at ILC in Dover was set up with a pressure control panel that provided all the breathing air and the cooling water for the liquid cooling garment. Also in the lab was the equipment needed to provide the interface with the suit to make sure it was carried out as it would be on the moon. During the later three lunar missions, ILC had a mockup lunar rover in the lab that proved to be more realistic than many of the other mockups NASA had. It became known as the Dover Rover.

ILC senior lab technician Ken Shane was put in charge of the suit subjects at ILC. He was also a suit subject when needed. The test subjects were often called upon when an issue came up that the engineers needed to study more closely in a pressurized suit that was being flexed and stressed to help duplicate problems. Shane recalls many instances when he or the other subjects would have to get into a suit and walk the treadmill for long durations so the engineers could evaluate a problem.

In August 1968, Shane traveled to the Manned Spaceflight Center in Houston to demonstrate to NASA that the new TMG design would work throughout an entire mission. ILC had integrated the new TMG onto a model A-7L suit and had him enter the command module and lunar module mockups and test all the interfaces with the systems inside. He also had to climb down the ladder of the lunar module and walk 100 feet across a lunar simulation surface complete with lunar-type rocks. He did various experiments that simulated what the crews would have to do on the moon. Shane had to demonstrate to everyone every scenario that NASA was planning for the upcoming lunar missions. Over four full days, testing proved that the new integrated Beta-cloth TMG would work. The exercise uncovered a number of issues related to premature wear on the Beta-cloth fabric, but that was not a complete surprise.[13] The wear issue was resolved by adding abrasion patches made of Teflon cloth. Eventually, ILC added a complete cover layer of Teflon fabric over the Beta cloth. For Shane, the most memorable part of the test was when he had to move through the tunnel that connected the command module and the lunar module. He confessed to me that he was always a bit claustrophobic inside the suit, and being asked to go inside the spacecraft simulators in such tight quarters made him very uncomfortable at times.

It was not uncommon for parts of the suit to fail in this laboratory testing. Typically, failure occurred well beyond any amount of cycles the suits would ever see, but NASA required double the amount of cycles expected just to have an adequate margin of safety. This testing also helped everyone understand the dependability of the suit based on the numbers of cycles that could be achieved. John McMullen told me about the time that the space suit was undergoing qualification tested in the ILC labs and a part of the suit broke in cycling as one of the missions was on its way to the moon. When one of the managers at NASA got word of the failure, his instant reply was "stop all testing immediately." He knew that many more cycles than were necessary were accomplished, and his fear was that this failure would drive others at NASA to question the reliability of the suits and the safety of the crew that was currently wearing the suits. (In this case, the testing had to do with the leg-restraint cables. Astronaut John Young was breaking them in the lab when he putting excessive force on them.) As Apollo advanced beyond the first few missions, everyone intimately involved knew that the suits were more than safe, including the astronaut who would wear them.[14] That became evident as the videos taken during many of the missions showed the crew jumping, falling, and hammering tools into the surface of the moon with their gloves.

Laboratory cycle testing revealed that certain areas and parts of the suit were more fragile than others. These parts included the pressure-sealing zippers, the convolutes, and the restraint cables. If any of these parts failed on a mission, it could have had serious consequences.

Inspecting Space Suits

All of work to make the Apollo space suits was accomplished by following written instructions known as tables of operations, or TOs. These TOs provided written details about how to sew parts together or bond the various bladder layers together, for example. At various intervals within the TOs, the work had to be inspected by the ILC quality inspector and or the in-house government inspector NASA employed and managed out of a local Philadelphia office. This was critical for many of the operations, particularly when the next steps would close up an area that would no longer be visible. After inspecting a step that called for this inspection, the inspector(s) would stamp off the corresponding step on the paperwork using a numbered ink stamp issued to them. Additionally, the ILC quality inspector and perhaps the in-house government inspector would witness the assembly of hardware or the testing of the suits. In these instances, no work could be done unless all the parties were together. If any of the individuals took a break, all work stopped until the group was back together once again.

All of the ILC inspectors were women. They were tough and took their jobs seriously. Women such as Clyde Wosylkowski and Madeline Ivory had to make sure that no broken stitching or seams not meeting strict dimensions got past them. They would often err on the side of caution and document any issues they found on a withhold tag, which then would have to be reviewed and dispositioned. Engineering and manufacturing supervisors would determine the disposition, which could include the instruction to "use as is" or to rework or scrap a piece. Scrapping a part could become expensive, so reworking was the best option whenever possible.

The inspectors got along well with the manufacturing folks for the most part, but tensions developed when the pressure was on and manufacturing was attempting to meet a tight delivery schedule. The inspectors could easily slow that process down when they discovered something they did not like, but their work was necessary to ensure the safety of astronauts when they were some quarter-million miles from home.

Inspectors were called upon at times to carry out special inspections because problems were discovered after production. In May 1969, Len Shepard issued an internal memo titled "PGA Thread Verification," upon discovery that a Beta-fiber thread had been accidently substituted for nylon thread

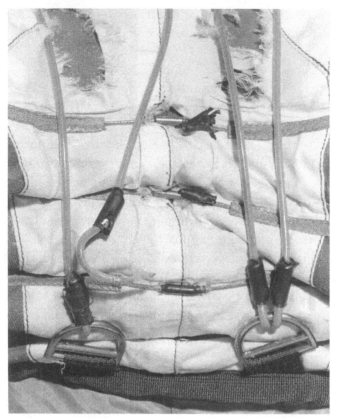

Figure 7.10. This photo, which was taken after many cycles of testing in the ILC Laboratory, shows the damage caused by the constant wear of the restraint cable against the abrasion layer of fabric in the rear crotch area. It also shows a failure of the cable itself where it pulled out of the swage fitting intended to secure it to the D ring. Courtesy ILC Dover LP 2020.

during the fabrication of some restraint material pieces. This restraint layer took all of the pressure loads within the suit and had to be structurally sound. Beta thread had a low tensile value and was much weaker than the required nylon. Once this problem was discovered, NASA ordered ILC to inspect all suits that had been produced to date to verify that the proper nylon thread had been used. The solution was to put all the suits under a blacklight in dark surroundings because the Beta thread would fluoresce under the blacklight but the nylon thread would not. This process verified that suit A-7L serial number 020 had some Beta threads in the restraint layer of the arm assembly.[15] Although the details of how the repair was done are not available, it is likely that the Beta threads were removed and the correct nylon threads were sewn into place.

X-Raying Space Suits

Sewing was not a task that could be rushed. Rushed sewing would have pushed the scrap rate through the roof. Rejected parts in the sewing operation could cost a lot of money and cause delays that no one could afford. One of the earlier problems came about because of pins. The seamstresses were using straight pins to hold the parts in place while they were being sewn together. It was soon evident that sometimes a seamstress could leave the pins inside a garment section that was completely closed after sewing. Pins that were captured within the layers of material of a pressurized space suit presented a situation that was not welcome, and this issue became the focus of attention for everyone.

Mrs. Eleanor Foraker, the lead seamstress on the program, made sure that her staff did their absolute best. She was a mother hen who defended her staff if they needed defending but was also quick to raise hell with them if they stepped out of line. She knew the importance of calling out the problems before they were found outside the production area. It wasn't long before a pin was found in a suit (reportedly in a lunar boot that was delivered to NASA). Soon after that, Mrs. Foraker made sure that each seamstress was issued a set of pins with a particular color of head. At the end of the day, each operator had to produce all of her pins. God forbid if one of the women was short a pin. Several women recalled that Eleanor would not hesitate to come up behind a seamstress who left her pin inside a garment and stick the pin in their behind as they sat at their machine, followed by a good chewing out that was long remembered by the offender.

To make sure the sewn garments were free of any foreign objects of any kind, including pins, a call was made one day to Kent General Hospital, located about five miles away. The question was asked about the possibility of placing an Apollo space suit in their X-ray machine to see if any metallic objects that shouldn't be there could be detected. The staff at Kent General was no doubt very intrigued by the request and certainly wanted to play a part in helping our nation get to the moon, so they agreed to give it a try. Before long, ILC was running the suits over to the hospital and the process was working out just fine until it was found that patients were getting bumped on occasion because the suits were getting priority. It wasn't long before NASA was able to supply a government-furnished X-ray machine to ILC and the patient backlog at Kent General cleared up.[16] NASA ultimately furnished ILC with a Picker brand X-Ray machine that was used to X-ray the TMG cover layers and the torso-limb suit assembly. ILC also used a

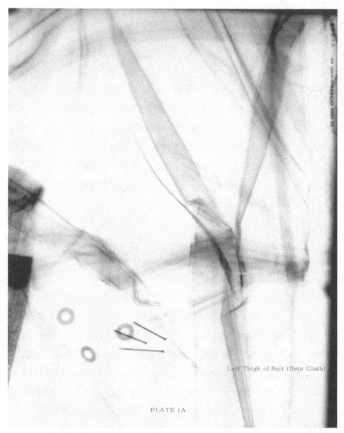

Left Thigh of Suit (Beta Cloth)

PLATE 1A

Figure 7.11. Image of early trial X-ray to see if pins implanted in the thermal micrometeoroid garment could be detected. This procedure is still used today on modern space suits to detect foreign objects such as broken needles or basting pins. Courtesy ILC Dover LP 2020.

smaller Norelco 200 X-ray machine to radiograph the boots for the pressure-garment assembly, the extravehicular gloves, and the lunar boots.[17]

Rubber Problems

The biggest problem with the Apollo space suit was the many reinforced rubber parts that all played a very critical role in its structural integrity. In the 1950s and 1960s, when this suit was being developed, rubber was the best material for the job; it was relatively easy to work with and provided plenty of soft flexibility. It did the job for short-term use, and the Apollo suits would be used under relatively short-term conditions. The problem was that ILC engineers were using the rubber-dipping facilities at Playtex,

ILC's former parent company, and assumed they knew everything there was to know about rubber properties. Unfortunately, this would prove to be incorrect.

On Friday, December 11, 1970, one of the most significant quality issues with the Apollo space suit was uncovered. Astronaut Ron Evans was suiting in the Kennedy Space Center in preparation for training associated with his role as backup command module pilot for Apollo 14. During the suit-up procedure, the technician could not pressurize the suit for an unknown reason. At the time, Evans was using a model A-7L suit, serial number 034, that had initially been made in February 1969 as a backup suit for Mike Collins for his Apollo 11 mission. It was common practice to pass the older backup suits around for training purposes since it saved NASA a lot of money.

Technicians and quality personnel on site at Kennedy immediately began inspecting the suit to find out why it would not hold the air pressure and soon discovered that the left boot bladder had a three-inch irregular-shaped hole. What was even more unsettling was that the rubber around the edges of the hole appeared to be reverting back to a soft, almost liquid-like rubber. NASA did not take the implications of this issue lightly. They were only six weeks away from the launch of Apollo 14, and finding this kind of rubber failure in a suit for reasons unknown was shocking.

The suit was hand carried back to Houston that evening so an immediate investigation into the cause could begin. NASA called in all their materials experts and their engineers to begin investigation work that lasted through the weekend. They also asked ILC engineers if they were aware of what the problem could be. ILC responded by saying that they had noticed a very similar issue with a convolute in the fall of 1969, but they had attributed it to exposure to a solvent followed by long-term storage of the finished convolute in a plastic bag. That Saturday, NASA called in Dr. B. Mosier from the Institute for Research in Houston, Texas. NASA had experience working with this organization for independent materials testing. As a starting point, Dr. Mosier recommended that his laboratory begin testing both the bad rubber from the boot and some known, good-looking rubber to compare results. Those tests were carried out the next day. There was no immediate answer based on the testing, but over the next year, NASA gave Mosier and his group contracts worth $26,197 ($180,000 in 2019 dollars) to help it and ILC solve the rubber issues.

Immediately after the problem was identified, NASA brought in ILC engineers to identify every lot of rubber made around the time of the Evans boot bladders so that potentially bad rubber parts could be isolated and removed from any other suit they might be located in. ILC began investigation work to document all of the convolutes in all the flight suits that were in line to be used. ILC linked the individual convolutes to each lot of rubber made in the Playtex facility.[18]

Initial reports followed numerous tests. One memo dated just a few days after the problem was identified suggested that an accumulation of unwashed salts from perspiration had gathered in the valleys of the convolute. The salts were thought to possibly contribute to the failure of the rubber, but this was just the beginning of the inquisition process. It turned out that salts were not the problem.

Soon, everyone realized that this was a significant problem and that it would take longer to solve. ILC began looking at the age of the rubber convolutes in each active suit, particularly the ones related to the Apollo 14 mission. Since many of the installed convolutes had been installed for several months, it was agreed that they would all be replaced with brand-new rubber. Following NASA approval, ILC employed a small jet aircraft that would fly the primary and backup Apollo 14 suits from Cape Kennedy to Dover so the convolutes could be replaced at the plant. To reduce risks, only primary or backup suits could be flown to and from ILC at the same time, never both together. Remarkably, ILC accomplished the task of switching out the numerous rubber convolutes in both the flight and backup space suits within two weeks of the mission. The Apollo 14 flight suits performed well and no rubber issues emerged.

As more experts were brought in, attention focused on the dipping facility where the Apollo suit convolutes were being made. This was done at the Playtex plant in Dover. The Playtex folks were considered the experts up to this point, as they had been involved in designing latex products for consumers since the founding of the company almost thirty years earlier. This capability and expertise was what had gotten ILC involved in the space suit business to begin with. That dipping facility is where the very first convolutes were made when Len and George began their suit-making journey in the 1950s.

The Playtex lab was using a rubber compound identified as XN4802 for the ILC convolutes. Although this process of dipping the convolutes was being performed at the Playtex plant, ILC personnel were involved in witnessing the mixing of each batch of the XN4802 latex and signing off on a verification memo. It can be assumed that the ILC personnel had a lot of respect for the chemists of Playtex and did not challenge the chemistry or the processing involved in the dipping of the convolutes. Later, when outside experts looked more closely at the process Playtex was following, they found that the ovens used to cure the products were not calibrated correctly and it became apparent that the rigor needed was not in place. Curing temperatures were a critical part of this process and had to be controlled, but these ovens were varying by as much as 15°F. This attitude would have to change, since the formulation and processing of latex for the Apollo space suits should have far more rigor than the process used for the latex for bras, girdles, and rubber gloves. It was clear that the ILC engineers would have to become experts in the chemistry of latex rubber dipping. ILC engineer Bob Wise and a few others played a big part in overseeing the fix to this problem.

On January 7, 1971, ILC held a meeting with John Carl, the author of a DuPont Corporation book titled *Neoprene Latex*. The memo that summarized the meeting noted that Carl felt that the aging of the failed bladder showed characteristics of ozone aging, during which stress contributes to degradation. The failure happened to be in an area where stress was present due to the shape of the convolute. He also said that ionic copper could catalyze the degradation process, but at that time it was not known how much copper could have been present in the dipping process. Carl recommended that ILC increase the antioxidants used in the curing process to counter any copper contamination. He specifically recommended that a compound called Agerite White be added at a rate of two parts per hundred of rubber. The Agerite White would act as a copper inhibitor, thus increasing the life of the product. Assuming that the presence of copper was a primary contributor to the issue, Carl projected that eliminating the copper threat would give the rubber a shelf life of ten years. Carl also offered a few other suggestions for improving the formulation.

ILC personnel met with NASA's Crew Systems Division personnel on March 15, 1971. Al Gross of ILC made the following points that summarized the lengthy investigation and zeroed in on the solutions to the problem:

Use the Agerite White antioxidant.

Eliminate the source of the copper contamination.

Run physical property testing on all older batches of rubber to identify poorly performing lots that could be represented by convolutes already installed in suits.

The change that played the most significant role in solving the problem was building a brand-new dipping facility at the ILC plant and gaining better control of the entire dipping process. Removing any trace of copper piping used in the facility was a good start. On April 14, 1971, NASA issued a contract change authorizing ILC to "prepare and release procurement, process, and control documentation to allow manufacturing of the dipping compound at ILC Industries instead of purchasing from Playtex Corporation." The details of the process change included adding Agerite White and removing Santo-white (an antioxidant) from the dipping process.[19] ILC was operating its new dipping facilities by late May 1971 and performing all of its own testing to verify the physical properties and the antioxidant properties. These new convolutes were used in all the new A-7LB suits.

It is a fact well known to many of us suit experts, including the staff at the Smithsonian in charge of the suit conservation, that the rubber convolutes of the post–Apollo 14 model A-7LB suits made with the Agerite White additive are almost as soft and flexible today as the day they were made. In contrast, the rubber in the earlier A-7L is essentially turning to dust. The model A-7LB suit that was built for Gene Cernan, which I was fortunate enough to get into not long ago, could only have been pressurized with the new convolutes; older suits are totally incapable of being pressurized ever again.

Although all the focus was on the latex rubber convolutes, questions were also asked about the expected shelf life of the bladder layer of the suit, which was made of neoprene-coated nylon. This material was purchased from Reeves Inc. They manufactured it using what they called Type W Neoprene, which contained antioxidant 2246 added at two parts per hundred of rubber. It also contained carbon black and an inorganic pigment. Reeves felt that the usable shelf life would be ten years if it was protected from oxygen and light. No issues were ever discovered with this material throughout the Apollo missions.

From Production to Mission Support

Sewing and Cementing Space Suits

The main pressure-garment assembly was made at the Dover facility on Pear Street. At that location, all the materials were laid out and cut to very close tolerances using the proper patterns. Quality inspectors were ever present throughout the process to make sure that all parts conformed to dimensions and lacked any defects. A group of around twenty seamstresses would sew the torso, the arms, the legs, and the boot assemblies and then sew all these parts together to form the single suit. Others would cement and install the neoprene-coated nylon layer that made up the pressure bladder that held oxygen inside the suit. A liner assembly followed that provided comfort, made it easier to don the suit, and helped prevent abrasion. At the end of the process, hardware assemblers would install all the mechanical hardware such as the wrist and helmet disconnects, the gas and electrical connectors, and the restraint cables and associated hardware pieces. Testing was also performed at the Dover plant.

Eleanor Foraker supervised a group of eighty-three seamstresses, assemblers, and pattern cutters who were assigned to the Frederica facility. Many others provided support with scheduling, quality, maintenance, and so forth. The challenge of making the TMG was to get the bulk of the garment under the machine to sew the final seams. According to Eleanor, many mistakes were made when it came to stitching through all the layers of the insulation. Often the problems emerged when an operator tried to stitch around the small pass-through holes cut into the TMG for various hardware pieces. Additionally, the scrap rate was high because operators had to

Figure 7.12. A view of the Apollo production floor at ILC's main Dover Plant in 1967. Dedicated women and their industrial-grade Singer sewing machines played a key role in the success of the Apollo space suit program. Courtesy ILC Dover LP 2020.

Figure 7.13. One of the two custom-made sewing machines used to make the oversized Apollo thermal micrometeoroid garment covers. Courtesy ILC Dover LP 2020.

bunch up the fragile glass-fiber fabric in the relatively small space between the needle and the body of the Singer sewing machines. ILC overcame this problem by having an industrial engineer named Paul Martin experiment with one of the older machines by extending the body on a long arm. This significantly opened the workspace so that operators could manipulate the TMG so the seam area lay flat. The seamstresses loved the new machine and named it Big Moe. After they had worked with Big Moe for only two months, the increase in quantity and quality was obvious. A second machine was made that the operators christened Sweet Sue. Because of these two machines, productivity soon increased by 80 percent and the product had greater integrity. In addition, the operators experienced far less physical and mental fatigue.

Eleanor Foraker is an excellent example of a hands-on employee who made it their mission to see that the Apollo suits went out of the ILC plant on time with the best possible quality. She transferred from Playtex in 1962, where she had been taught to sew many of the Playtex apparel items over the ten years she had worked there. She learned about the space suit business from engineers such as Mel Case, Bob Wise and George Durney. As

Figure 7.14. Eleanor Foraker was the lead supervisor at the Frederica facility that made all the Apollo thermal microme- teoroid garments. Courtesy ILC Dover LP 2020.

she became more proficient at her job, the tables turned and she became the one who taught the engineers details about how to sew space suits together.

The manufacturing process began at both facilities with the precision cutting of all the pieces of the suit that would ultimately be taped, sewn, or glued together. Francine (Fran) Burris. who oversaw women in the Fred- erica plant where the TMG was cut, recalled how tough that job was. She would typically work up to twelve hours per day and on many occasions she worked seven days a week. Much is made of the seamstresses that did the critical sewing, but the women who did the precision cutting of some very delicate materials also deserve a lot of credit for their skill. Fran and the others were under constant pressure to provide the parts needed for the suit and mistakes were frowned upon. She worked for Eleanor Foraker, who was a fair supervisor, but you did not want to make Eleanor look bad. Francine recalled that Eleanor's boss, Ron Tenaro, jokingly said that the two of them worked well together because Eleanor was always after Francine to avoid mistakes and Francine was always afraid of making mistakes because of Eleanor's threats. In looking back on it, though, Francine declared that it was all worth it in the end. She recalled that the seamstresses were the ones who received much of the glory and attention because of the unique- ness of what they were doing and the skills they had to possess. When the astronauts would come to ILC for their fit checks, it was the sewing rooms that they would visit, not her cutting operation. When awards were handed out, such as the all-expense-paid trips to the Kennedy Space Center to attend pre-launch parties and then see the launch the next day, typically they went to engineers and seamstresses. She said that she totally understood it. Francine never lost sight of the important role she played, and to her surprise, ILC did eventually honor her for her hard work and dedication. In January 1971, ILC flew her and her husband to Florida to get a closeup tour of the Space Center and to witness the launch of Apollo 14. Being there to hear and feel the Saturn rocket lifting the three astronauts and the space suits she had helped make to the moon made it all worthwhile.

Some of the details of Francine's job involved the outer TMG layer of the suit. She or the women who worked under her would lay the pattern piece over the Beta cloth, mark the outline, and then very carefully cut the part along the lines. They would then brush on a chemical known as Kel-F, a high-temperature, fireproof sealer that would keep the edges from unraveling along the cut edges, which the Beta cloth was very prone to do. The parts with the wet Kel-F sealant were dried in a large oven for a short time and then inspected to make sure they had "made the cut," so to speak.

By early 1969, as ILC was running hard just to keep up with production of the model A-7L suit, another group was already developing the next-generation TMG for the model A-7LB suit. ILC materials engineer John Scheible wrote a memo to Bob Woods that stated that approximately 256 patterns had been made to cut approximately 454 pieces to make the integrated TMG. By the time the final design was established, these numbers may have changed, but the memo provides some insight into how much effort had to be put into this garment.

One of the more difficult jobs was the marking and cutting of the numerous layers of the very thin aluminized mylar film, which was extremely difficult to work with. Francine recalled the day when one of the engineers thought they were helping her out by offering to provide her and the other ladies with electric scissors to speed up the cutting process. She knew at that moment that these smart engineers had no idea how tedious a job it was to cut these thin film materials and that electric scissors would increase the scrap rate by as much as 100 percent. She recalls that the engineer who offered the idea was not happy with her response even if it was the correct one. She continued with mechanical scissors. After the Apollo 12 mission, the cover layer of Alan Bean's suit was dissected to study the effects of the wear and tear on the garment. It was discovered that the aluminized mylar had ripped and broken apart, due primarily to the fragile nature of the film.

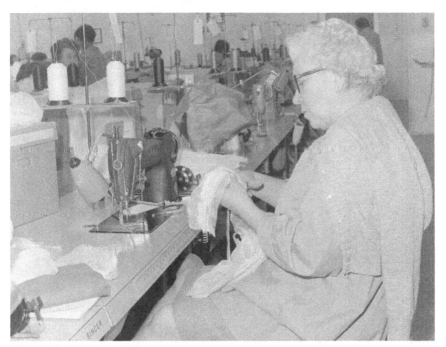

Figure 7.15. One of the Apollo seamstresses assembling a section of the ventilation system to be installed into an Apollo suit. Courtesy ILC Dover LP 2020.

Figure 7.16. This sewing operator is assembling the thermal micrometeoroid garment under the long arm of the specially made sewing machine designed and built by ILC. You can see by the size of the garment that it would not fit under the standard sewing machines. Courtesy ILC Dover LP 2020

Figure 7.17. One of the very talented ILC seamstresses, Mrs. Iona Allen, who personally sewed Neil Armstrong's lunar boots. Courtesy ILC Dover LP 2020.

For the new suits for all the following missions, ILC spread the fresh, uncut mylar film out on a table and bonded thin strips of adhesive-backed Kapton film to it to form grids that provided structural strength and prevented tearing across larger areas of the film layers. This is known as ripstop in the business.

The seamstresses did the bulk of the fine, very detailed work. Their skills were pivotal to the success of the program and the safety of the astronauts. Each pressure-garment assembly was made of over 800 pattern pieces. Many of these pieces needed to be sewn together to make the suits and few mistakes could be made. The main room in Dover where the sewing was done consisted of about twenty heavy-duty Singer sewing machines, each of which was set up to do different tasks.

Suit engineer Mel Case had to provide sketches to get his ideas across to the seamstresses during the very early stages of development. He understood how certain types of sewn seams would provide the strength required to hold the most critical parts of the suits together. The seamstresses would then take his ideas and make sure that they could carry them out on their machines. They would often give feedback to the engineer that helped establish the proper procedures to follow for all future work. Once it was clear that the design and the process would work, the details had to be put into an approved table of operation. Model makers would rely on blueprint drawings.

In these earlier years, many women picked up the sewing trade from their mothers or in high school. The United States was not competing in the

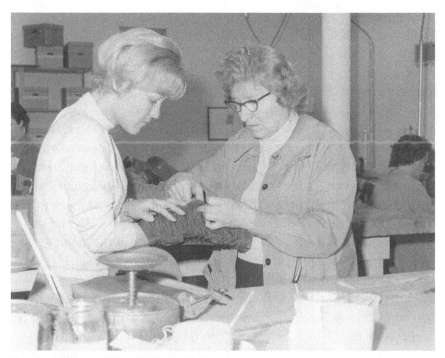

Figure 7.18. Evelyn Kibler (*left*) and Thelma Breeding (*right*) cement a convolute section for assembly in an Apollo suit. Courtesy ILC Dover LP 2020.

global marketplace with China or Indonesia, so much of the sewn goods consumed in this country were made in this country. Several of the seamstresses were recruited from within the Playtex division while others came from Leeds Luggage manufacturing in Smyrna, Delaware, and another local sewing factory in Greensboro, Maryland. Seamstresses were given a test of their capabilities. It was not good enough that they could sew a straight line; they also had to have the temperament to sit at a sewing machine and stitch parts together within 3/32 of an inch and at a very slow speed so as not to puncture very expensive rubber convolutes, for example. The concentration that was needed was incredible. This was not something the typical seamstress was cut out for. Many did not have the patience or were too nervous to give it a try.

Roberta Pilkenton, or Bert as she was known, was one of the excellent seamstresses brought on to sew the Apollo suits. Bert was a great example of the tough, working-class ladies who made the Apollo suits. She had been born in West Virginia and eventually moved to southern Delaware. She was taught to sew at a young age by her mother. When she entered the workforce, she bounced around at a couple of odd jobs before ending up at Playtex in 1955, where she sewed girdles for a short period. Sometime around 1963, she transferred over to ILC Industries and began working on the space suit line. She started sewing various portions of what was called the restraint material for the pressure suit. This was the layer of the suit that basically held the suit together structurally and prevented a catastrophic failure when it was pressurized. She had to make many of the seams with what is known as a blind stitch, meaning that she could not see what she was sewing and could only tell by feeling the results of the seam from underneath. This was grueling work. It was not long before she comprehended the immensity of what it was she was involved in, as would many others. She said that when she went home at the end of the day to her mobile home in the outskirts of Denton, Maryland, and looked up at the moon when it was shining its brightest, an intense feeling would overcome her. The idea that her sewing would keep an astronaut alive on that distant moon was hard to fathom, but it hit home with her. That memory from the night before would occupy a corner of her mind the next morning as she sat at her Singer machine, sewing the suit that could be next on the moon.

In one of the last of many TV interviews she gave, Bert told the reporter "When we built the Apollo suit, the one that's in the Smithsonian Institute, some of us girl's names are inside that suit." She later told me that she and others had marked their names on the inside mylar layers of the Neil Armstrong's suit. Enough time had passed that she felt she could tell that story. The workers at Grumman who built the lunar lander also later confessed that they had added their names and personal effects inside the insulation layers attached to the skin of the lander.

An extraordinary amount of labor was expended on the cementing of the various Neoprene bladder sections and the taping of the mylar film layers. It was quite a labor-intensive process that took a lot of training and attention to detail. An incorrectly cemented seam could spell just as much trouble as an incorrectly sewn seam. The urethane adhesive that sealed the neoprene materials had to be applied just right. Many of the women who did this job were trained in several areas of assembly, including sewing and cementing.

Mrs. Ruth Embert worked as a seamstress in the Frederica plant, where she made the outer-garment cover layer. She took great pride in the work she did, which she expressed in the form of poetry. She probably penned the following verse after the Apollo 8 mission circled the moon and before the Apollo 11 mission landed on it—meaning sometime between January and July 1969.

"Reaching for the Moon," by Ruth Embert

Since I was just a little child
I've loved to gaze on high
To watch the moon on clear sweet nights
A'riding through the sky.

Its face would seem to smile at me
It seemed to wink an eye
At all the white and billowing clouds
That drifted through the sky.

Sometimes, it seemed so close to me
Sometimes, so far away-
On stormy nights, 'twould disappear
As dawn would turn to day.

But then, when I would fall asleep
Its rays would shine through windowpane
And drowsily, as I closed my eyes
I'd think, the moon is still the same!

But now there's been a turn-about
America's reached the moon!
And through *my* help and more like me
We're going to land there soon!

And deep within my heart I know
I'll feel a happy thrill
To me, it's been a life-long friend
I hope it always will . . .

Repairing Space Suits

One morning when Bert Pilkenton reported to work, she was asked to take a plane immediately to the Kennedy Space Center because they needed her to replace a knee convolute on one of the flight suits that was only days away from going to the moon. (It was determined that one of the knee convolutes already in the suit was from a batch of rubber that was in danger of failing due to premature aging that was causing several problems around the time of Apollo 14.) Bert understood the need, but she had never flown before and just the thought of flying petrified her. She was aware of the impact this had on the Apollo mission, and now the importance of what she was doing was really sinking in. There were other good seamstresses, but for a task this critical, she was the one they relied on.

Bert had just gotten married months before and had to call home to let her husband know that she had to get on a plane right away and fly to Florida. Her husband told her there was no way she was flying to Florida. After a lot of back and forth conversation that escalated at times, she hung up the phone and let her boss know that she would do it. Within an hour, her boss came to her with an executive-style leather briefcase containing the replacement suit part. She thought the briefcase idea was a bit odd, but the best part was yet to come. As he handed her the briefcase, she noticed that he had a pair of handcuffs. He fastened one end to the handle of the briefcase and then reached out and attached the other end to her right wrist. This all happened in a split second and she had no time to react. She was told that the fate of this Apollo mission was now handcuffed to her wrist. The key was given to her, but she was told that under no circumstances was she to remove the handcuffs until she was in the suit laboratory at the Space Center where she would swap out the part. She ended up on a flight that afternoon that had a changeover in Atlanta, and with time on her hands—and a briefcase—she found the nearest bar and had a stiff drink. She told me that she had some interesting conversations with the others in the bar that afternoon in Atlanta.

Bert finally made it to the suit lab that night, where she proceeded to take the suit apart to the point where she could remove the convolute and sew in the new one. If she made a mistake and the sewing needle punctured the new rubber convolute just fractions of an inch from the intended area, it would have to be scrapped and she had no other replacement. She must have felt as though the weight of the world was on her. She completed the job early the next morning and the suit passed all testing. Thanks to the skills and determination of Bert Pilkington, the suit was on its way to the moon within a few days. Her marriage also survived.

Production Schedules and Fit Checks

The time from planning to final acceptance of an Apollo suit was approximately 120 calendar days, but of course that varied for many reasons, including personnel resources, rework issues, or design changes that took place during the build process. The build process alone took about 40 calendar days. It was not uncommon for changes to be made on an almost daily basis, since the suits were customized and were constantly evolving. The fit checks were typically performed no less than seven months before the scheduled launch date. This allowed time to send the suits to Houston for vacuum-chamber testing with the astronaut.

Before any of the Apollo missions, NASA flew the first crews to Dover Air Force Base on a specific date for their fit checks. Astronaut Fred Haise once told me that it was an exciting moment for an Apollo astronaut to hear that they were scheduled to fly to Dover AFB because it meant they were selected for a mission. From the start of the first manned flights, NASA kept the names of the astronauts secret, so code names were provided for each astronaut (see Appendix B.) NASA did not want the names revealed in case crew changes were made during the process that took place before the official release of the crew members' names. NASA would provide the coded list only to Len Shepard of ILC, who would then decide who else at ILC could have access to the names. That group was very small.

There was a lot to consider when making a new Apollo space suit. The process would begin when NASA identified an astronaut for a mission. They would then need to be scheduled to report to ILC for measurements. Sometimes the measurements had to be taken in Houston because of the astronaut's tight schedule. Sixty-six measurements were taken of various body parts, from the head circumference to the width of the foot and everything in between, including measurements such as the suprasternal height and the gluteal arc length. Richard Ellis was one of the key ILC employees responsible for taking the astronauts' measurements. Richard performed many roles at ILC in addition to his job as a model maker.

Richard once told me the story of how he had to meet with Neil Armstrong in a hotel in some forsaken and forgotten part of the country at a pre-arranged time and date so that he could get Armstrong's dimensions. Time was running out and there did not appear to be many openings in Armstrong's schedule to permit him to visit ILC. The measurement process itself took only about one to two hours, so carving out a full day from an astronaut's schedule for travel seemed like a poor use of very valuable time. It was efficient to send Richard Ellis to all parts of the country if need be to corner the astronauts. On the measurement form, Richard entered the location of the fit check as "MSC" (for Manned Spaceflight Center), likely because it was just easier to do so.

2 OCT 67 ILC

MEDIUM LONG

Subject	NEIL ARMSTRONG	Location	MSC
Date	10/3/67	ILC Tech.	R. Ellis

Measurement Location	CM.	IN.	Measurement Location	CM.	IN.
Weight	173	LBS.	Upper Thigh Circumference		23 3/4
Height		70 3/8	Mid Thigh Circumference		21.0
Cervical Height		60 1/2	Lower Thigh Circumference		15 1/2
Mid Shoulder Height Right		59 7/8	Knee Circumference		15 1/8
Mid Shoulder Height Left		59 7/8	Calf Circumference		15
Shoulder Height Right		57 3/4	Lower Leg		9 1/8
Shoulder Height Left		57 3/4	Ankle Bone		10 1/4
Suprasternale Height		57 1/2	Scye Circumference Right		18 1/4
Nipple Height		51 3/4	Scye Circumference Left		18 1/2
Waist Height (Back)		43 1/4	Axillary Arm Circumference		12 7/8
Trochanteric Height		36 3/8	Biceps Flexed Circum.		13
Knee Cap		21 1/8	Elbow " "		12 3/8
Center Knee-Floor Height		20	Forearm Flexed Circum.		11
Crotch Height		33 1/2	Wrist Circumference		7
Shoulder-Elbow Length		14 1/8	Sleeve Inseam Right		19
Inter scye Breadth		14 3/4	Shoulder-Elbow Pivot		12 7/8
Biacromial Breadth		16 1/2	Elbow Pivot-Wrist		11 7/8
Shoulder Breadth		19 3/8	Wrist for Finger Tip		7 1/2
Chest Breadth		15	Vert Trunk Circ. Right		68
Waist Breadth		12 1/2	Waist, Front Length		15 1/2
Hip Breadth		13 3/8	Anterior Neck Length		4 1/2
Vert Trunk Dia. Right		25 7/8	Posterior Neck Length		4
Vert Trunk Dia. Left		25 1/2	Waist Back Length		18 5/8
Head Circumference		22 5/8	Gluteal Arc Length		10 1/2
Neck Circumference		15 1/2	Crotch Length		30
Shoulder Circumference		47	Span		71 5/8
Chest at scye		39 1/2	Span Free		—
Chest at nipple		38 1/2	Metacarple 2		8 1/2
Waist Circumference		34 1/2	Extended Arm Length	LEFT 76.7 RIGHT 79.4	
Buttock Circumference		39	Mid-Shoulder/Top of Head		10 1/4

FOOT	Right	Left	
			WEARS B SHOE 9 1/2
Length	9 1/2	9	
Instep Length	10	9 1/2	
Width	B	B	

Figure 7.19. A copy of the physical sizing data ILC's Richard Ellis gathered from Neil Armstrong on October 2, 1967. Ellis recorded sixty-six different measurements that ILC engineers used to make the patterns for the suit. Courtesy ILC Dover LP 2020.

Once the dimensions were made available to ILC, engineers such as Mel Case and George Durney would review what size components would be needed to assemble a complete suit to properly fit the astronaut. There were some basic sizes to select from for the arms and legs, but the torso was patterned from the dimensions of each astronaut, since this was a critical element of the suit sizing. Finally, the pressure boots would be chosen based on shoe size. These would all be assembled together to form one pressure-garment assembly. The gloves were a separate matter, since a hand cast would have to be made of each hand. The model makers would then set out to form the mold that would be used to dip the bladder for each glove based on very precisely measured dimensions from the hand cast.

Production was laid out based on the major suit components that were being assembled in separate areas and then brought together at the final assembly levels. The pressure-garment assemblies were made at the Dover facility while the TMG was made in Frederica and the schedule had to be such that they came together at about the same time. Some seamstresses were trained to make the boot assemblies, while others were trained to make the gloves. The better seamstresses who had a lot of experience could move to cover different areas if the need existed. Other seamstresses concentrated on modifying suits, for example when suits were returned to ILC to have a set of gas connectors added or removed based on whether NASA wanted to change the suit for intravehicular or extravehicular use. Such modifications saved money by avoiding the manufacture of a completely new suit. Such conversion was common for the training suits that were moved around between missions.

Once a completion date for each suit was forecast, ILC would coordinate with the NASA astronaut office and schedule the respective crew members to come to Dover for their fit check sometime around that completion date.

The fit check typically took a full day to complete, but two astronauts could be fit checked in one day if necessary. The astronauts would ordinarily fly a NASA-owned T-38 jet from Houston or some other city they were visiting into Dover Air Force Base. A representative from ILC would drive the eight miles to the base to retrieve the astronaut(s) and bring them back to the plant. This was always a special time for the employee who was fortunate enough to pick them up, since they had a good twenty minutes in each direction to get to chat about whatever came up. They always hoped that would include some inside story of what it was like to train for going to the moon. John McMullen, who was fortunate enough to pick up Neil

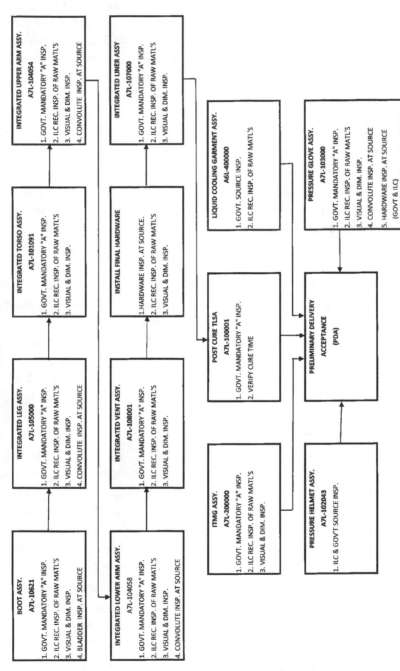

Figure 7.20. This flow diagram depicted the major components of the Apollo suit and how they were all brought together at different stages to form one space suit assembly ready for the preliminary delivery acceptance testing that was the government's official buy off. Courtesy ILC Dover LP 2020.

Armstrong on one occasion, recalled that all Neil wanted to talk about was the stock market, and more specifically, the stock price of ILC Industries.

On December 15, 1970, Harrison Schmitt flew his T-38 to Dover Air Force Base and had the fit check performed on his Apollo 17 training suit. Upon leaving the plant, he asked a couple of ILC employees if they would mind picking up a Christmas tree from the area since he could not obtain a fresh pine tree in Houston, Texas. ILC employee Ralph Armstrong agreed and picked out a nice fresh tree. He delivered both the tree and Schmitt to his T-38 for the trip back to Houston. Ralph recalled shoving the tree under the belly of the plane in the unpressurized cargo hold. Many years later Schmitt told me that plan failed miserably because all the needles fell off the tree days after getting it home. He assumed that the extreme cold at altitude did it in.

The fit checks were conducted in a special room called Astronaut Heaven that was upstairs and away from the many bustling areas at ILC. Even as the astronauts moved between the front entrance and this room, they rarely met any ILC employees. This was intentional early in the program, since NASA wanted to keep crew selections secret until a date when they decided to make the announcement. The initial fit check was conducted while an astronaut was wearing only the pressure-garment assembly that was pressurized to 3.7 pounds per square inch. This initial check would be done in the presence of the ILC design engineer, typically Mel Case. The astronaut would provide feedback on the fit and the engineer would record the comments on an ILC Form number 1031. It included a block titled "Summary of Problem Area." In most cases, there was an entry of one sort or another in this block. The engineer would then have to figure out the cause and how to resolve the fit or comfort issue and update their recommended action under the next two blocks, which were titled "Reason for Problems" and "Corrective Action." One of the primary focuses of the fit check was to get the proper lengths of the metal restraint cables throughout the suit. Many of the cable ends were not swaged or crimped tightly in position before the fit check because the lengths of the cable were established based on the astronaut's feedback about comfort and range of motion. It was important to do this when the suit was pressurized. The fit-check engineer would use an adjustable U-shaped clamp to temporarily secure the cable until it could be permanently crimped.

Following the initial fit checks at pressure, the suit was removed and taken back to the production area, where the hardgoods technician would crimp the restraint cables in place. If necessary, the women would have to re-size the suit based on the fit and comments. A series of lacing cords in the legs and arms could be let out or taken in to lengthen or shorten the appendages and (hopefully) satisfy the crew member. The astronaut would then be resized. While the suit was once again given to the women in production so they could lace on the bulky TMG cover layer, the astronaut would have a long break. Once that was finished, the entire suit would then be brought back to Astronaut Heaven, where it would have one final fit check. The astronaut could be in the suit for up to two hours while the ILC engineer asked more detailed questions so that all concerns were addressed. This also gave the astronaut time to get familiar with their suit. At the end of the fit check, the astronaut would sign at the bottom of the form as evidence that they agreed with the documented results. There was also a block at the top of the Form 1031 that had check boxes for "Status: Fit Satisfactory" or 'Modification Required." If any issues were not resolved at this point, ILC was responsible for ensuring that the suit was taken care of to resolve any issues. This could mean rescheduling the fit-check process if necessary. This entire routine was repeated for the backup suit. There was also a sizing exercise for training suits that frequently required some adjustments, since the suits were not custom made for the astronaut it would be assigned to.

One of the ILC engineers told me a story about two crew members who were fit checked together on the same day. The commander was quick to accept his suit after only minor adjustments. He impressed everyone in attendance as a get-it-done kind of guy who didn't mind a pressure point in the suit here or there. He liked his suit and although a few areas still had some slight issues that were not simple to fix, he agreed that he was very comfortable and could wear it on the moon with few problems. His lunar module pilot, who followed him, had many complaints that seemed to be relatively minor. After production had made several adjustments, he continued to want further modifications. The few folks gathered in the test room could see the commander getting a bit irritated, and he finally told his crewmate half-jokingly, "Listen, if you want to fly with me, you'd better sign off on this suit now or you're off my mission." It wasn't long before the suit seemed to be fitting much better. This was an example of the challenges the ILC folks faced related to suit sizing. It was not an exact science and the subjectivity of the fit was what ILC had to deal with, along with the mantra they accepted that "the customer is always right." Aside from the fit itself, the mobility and range of motion of the suit was of concern to some crew members, while others were fine and never complained. It was

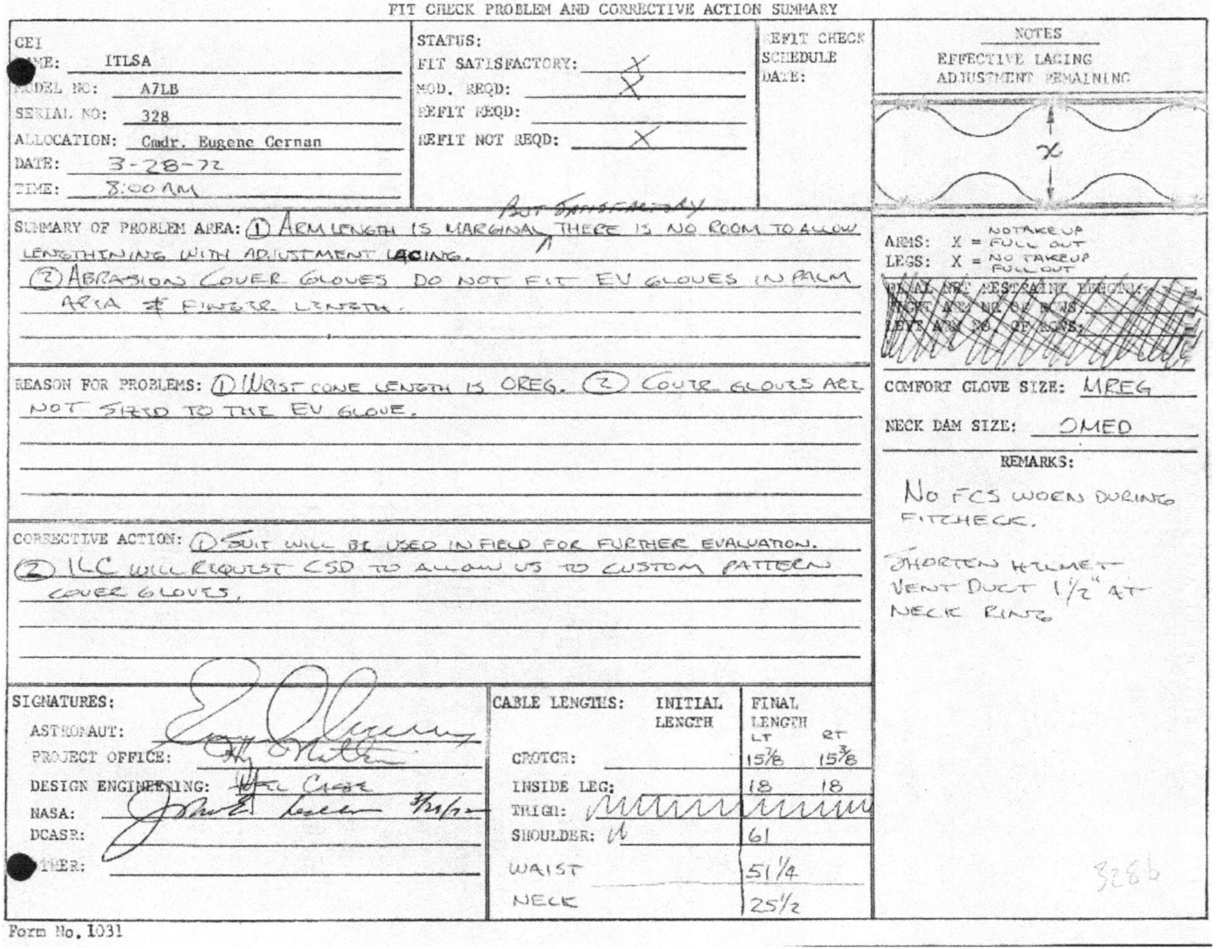

FIT CHECK PROBLEM AND CORRECTIVE ACTION SUMMARY

CEI
NAME: ITLSA
MODEL NO: A7LB
SERIAL NO: 328
ALLOCATION: Cmdr. Eugene Cernan
DATE: 3-28-72
TIME: 8:00 AM

STATUS:
FIT SATISFACTORY: X
MOD. REQD: X
REFIT REQD:
REFIT NOT REQD: X

REFIT CHECK SCHEDULE DATE:

NOTES
EFFECTIVE LACING ADJUSTMENT REMAINING
x

ARMS: X = NO TAKE UP FULL OUT
LEGS: X = NO TAKE UP FULL OUT

SUMMARY OF PROBLEM AREA: ① ARM LENGTH IS MARGINAL BUT SATISFACTORY THERE IS NO ROOM TO ALLOW LENGTHENING WITH ADJUSTMENT LACING.
② ABRASION COVER GLOVES DO NOT FIT EV GLOVES IN PALM AREA & FINGER LENGTH.

REASON FOR PROBLEMS: ① WRIST CONE LENGTH IS OREG. ② COVER GLOVES ARE NOT SIZED TO THE EV GLOVE.

COMFORT GLOVE SIZE: MREG
NECK DAM SIZE: OMED

CORRECTIVE ACTION: ① SUIT WILL BE USED IN FIELD FOR FURTHER EVALUATION.
② ILC WILL REQUEST CSD TO ALLOW US TO CUSTOM PATTERN COVER GLOVES.

REMARKS:
NO FCS WORN DURING FITCHECK.
SHORTEN HELMET VENT DUCT 1/2" AT NECK RING

SIGNATURES:
ASTRONAUT:
PROJECT OFFICE:
DESIGN ENGINEERING:
NASA:
DCASR:
OTHER:

CABLE LENGTHS:	INITIAL LENGTH	FINAL LENGTH LT	FINAL LENGTH RT
CROTCH:		15 7/8	15 7/8
INSIDE LEG:		18	18
THIGH:			
SHOULDER:		61	
WAIST		51 1/4	
NECK		25 1/2	

328b

Form No. 1031

Figure 7.21. The fit-check summary form that ILC used to verify the fit of each suit with the astronaut it was made for. Here is a copy of Form 1031 for the primary flight suit Gene Cernan used on his Apollo 17 mission. Courtesy ILC Dover LP 2020.

a tough balancing act and the stakes were high. After one of his fit checks at ILC, astronaut Jim Lovell left a note behind for the seamstresses that said, "Please sew straight and careful. I'd hate to have a tear in my pants on the moon." As humorously as he intended the comment, he was also serious.

Pre-Mission Testing and Training

The new suits were typically delivered to Houston's Manned Spacecraft Center. The crews would wear these suits in a man-rated vacuum chamber, where they would be thoroughly tested to make sure everything worked. It was also a good familiarization for the astronaut. Following the chamber testing in Houston, the primary and backup suits were flown to the Kennedy Space Center. So as not to wear the flight suits out, most of the crew training was performed with the assigned training suit, but it was not uncommon for the crews to use their primary suits on a limited basis. At the Kennedy Space Center, there were two vacuum chambers, Chamber A and Chamber B. The command module mockup was in Chamber A and the lunar module mockup was in Chamber B. These chambers could pull the appropriate vacuum pressures to simulate the altitudes anticipated, including the full vacuum in Chamber B to represent the lunar surface. Between training both the primary crews and the backup crews, the ILC suit technicians including Ron Woods were kept busy turning the suits around, which involved cleaning, repairing, and readying them for the next day. Recall that a total of fifteen suits were assigned to the primary crew and the backup crew. Training suits were marked as Class III suits, which meant that they could support basic nonvacuum environment training, while the primary and backup suits were marked as Class I, which meant that they were maintained clean and in the current configuration and were suitable for vacuum chamber work.

As the missions advanced beyond Apollo 11, the crews spent much more time training for their extravehicular activity on the lunar surface, so the training suits got quite a workout and were often being repaired as time permitted. Much of this training was conducted in Houston in areas set up to represent the lunar surface.

Approximately two months before the scheduled launch, the crews would take part in what was called the Crew Fit and Crew Function Exercise (also called CF^2). This exercise was conducted with all the primary space suit gear and the accessories assigned for the mission so that the crew could walk through the checklist and make comments about any issues. Quality inspectors were assigned to the exercise to record all comments from the astronauts and verify that the procedures were accurately detailed. Finally, just two weeks before the scheduled launch date, the flight suits would be used for the countdown demonstration that took place at the same time as the scheduled activities on the launch day so that the astronauts and the launch team could benefit from rehearsing the full countdown procedures.

Finally, not long after the countdown demonstration, a final fit check would be done to verify that the suits were properly sized. At this time, there was a focus on the sizing of the arms and the legs, where final adjustments could easily be made. Following this fit check, the suits would be taken away and stored in prepared for the launch day.

Field Support and Suiting Up Astronauts

Throughout Apollo and Skylab, ILC Industries located and hired both engineers and suit technicians at the Manned Spacecraft Center in Houston and the Kennedy Space Center in Florida. NASA issued several Apollo suits to North American Aviation and Grumman Aviation to support command module and lunar module interface testing and training. When ILC was asked to provide support at these two sites, they found a great candidate for the job in Ray Winward. In 1962, Ray had done a brief but intense stint in the air force working in the Physiological Training Units at Edwards Air Force Base, which was the prime location to be if you wanted to work around the cutting edge of aerospace technology. Ray had the enviable good fortune to fly with the likes of Neil Armstrong and Mike Collins in the back seat of their F-104 chase planes during various missions at Edwards. Ray later learned that it was Mike Collins who had recommended that ILC hire him because of his background. Ray and his family were brought to Dover to start with ILC in September 1965, when ILC was in dire need of skilled workers as they geared up for the contract. Ray became the space suit interface engineering representative at Grumman in 1966 and 1967. Ray had a suit that was assigned specifically to him that he wore inside the lunar module any time a change was made to the module that might impact the interface between the crew member and the part of the module that was changed. He was also responsible for suiting up the astronauts who travelled to Grumman for lunar module evaluations. By early 1967, it was determined that future astronaut training would be carried out at the Kennedy Space Center in Florida or at the Johnson Space Center in Houston. Ray and his family made the trip to Houston, where they stayed for the next eighteen months. This was followed by a tour of duty as the space suit interface engineering representative at North American Rockwell in

Downey, California. His duties there were similar to the ones he had when he worked at Grumman. Finally, by the end of 1968, Ray and family moved back to ILC Industries in Dover, where he completed the last of his five years working as part of the team that did the final suit fit checks for the astronaut crews.[20]

One of the ILC engineers, Al Gross, was hired to work out of the Kennedy Space Center and later transferred to the ILC office in Houston. There he joined others such as Walt Sawyer and Ron Minks. This team became the right hand of NASA's Jim McBarron and others when they needed help understanding issues and providing solutions related to the Apollo space suits. NASA's chief of the Crew Systems Division wrote a letter of appreciation to ILC's Apollo manager, Len Shepard, that thanked Minks and credited him with working after hours and on weekends and holidays, when he was not compensated, to do design studies on a proposed redundant loop for the cooling garment. The NASA chief also thanked Walt Sayer for his work on the design of a drink bag that would eventually be used in the Apollo suits.[21] NASA managed the program but had full confidence in the abilities of ILC engineers and trusted their work. Everyone had to come together as a team to make things work, and it took a good deal of interface with the ILC engineers to carry it through. Eventually, Al Gross was transferred to the main plant in Dover and became involved in the systems engineering group, where he frequently helped with the fit checks of the flight, training, and backup suits of the astronauts as they came through Dover.

At both the Kennedy Space Center and in Houston, the ILC suit technicians were involved in many activities related to the training of the astronaut crews. They prepared all the suits and related gear for simulator and vacuum chamber training and made updates and repairs to the suits as needed. At the Kennedy site, the ILC technicians supported the suit up of the astronauts for their missions. After the suit up, the ILC technicians would even accompany the astronauts to the command module atop the Saturn rocket, a full thirty-eight stories above the Atlantic Ocean. Ron Woods was one of these fortunate ILC suit technicians. ILC Industries employed him through most of the Apollo era as one of the Kennedy Space Center technicians. Although ILC originally hired him for its Houston office, it wasn't long before he was asked one Friday afternoon if he wanted to transfer to the Kennedy Space Center, where much of spacecraft interface training and testing was taking place, not to mention the launch activities.

It didn't take him long to make his decision and the next Monday he reported for duty in Florida.

Ron was fortunate enough to suit up Buzz Aldrin for his Apollo 11 mission and support the Apollo 15 and Apollo-Soyuz Test Program missions. In between those missions, he constantly worked to support the other missions during training. Turning the suits around required more work that many can appreciate. Ron's career path began with the army's elite 101st Airborne Division, where he learned how to take care of the survival gear that would keep the soldiers alive. One of the skills he learned was sewing. When he left the army and went home to Texas, his mother passed on a job advertisement she had seen in a local paper from Brown and Root Northrop, which was looking for people to work on a space suit–related contract involving the physical cycle testing of space suit arms and other parts. Not long after that, Ron made the switch over to ILC Industries after making friends with some of the ILC technicians assigned to the Houston office. An ILC employee asked if he'd be willing to come over to ILC because there was a real need for suit technicians. Aside from the military life-support background Ron had, he also knew a thing or two about sewing. Suit technicians were frequently called upon to make modifications and repairs to suits in the field that were supporting the various training operations and the actual missions. For example, pressure or restraint zippers might need replacement, or crew options such as comfort pads would be sewn into the suits to provide relief from pressure points. It was not at all unusual for a suit problem to emerge during training exercises that would require Ron or the others to stick around late into the evening or even the early morning hours just to get the suit(s) repaired, cleaned, and prepped for the next day's activities. Ron also recalled frequent trips away from the Cape when he had to hand carry space suits or support items such as gloves back to ILC in Dover for repairs. In those days, the suits were considered top secret, so suit technicians had to hand carry the items at all times or in the case of the larger suit containers, they would have to watch those containers being loaded into the cargo hold of the plane and be the last on before takeoff and the first off when they arrived at their destination to gather the case. He or the others would often have to wait several hours at the Dover plant for the repairs or upgrades to be made so that they could hand carry the item(s) back to Cape Kennedy or to Houston.

Ron and the other technicians were trained in the skilled art of helping the astronauts with their suits by Joe Schmitt, one of the best technicians

NASA had. They all felt honored to be trained by and work with Joe. It was not just a simple job of shoving an astronaut into their suit. There was an art to performing the many tasks associated with that activity. Throughout the process, it was critical that the technician display unwavering confidence and demonstrate their ability to perform the functions without missing a beat. The crew counted on that and expected it. The technicians had to provide 100 percent focus on the details so that no problems were overlooked. These guys were the last to see the suits before they left Earth on their mission and it was up to them to spot any problems and get them corrected before the mission or before training.

NASA's Joe Schmitt had worked his way through many jobs related to early aviation in the 1930s. During the war years he worked as a mechanic. He eventually ended up at NACA, the forerunner of NASA. While he was a mechanic at NACA, he supported and witnessed many of the early record-breaking flights by pilots such as Chuck Yeager. During a short stint at Langley Air Force Base in Virginia, he took classes in parachute rigging and aircraft clothing repair, which set the stage for him to be a natural fit in the suit business. When NACA formed the Space Task Group, which was charged with managing the initial manned space missions, Joe was selected as the first space suit mechanic based on his experience. He suited up Alan Shepard for the first manned flight into space. He suited John Glenn for his historic orbit around the earth. He and Ron Woods participated in suiting up the Apollo 11 crew. Ron suited Buzz Aldrin while Joe Schmitt suited Mike Collins and ILC's Troy Stewart suited Neil Armstrong. It was a day they would never forget.[22]

When the Apollo program concluded, Ron's value to the manned space program was recognized: NASA was happy to hire him to support the space shuttle mission from beginning to end.

Field Optional Items

One of the challenges ILC faced during the fit checks and comfort evaluations with the crew members related to what were called field optional items, or FOIs as they became known. During the 190 fit checks performed at ILC Industries and the resulting modifications to improve comfort and fit, a total of thirty options were arrived at. This included such things as adding comfort pads to the liners inside the suit and rotating the gas connectors so the crew member could see the tabs more easily. After a short time, NASA requested that the changes and additions be routed through NASA for review and for initiation of configuration change control board directives, specification change notices, and subsequent engineering change orders. This evolved into a master list that outlined all possible FOIs and provided a menu of options to review with each crew member for their possible selection. One example of the options involved the gas connectors and was outlined as follows:

> The crewman had the option of positioning the "lock-locks" in the orientation most comfortable to him when attaching O2 and exhaust lines. This item was noted in the CEI specification as "orientation of gas connectors locks" for all Apollo, Skylab and ASTP suits. This particular FOI required different clocking configurations for the command module pilot as compared to clocking for the EV crewman's suits since the gas connector location on the pressure garment assemblies were different.[23]

At the start of the program, there was little or no control over the addition of these changes or additions because they were seen as minor things that were done in the interest of pleasing the crew members. However, at the list grew, more discipline was needed so that changes could be made to drawings and more rigor was followed, such as the addition of the changes to the acceptance data pack that accompanied each suit.[24]

Not long after the Apollo program started, crew members began requesting additional changes after their suits had left ILC and were being used to support tests or training in Houston or Florida. ILC technicians in the field would make additional changes based on sizing and comfort issues. If an arm or leg sizing was changed, for instance, it might result in other problems that required additional comfort pads in other locations.

At the end of the Apollo program, an ILC Program Management Study report dated April 30, 1974, recommended that on all future programs, the suits should undergo all fit checks in the field, where all optional equipment would be installed and the configuration would be updated and maintained in the data pack that accompanied the suit and contained all written records of the suit configuration.[25]

Main Components of the Model A-7L Suit

The following sections describe the major subassemblies of the model A-7L suits. The space suits and all support equipment that was used throughout Apollo were continually evolving; slight design changes were made and different manufacturing techniques were used from mission to mission.

This overview describes how the suits were made and how the various parts came together and functioned. It also describes the more significant changes to the suits under the specific missions outlined below.

Torso Assembly

The torso assembly consisted of everything minus the arms and legs. It included the neoprene bladder and nylon restraint material, the pressure and restraint zippers, and all the pass-through openings for the oxygen, exhaust gas, water cooling (in the extravehicular activity suits), and electrical systems.

The torso assembly was made from patterns that formed the basis for each astronaut's torso. Modifications to the patterns were made based on the exact dimensions of the astronaut the suit was being made for. Some astronauts had a large chest diameter and a smaller waist, for instance, and that all had to be considered. All pass-throughs for the gas and electrical connectors were marked on the torso pieces during the marking and cutting process. A reinforcement backing layer was also added to both sides of the chest. The cuts for the pass-throughs on the command module pilot suits were made only on the right; the left side was not cut but was ready to have pass-throughs installed later if the suit was selected for use as an extravehicular activity (EVA) training suit. It was less costly to make all suits with this option rather than modify a suit that did not have the reinforcement areas and locations marked.

Astronauts entered and exited the suits by way of the pressure-sealing and restraint zippers located in the torso. The pressure-sealing zipper was bonded onto the neoprene-coated nylon inner bladder layer; the restraint zipper was sewn onto the blue nylon restraint layer. Both zippers had webbing lanyards that were clipped onto the pull tab to aid the user when closing or opening them. This was necessary since it would be virtually impossible to reach the pull tabs in the back of the suit on the A-7L models. Red webbing was used on the restraint zipper while the blue lanyard was used on the pressure-sealing pull tab. When the lanyards were not in use, they were removed and stowed in the utility pocket of the left thigh on the first models, but that location varied with the later A-7L suit models.

Arm Development and Modifications

Around February 1968, many suit-related tests and fit checks were being run at NASA and numerous reports were being circulated that indicated

Figure 7.22. The Apollo A-7L command module pilot (CMP) suit. Note the three circular markings and reinforcement patches on the left chest area. It was common practice to fabricate all CMP suits with these locations in case the suit was later selected to be modified as an extravehicular training suit that required the dual set of connectors. Courtesy ILC Dover LP 2020.

the problems astronauts were experiencing with the arms and shoulders of the EVA suits related to comfort and range of motion. That month, a letter from Richard Johnson, chief of the Crew Systems Division at NASA, to Len Shepard outlined the complaints of astronauts Jim McDivitt and Rusty Schweickart.[26] The letter noted that the arm lengths had increased to the point that the bearings were interfering with elbow mobility. The cable turnaround ring was digging into the shoulder when the arm was raised and the astronauts were not able to raise their arms above their shoulders without using excessive force. These issues were related to the model A-6L suits, but there was no magic formula at that time that was going to fix them with the upcoming model A-7L suits. Aside from the fit and comfort problems, there was the nagging issue of interference between the suits when three astronauts were seated side by side in the command module. The width of the arms and shoulders of the suits resulted in little room to maneuver, particularly for the command module pilot in the center seat.

One of the earliest documented fixes to the shoulders was in May 1968, when a solution was found to the fact that the shoulder convolute was not collapsing uniformly, thus impacting comfort and range of motion. This was caused by the cable turnaround ring on the outside of the shoulder that pinched the shoulder convolute and did not allow it to flex naturally. The solution was to add an oversized polycarbonate sheet over the hardware and reinforcement patches to the inside surfaces of the shoulder convolute. The polycarbonate sheet was slightly larger than the turnaround hardware that was on top of it, thus spreading the loads out so as not to collapse the convolute directly under it.[27]

As NASA and ILC engineers were carefully evaluating the A-7L suit in the early part of 1969 in preparation for the lunar landings, several observers continued to express concern about the deficiencies of the EVA suit, particularly the arms. It seemed to be a recurring theme. Astronaut Ken Mattingly wrote a memo on March 12, 1969, that raised many issues about a suit he was evaluating during a series of lunar-surface qualification tests. The arms were apparently not properly sized, since his elbow was not located in the convolute where it should have been. However, it was supposed to be sized for him. His observation was that the distance from the fingertip to the top of the arm assembly wrist-cone was fixed, which meant that all the sizing was done between the wrist cone and the shoulder cone. Mattingly felt that this was a design issue. In the concluding comments section of the memo, he wrote, "I'm more convinced than ever that every effort should be expended to ensure that improved arm mobility (force requirements) is provided for the G1 [Apollo 11] mission."[28]

Primarily because of the limitations in the mobility of the A-7L suits, NASA began internal discussions about suit improvements in late 1968 and early 1969. Two actions came out of the discussions. The first was to issue a two-phased procurement plan for advanced hard suits, also known as advanced extravehicular suits, that Litton Industries and Garrett AiResearch were developing.[29] This program was being administered by the Office of Advanced Research & Technology within NASA. The plan under discussion was to use these hard suits for Apollo 17 and later missions, where longer-duration lunar exploration was anticipated. Phase I of the plan was to have each contractor develop and deliver two pre-qualified hard suits. Phase II would provide the contract award for mission-ready suits based on the evaluations of the Phase I efforts.[30]

The second action was to have ILC modify the arm assembly of the A-7L EVA suits. This was driven by feedback from Ken Mattingly and other astronauts and test subjects. By mid- to late 1968, ILC was already hard at work on the next-generation suit, called the Omega III. The design of the arms was critical to the success of any future suit model. In September 1968, when ILC demonstrated the new suit to NASA, engineers there immediately began looking for improvements that could be designed into the A-7L suits as soon as possible, ideally before the first lunar mission. (More details about the ILC Omega suit and the advanced extravehicular suits are provided in a later chapter.)

After the Omega III suit demonstration in September 1968, NASA began serious discussions with ILC about putting the new arm into the Apollo 11 suits. NASA asked ILC to press on with the development of the arm to support this first lunar mission, since the Apollo 11 crew needed all the mobility they could get. An abundance of data on the tests that were performed on the new arms and their bearings prove their structural integrity and ability to retain air. One set of tests subjected three sets of arms and bearings to increasing air pressure until failure occurred at 20, 53 and 49 pounds per square inch.[31] On average, this was ten times the pressure the arm would operate at. Much more formal testing was done to verify that the arm would do the job.

By April 1969, ILC had completed design verification testing on the new arm design. This confirmed that it would meet all the mission requirements. Testing on the new bearing design was also run at Wyle Laboratories in Huntsville, Alabama, to verify that it would pass environmental design requirements related to thermal vacuum exposure, vibration, impact shock, and sand and dust compatibility. This testing revealed shortcomings in the urethane pressure-retaining seal ILC had used in the bearing. ILC responded by replacing it with a Fluorel-silicone seal. Retesting with the new seal yielded passing results. In the end, the test findings were favorable. The key improvements included

New upper- and lower-arm cone sections that were reduced in the cross-section to shrink the overall bulk.

An upper-arm bearing (part number A7L-104050) that ILC had designed and manufactured that permitted low torque rotation. The length of the bearing was reduced from 1¾ inches to ⅝ inch. The bearing was also moved up the arm 1 inch to assure that the elbow was located in the center of the elbow convolute.

A new elbow convolute that was increased 2 inches in length to permit 135° of elbow flex, as opposed to only 70° on the A-6L style arm. This also put the wearers' elbow within the center of the convolute and fixed the problem Ken Mattingly had.

A shoulder-cable turnaround ring that was raised higher on the arm to take advantage of the new cone design and provide improved mobility.

New liner and abrasion prevention scuff patches to reduce wear issues experienced in the old arm design.[32]

Records show that work began on April 21, 1969 to remove the old arms on Neil Armstrong's serial number 056 suit and replace them with the new configuration arms. That work was completed on June 3, 1969.[33] Because of the urgent need to get this suit back to NASA for the mission, the ILC workers had no weekends off between the start and completion dates. Buzz Aldrin's suit was reworked around the same time, including the installation of a newly designed six-inch elbow convolute because of sizing issues Aldrin had had during his fit check at ILC in Dover. Because of all these changes and updates to the primary flight suits, there was no time to make the changes to either of the backup suits that contained the old arm design. Fortunately, the astronauts did not have to use them.

The arms used for the intravehicular suits consisted of a restraint-layer system of interwoven cords known as the link-net system that the David Clark Company had invented. It was good for restraining the shape of a pressurized cylinder such as an arm assembly and provided adequate flex and range of motion, particularly for a crew member inside the capsule. While this system was not acceptable for the crew member performing extravehicular activity, it was quite acceptable for the command module pilots and helped reduce the width and interference problems between the two crewmen on either side. ILC hired a local woman named Joanne (Jo) Thompson, who became one of the best Research & Engineering seamstresses the company had. Mrs. Thompson had come to ILC one day with a

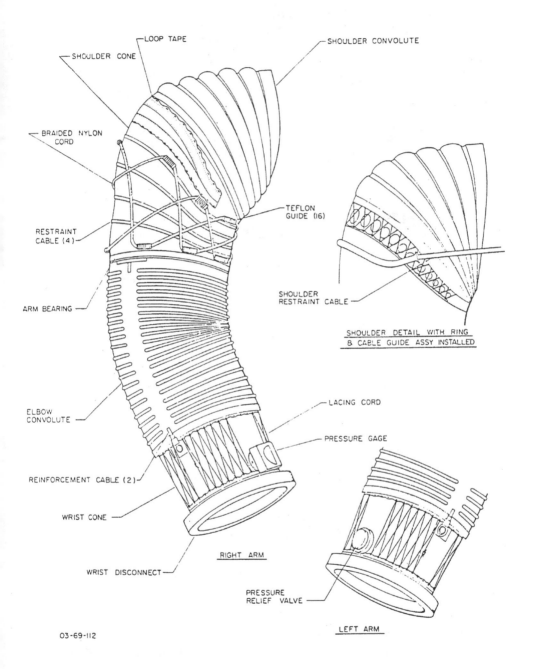

LOOP TAPE

SHOULDER CONE

SHOULDER CONVOLUTE

BRAIDED NYLON
CORD

TEFLON
GUIDE (16)

RESTRAINT
CABLE (4)

SHOULDER
RESTRAINT CABLE

ARM BEARING

SHOULDER DETAIL WITH RING
& CABLE GUIDE ASSY INSTALLED

LACING CORD

ELBOW
CONVOLUTE

PRESSURE GAGE

REINFORCEMENT CABLE (2)

WRIST CONE

RIGHT ARM

WRIST DISCONNECT

PRESSURE
RELIEF VALVE

LEFT ARM

03-69-112

Figure 7.23. The extravehicular arm assembly used on Apollo 11. Courtesy ILC Dover LP 2020.

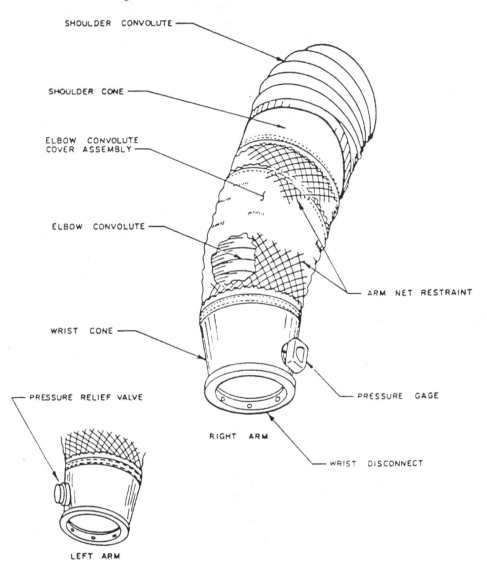

SHOULDER CONVOLUTE —

SHOULDER CONE —

ELBOW CONVOLUTE
COVER ASSEMBLY —

ELBOW CONVOLUTE —

—— ARM NET RESTRAINT

WRIST CONE —

— PRESSURE RELIEF VALVE

—— PRESSURE GAGE

RIGHT ARM

— WRIST DISCONNECT

LEFT ARM

Figure 7.24. The Apollo A-7L intravehicular link-net arm assembly for Apollo 11. Courtesy ILC Dover LP 2020.

friend who was interested in taking the sewing test to see if she could get a job at ILC. While she was there, someone asked Jo if she was interested in taking the test and she somewhat reluctantly agreed. Although her friend failed the sewing test, Jo passed and was offered the job.

Jo was eventually given the task of understanding how to make the link-net arrangement at ILC. No machine existed that could make the tube of

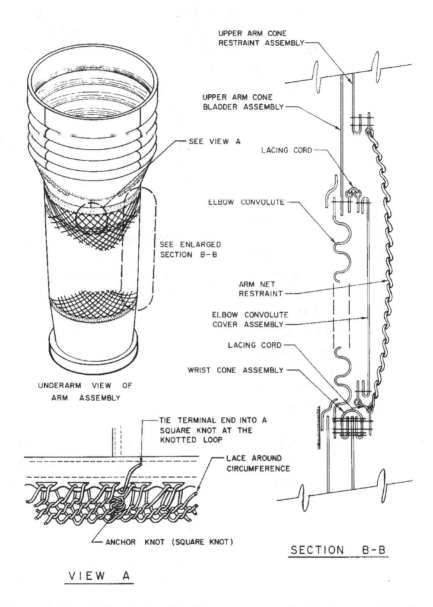

UPPER ARM CONE
RESTRAINT ASSEMBLY

UPPER ARM CONE
BLADDER ASSEMBLY

SEE VIEW A

LACING CORD

ELBOW CONVOLUTE

SEE ENLARGED
SECTION B-B

ARM NET
RESTRAINT

ELBOW CONVOLUTE
COVER ASSEMBLY

LACING CORD

WRIST CONE ASSEMBLY

UNDERARM VIEW OF
ARM ASSEMBLY

TIE TERMINAL END INTO A
SQUARE KNOT AT THE
KNOTTED LOOP

LACE AROUND
CIRCUMFERENCE

ANCHOR KNOT (SQUARE KNOT)

SECTION B-B

VIEW A

Figure 7.25. The detailed sketch of the link-net arm assembly. The lacing cord (labeled View A) would be let out or taken in on both the wrist cone and the upper arm cone to lengthen or shorten the arm assembly based on the fit-check sizing of the astronaut. Courtesy ILC Dover LP 2020.

link-net cords, but before too long she had it figured out and was making enough of it for the arms of the intravehicular suit. A total of 200 yards of the cord was required to make a single elbow section. The link-net system was fabricated with nylon until Apollo 10, after which Nomex was used.[34]

The cost of arm development totaled $56,146 ($389,000 in 2019 dollars). The cost to retrofit an estimated nineteen space suits that had already been manufactured cost $80,294 ($557,000 in 2019 dollars). Changes to costs for the production of thirteen new suits on the contract totaled $22,217 ($154,000 in 2019 dollars). With tooling and qualification testing, the grand total was $168,015 ($1.17 million in 2019 dollars) for the new arms, but it turned out to be money well spent.[35]

NASA directed ILC to make one suit available per week immediately following the qualification testing of the new arm design that was to conclude on or about May 7, 1969. It issued a contract change authorization on April 28, 1969, to remove the arms in the following suits and replace them with the new arms with the low-profile bearings: "A-7L-045, 046, 047, 049, 053, 057, 060, 061, 062, 063, 064, 065, 067, 068, 070, 071, 073, 074, 076, 077, 078, and subsequent EV suits."[36] ILC replied that it could not replace the arms on suits scheduled to support the Apollo 10 mission, since the suits were being used to support a launch schedule that was less than one month away. That would include suits A-7L-045, -046, -047, and -049. Those suits would retain the old-style arms with the link-net restraint.[37]

The wrist disconnect was secured to the suit by wrapping a cord around the bladder and restraint layers of the lower arm after it was bonded onto the disconnect to prevent leakage. The cord provided the structural strength to hold it in place. This was then covered with a silicone wrist band to provide protection.

Leg Assembly

The leg assemblies were sewn onto the torso assembly. Most of the pressure and man-loads were carried through aircraft-grade restraint cables. Cables located on both the inner and outer sides of the legs carried the loads from the waist down to the boots.

The leg-thigh restraint cable system was an ingenious design by George Durney. Homer Reihm described it this way:

> The thigh convolute, which controlled step height, used vinyl coated stainless steel cables on the outside and inside of the hip to control pressurized length. Both ends of both cables were attached to sections

of the suit. George Durney was searching for a way to add range to the joint without increasing standing length, which caused other problems in sizing. George was aware that the cables could operate under small radius of curvature, a situation he had experienced as both a pilot and air craft mechanic. He designed a small pulley that retained each cable at the bottom of the thigh. During stepping, the pulley effectively rode up the cable which allowed for improved step height thus alleviating a clear suit short coming without altering the natural anatomy of the knee, thigh and hip. The pulley was thin so as to not create a pressure point on the wearer, yet also large enough in diameter to exceed the minimum bend radius of the cables, hence assuring maximum flex life of the cable.[38]

Figures A.3a and A.3b in appendix A show the details of this design.

Although every effort was made to size the suit to each astronaut during the build, some adjustment had to made so that each crew member's suit fit almost perfectly. For example, a sizing adjustment was made to lengthen or shorten the legs so that the fit between the shoulders and the heels of the feet was snug but not tight. This was done with sizing cords that were laced through loop tapes sewn onto the bottom of the leg restraint. Tightening these cords or letting them out adjusted the length. During the Apollo 12 mission, Alan Bean had to lengthen the sizing cords on Pete Conrad's suit because it was made too short before the mission. Pete was in agony because the suit was squeezing his shoulders when his heels were forced tight into the boots as he stood up. This change was not meant to be done during the mission, and Alan later noted how hard it was to do in the cramped lunar module.

Pressure Boots

The A-5L boots did not have any convolutes in the ankles. Because the Apollo suit was the very first walking space suit, its entire lower torso had to provide the best mobility possible. As part of the redesign of the suit, ILC added a convolute and began evaluation. Richard Ellis, who served as both a premier model maker and a suit subject, had begun evaluating new convolute ideas by February 1966. He did this by walking the treadmill in the pressurized suit. ILC engineers and seamstresses installed a new A-6L convoluted boot on the right side of a model A-5L suit identified as Design Mobility Unit suit #1. They retained the old-style A-5L boot on the left. After Richard had walked on a treadmill for fifteen minutes, it was clear that

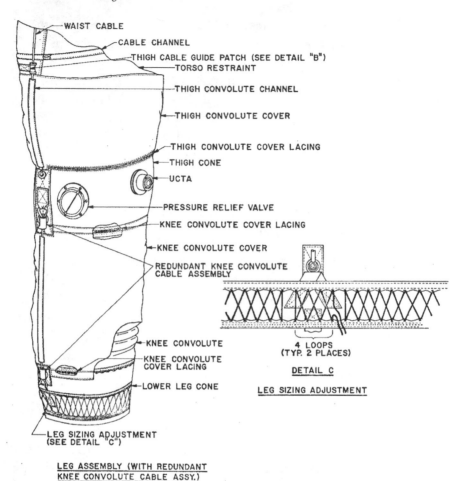

WAIST CABLE
CABLE CHANNEL
THIGH CABLE GUIDE PATCH (SEE DETAIL "B")
TORSO RESTRAINT
THIGH CONVOLUTE CHANNEL
THIGH CONVOLUTE COVER
THIGH CONVOLUTE COVER LACING
THIGH CONE
UCTA
PRESSURE RELIEF VALVE
KNEE CONVOLUTE COVER LACING
KNEE CONVOLUTE COVER
REDUNDANT KNEE CONVOLUTE CABLE ASSEMBLY
KNEE CONVOLUTE
KNEE CONVOLUTE COVER LACING
LOWER LEG CONE

4 LOOPS (TYP. 2 PLACES)
DETAIL C
LEG SIZING ADJUSTMENT

LEG SIZING ADJUSTMENT (SEE DETAIL "C")

LEG ASSEMBLY (WITH REDUNDANT KNEE CONVOLUTE CABLE ASSY.)

Figure 7.26. The leg assembly showing the details of the cable restraints and the sizing adjustments. The lower leg was zipped onto the top of the boot. Pressure was maintained because the pressure bladder built into the leg assembly was tucked into the boot assembly. This sketch shows the redundant knee cable that was proposed at one time but was never used on an Apollo mission. Courtesy ILC Dover LP 2020.

the convolute made a significant difference. As a result, all suits from that point used ankle convolutes.

The pressure boot was a two-part assembly. The outer, blue colored, nylon restraint layer of the boot was zippered onto the restraint assembly of the lower leg. The neoprene-coated nylon pressure bladder was one piece that was connected to the leg assembly and was simply tucked down inside the boot restraint as the two were assembled with a zipper that joined the leg restraint to the boot restraint. The pressure bladder was properly

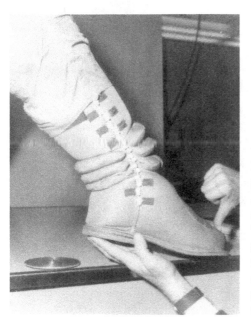

Figure 7.27. This engineering photo best illustrates the ability of the Apollo boot to flex under the pressure of the 3.75 pounds per square inch. You can easily see how the convoluted sections permitted the air to move from one side to the other without compressing the gas, thus making flexing the foot nearly effortless. Courtesy ILC Dover LP 2020.

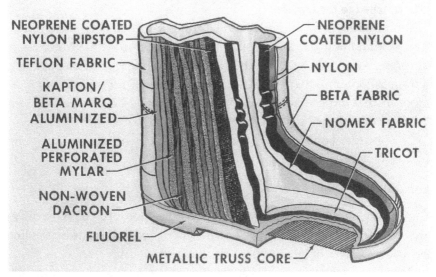

PGA/ITMG BOOT

NEOPRENE COATED NYLON RIPSTOP

TEFLON FABRIC

KAPTON/ BETA MARQ ALUMINIZED

ALUMINIZED PERFORATED MYLAR

NON-WOVEN DACRON

FLUOREL

METALLIC TRUSS CORE

NEOPRENE COATED NYLON

NYLON

BETA FABRIC

NOMEX FABRIC

TRICOT

Figure 7.28. This NASA diagram shows the layers of the Apollo pressure boot. Courtesy NASA.

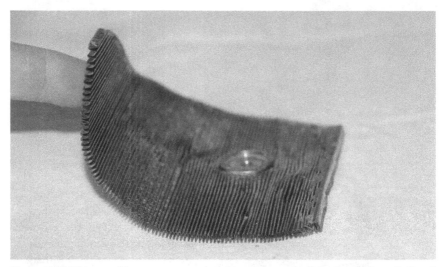

Figure 7.29. This metallic truss core was used in the bottom of the Apollo boot. It flexed when an astronaut was walking but prevented side-to-side bowing that could occur when the suit was pressurized. Photo by Bill Ayrey.

positioned and held securely in place with Velcro that joined the bottom of the pressure bladder to the sole of the boot. The vent system ducting was also attached to the bladder.

In the earlier A-5L and A-6L model suits, when the boot was pressurized, it would bow out along the bottom sole from toe to heel because of the pressures within the suit and the lack of anything rigid enough to keep it from ballooning outward. ILC engineers came up with a simple but ingenious fix that incorporated a metallic truss core that flexed when walking but prevented the width of the boot from bowing.

The pressure boots came in five sizes that adequately covered the foot sizes of the astronauts in the program.

Lunar Boots

Some of the most iconic and thought-provoking photos taken on the moon are the footprints the Apollo crews left. These footprints provide the first permanent record of human space travel beyond Earth.

There were many unknowns at the start of the Apollo program when it came to the makeup of the lunar surface and what it would take to provide the best protection for the astronauts as they ventured about the surface. The boots would be the one item of the space suit in constant contact with the lunar soil, so tremendous focus was placed on their design.

Figure 7.30. The lunar boot making a footprint that will be there for millions of years if it is not disturbed by future visitors or meteor impacts. Courtesy NASA.

The lunar surface temperatures were of concern to NASA, and they made it clear in the design requirements that the final product must address the issue. The challenge was to design a boot that would perform on a surface with temperatures up to +300°F (150°C). That was the temperature that NASA tested to. There was the unknown about the makeup of the lunar soil and its ability to support the weight of the astronauts without them sinking in too far. Traction was another concern. All these factors needed to be taken into consideration when it came to the design of the lunar boot.

As work started on the model A6L suit, Richard Ellis was busy working on one of the first lunar boot models. The engineers envisioned a fiberglass sole plate with silicone bonded onto the bottom surface. It would be connected to another fiberglass sole above by way of insulating foam that provided thermal isolation by limiting heat conduction between the two surfaces. These lunar shoes would be secured onto the pressure boot assembly by way of Velcro straps just before going out on the lunar surface. (See figure 6.5, which shows one of the early A6L suits with the slip-on TMG.) This style of boot never progressed further for a few reasons, one of the primary ones being the inability to flex the stiff fiberglass composite.

Many of the items designed for the space suit started with ideas that the engineer working on the project initiated. Then the draftsmen at ILC put

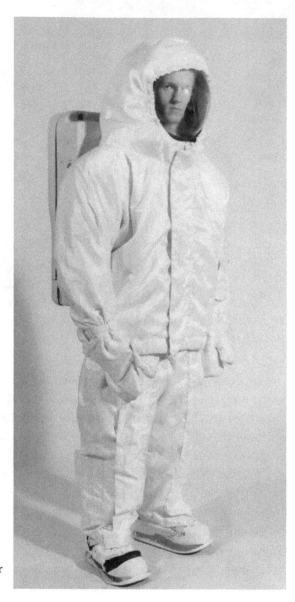

Figure 7.31. The model A6L suit with the first generation of the fiberglass lunar boot concept. Courtesy ILC Dover LP 2020.

them to paper in the form of sketches or formal drawings. For example, when the new lunar boot was designed, sketches of the tools to make the boot sole were provided to Kenny Dennis, one of ILC's very talented model makers. Kenny carved a prototype in plaster to form a mold. It was true artistry combined with model-making craftsmanship. What came out of his shop was a mold for a silicone tread that had a flange around the perimeter that was large enough that the outer protective cloth layers could be sewn to the body of the boot. It had a tall, ribbed front section that provided protection in case an astronaut stubbed their toe into a sharp rock while on the lunar surface. The sole was wide enough and long enough to likely prevent the wearer from sinking into the soft lunar soil, but how far down they might sink was an unknown at that time. This light blue-gray silicone boot sole is what formed the iconic tread pattern that became the symbol of the Apollo program. After the astronauts returned to the safety of the lunar module following their extravehicular activity, they would throw the boots out on the surface; not only were the boots no longer needed, but astronauts also had to reduce the take-off weight because they were bringing the lunar rocks back with them. However, Gene Cernan and Harrison Schmitt brought their lunar boots back with them; they are now at the Smithsonian National Air and Space Museum in Washington, DC. When I asked Cernan several years ago if the mission plan included the idea of bringing boots back, he admitted that it was his decision to bring them home in opposition to the plan. Ten pairs of lunar boots are still on the five landing sites of the moon's surface today and will be for perhaps millions of years to come.

When the boots were first being designed, the plan was to have a Nomex outer layer for the A6L block II suit. This gauntlet of this boot was cut quite high, about 14 inches above the boot sole, to provide added protection. It contained a plastic zipper that would close the gauntlet after it was slipped on over the pressure boot.

Studies had shown that the temperature on the lunar surface varies widely, from -200°F (-129°C) to +300°F (+150°C) depending on the exposure to the sun. To make the mission more manageable, all the landing sites NASA chose were exposed to lunar dawn, when the sun angle was not at its hottest nor its coldest.

A silicone made by GE had the properties to meet the requirements for the missions. It provided the flexibility needed and good strength and thermal isolation. ILC selected the GE silicone RTV-630. It would work in a range of temperatures from -75°F (-59°C) to +400°F (204°C).

The soles of the boots were made by mixing the two parts of the RTV-630 together and then pouring it into the boot mold, which was then cured in an oven until the silicone began to gel. Once it gelled, it was removed from the oven and pulled from the mold. An insert was placed inside the sole that consisted of a two-layer ply of Beta cloth stitched together to take the shape of the inside of the boot. The tackiness of the still-wet RTV helped secure the material in place. The Beta-cloth layer added structural

Figure 7.32. Details of the A6L thermal micrometeoroid garment pants being zipped. Note how the pants covered the thermal micrometeoroid garment of the lunar boot. Courtesy ILC Dover LP 2020.

Figure 7.33. ILC Engineers Richard Pulling (*left*) and Bob Wise look over drawing of the lunar boot Pulling is holding. Courtesy ILC Dover LP 2020.

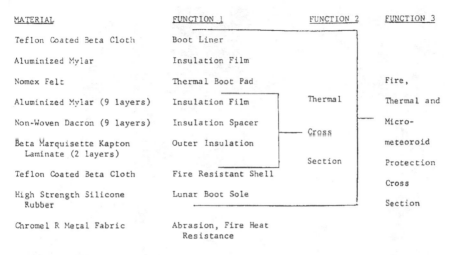

MATERIAL	FUNCTION 1	FUNCTION 2	FUNCTION 3
Teflon Coated Beta Cloth	Boot Liner		
Aluminized Mylar	Insulation Film		
Nomex Felt	Thermal Boot Pad		Fire,
Aluminized Mylar (9 layers)	Insulation Film	Thermal	Thermal and
Non-Woven Dacron (9 layers)	Insulation Spacer		Micro-
Beta Marquisette Kapton Laminate (2 layers)	Outer Insulation	Cross	meteoroid
Teflon Coated Beta Cloth	Fire Resistant Shell	Section	Protection Cross
High Strength Silicone Rubber	Lunar Boot Sole		Section
Chromel R Metal Fabric	Abrasion, Fire Heat Resistance		

Figure 7.34. This table shows the layers of the materials used in the Apollo lunar boot from the inside (*top*) to the outside (*bottom*) and the function that each one serves. "Familiarization & Operations Manual," Change 5, ILC document no. 8812700149B, October 1969.

strength. Then another layer of the RTV-630 was painted over the cloth to completely coat it. Following this process, a bag was inserted around the boot and a vacuum was pulled to press all materials together tightly. The boot was then placed under heat lamps until further gelling took place. Finally, the bag was removed and the boot was placed in a +93°C (+200°F) oven for two hours.

The inside liner of the boots was made of Teflon-coated Beta cloth, aluminized mylar, and Nomex felt to add insulation. That was followed by nine alternating layers of mylar and Dacron. The final layers consisted of Beta-marquisette Kapton film, Teflon-coated Beta cloth and Chromel-R fabric was inserted around the areas susceptible to abrasion. Chromel-R is a chromium steel cloth that cost $2,500 per yard in 1968 ($18,000 in 2019 dollars). This metal fabric resisted any punctures and abrasions because of its tough properties.

After the Apollo 1 fire and the findings of the NASA review boards, the materials used above the silicone sole were changed significantly. The A-7L boots used Beta cloth on the tongue section and multiple layers of aluminized mylar below. Most of the boot was covered with Chromel-R. The A-7L boot opened wide at the top so the pressure boot portion of the garment assembly could be slipped inside. A strap and buckle across the front of the boot secured it to the wearer.

Figure 7.35. Buzz Aldrin wearing the first model of the lunar boot in training. Because this boot did not include a top strap, the two inside snaps securing the top section came undone. Tape was used to keep the boot closed for this training exercise. Courtesy NASA.

Homer Reihm, the Apollo engineering manager at the time, said years later that he wished that they had molded the letters "ILC" into the bottom of the silicone boot soles as a form of shameless self-promotion but admitted that NASA probably would have asked that it be removed.

In an ILC memorandum on integrated lunar boots dated April 28, 1970, Apollo program manager Len Shepard addressed an issue NASA had raised with ILC engineers. NASA wanted ILC to investigate making pressure boots that would also serve as a lunar surface boot. NASA likely wanted to reduce weight and eliminate one more step in the process astronauts underwent as they suited up before opening the hatch and stepping out onto the moon. In the memo, Shepard pointed out that NASA required pressure boots to have Velcro on the bottoms to secure the feet to the command module couches and the floor of the lunar module. He also emphasized that if astronauts could not remove dirty boots and throw them onto the lunar surface, excessive dirt would be brought into the lunar module. The memo concluded by saying that since NASA did not appear interested in removing the Velcro requirement or providing further funding for redesign studies, ILC would remove this item from the engineering problem checklist.[39] The fact that no further push-back from NASA exists indicates that NASA agreed it was a dead issue and pursued it no further.

The first model A6L version lunar boot design (with the tall Nomex upper fabric) was given the part number A6L-206000. After the Apollo 1 fire and the suit redesign, the lunar boot was given part number A7L-106015-01/02. That boot included the fire-resistant Beta-cloth material. This boot design used a stud snap on each side of the Chromel-R upper flaps of the boot. This permitted the top of the boot to open wide enough to slip the pressure boot inside. Once the pressure boot was in the boot, both snaps were used to pull the top closed. However, Buzz Aldrin discovered during training for his Apollo 11 mission that rubbing against any object caused the snaps to come undone. The problem was remedied during the training by applying duct tape around the top of the boots to keep them closed. ILC looked for a solution to this problem immediately after it was discovered (duct tape not being one of them). A simple modification was made that consisted of adding a strap across the top that fastened with a snap on the other side. That in combination with the two other snaps proved to work. Lunar boots serial numbers 001 through 028 were either modified at ILC or ILC sent modification kits to Houston and the Kennedy Space Center so that upgrades could be made in the field. Serial numbers 001 through 028 of these boots were used for training. The next serial number, serial number 029, used part number A7L-106043-01/02 and was considered the first flight version. The documentation I have is not clear about what changes (if any) were made to the boots in terms of design or materials. Any changes that may have been made were very minor. They likely addressed only the insulation materials in the TMG layers and may have included the addition of the stitched letters "R" and "L" on the inside sole to denote right or left boot.

Following that, a boot with the part number A7L-106043-05/06 was made that included the crewman's name tag, which was stitched inside so the boots could not be confused during suit-up in the cramped lunar module. The lunar boots were made in two sizes: OMED (medium) and OLGE (large). The size OMED silicone soles were approximately 13¼ inches by 6 inches and had eight ribs across the width; the OLGE soles were approximately 14½ inches by 6½ inches and had nine ribs. The contract called for the lunar boots to be manufactured and delivered as an assigned item to a suit or in some cases to an astronaut.

Table 7.1 is taken from ILC documents generated sometime around August 1969 that documented lunar boot allocations as they were being manufactured. It is not updated to reflect the final versions, since there are examples where the part number A7L-106043 was changed to part number

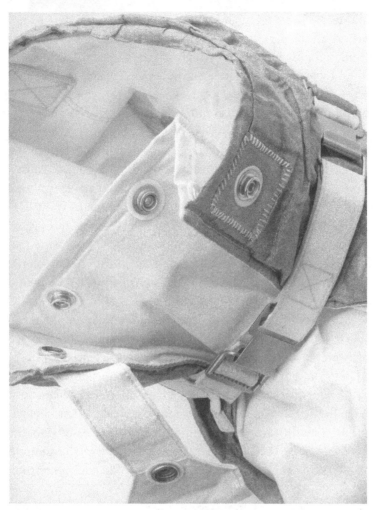

Figure 7.36. The top strap and snap assembly that was retrofitted onto the first twenty-eight boots made after engineers realized that the two inside snaps came undone too easily. This addition worked and was used in the remainder of the lunar boots ILC produced. Courtesy David Mather.

A7L-106062-01/02 and was described as essentially being of the same configuration as the model A-7L boots. For example, the lunar boots with serial numbers 055 and 062 were used on Apollo 15, and for that mission, the list of flown items records them both as having part number A7L-106062-03/04. However, the earlier list below has them identified as part number A7L-106043. It is fairly safe to conclude that this list of seventy-one pairs was likely the final count of boots manufactured for the program minus perhaps a few that were made for development purposes.

Table 7.1. Lunar boot allocation

Boot S/N	Original allocation	Production completion	Part number	Reallocation
001	A7L-001	5/3/1968	A7L-106015-01/02	Suit pool[a]
002	A6L-005	5/21/1968	A7L-106015-01/02	Suit pool
003	A6L-014	5/23/1968	A7L-106015-01/02	Suit pool
004	A7L-002	3/22/1968	A7L-106015-01/02	Young
005	A7L-003	5/31/1968	A7L-106015-01/02	Suit pool
006	A6L-013	6/26/1968	A7L-106015-03/04	Suit pool
007	A6L-012	8/6/1968	A7L-106015-03/04	Suit pool
008	A6L-024	7/1/1968	A7L-106015-03/04	Shepard
009	A6L-006	6/26/1968	A7L-106015-03/04	Shepard
010	A7L-024	7/1/1968	A7L-106015-03/04	Lovell
011	A7L-025	7/17/1968	A7L-106015-03/04	Haise
012	A7L-027	7/18/1968	A7L-106015-03/04	Armstrong
013	A6L-019	7/25/1968	A7L-106015-03/04	Shepard
014	A7L-014	7/26/1968	A7L-106015-01/02	Lind
015	A7L-020	7/29/1968	A7L-106015-01/02	Young
016	A7L-015	8/28/1968	A7L-106015-01/02	Suit pool
017	A7L-021	8/8/1968	A7L-106015-01/02	Suit pool
018	A7L-016	8/14/1968	A7L-106015-01/02	Conrad
019	A7L-018	8/19/1968	A7L-106015-01/02	Bean
020	A7L-029	8/16/1968	A7L-106015-01/02	Engle
021	A7L-030	8/28/1968	A7L-106015-01/02	Irwin
022	A7L-031	9/12/1968	A7L-106015-01/02	Suit pool
023	A7L-045	9/23/1968	A7L-106015-01/02	Suit pool
024	A7L-042	9/3/1968	A7L-106015-01/02	Suit pool
025	A7L-044	9/16/1968	A7L-106015-01/02	Aldrin
026	A7L-046	9/16/1968	A7L-106015-01/02	Scott
027	A7L-047	10/25/1968	A7L-106015-01/02	Mitchell
028	A7L-049	10/22/1968	A7L-106015-01/02	Cernan
029	A7L-039	12/6/1968	A7L-106043-01/02	A7L-039
030	A7L-050	12/11/1968	A7L-106015-01/02	A7L-050
031	(S/N not used)			
032	A7L-056	?	A7L-106043-01/02	KSC[b] spare
033	A7L-061	12/27/1968	A7L-106043-01/02	Unassigned
034	A7L-060	12/13/1968	A7L-106043-01/02	KSC spare
035	A7L-074	12/20/1968	A7L-106043-01/02	Haise
036	A7L-076	1/3/1969	A7L-106043-01/02	Suit pool
037	A7L-065	1/14/1969	A7L-106043-01/02	Conrad
038	A7L-067	?	A7L-106043-03/04	Shepard

(*continued*)

Table 7.1—*Continued*

Boot S/N	Original allocation	Production completion	Part number	Reallocation
039	(S/N not used)			
040	(S/N not used)			
041	(S/N not used)			
042	(S/N not used)			
043	A7L-051	1/15/1969	A7L-106043-03/04	Aldrin
044	A6L-065	1/20/1969	A7L-106043-03/04	KSC spare
045	S/N not used			
046	S/N not used			
047	S/N not used			
048	S/N not used			
049	A7L-060	2/7/1969	A7L-106043-03/04	KSC spare
050	A7L-057	4/15/1969	A7L-106043-05/06	Armstrong
051	COOPER	4/15/1969	A7L-106043-05/06	Duke
052	A7L-068	4/16/1969	A7L-106043-05/06	Engle
053	A7L-070	4/22/1969	A7L-106043-05/06	Bean
054	A7L-074	4/24/1969	A7L-106043-05/06	Lovell
055	A7L-080	5/5/1969	A7L-106043-05/06	Irwin
056	A7L-078	5/17/1969	A7L-106043-05/06	
057	A7L-084	6/5/1969	A7L-106043-05/06	
058	A7L-082	6/10/1969	A7L-106043-05/06	
059	S/N not used			
060	S/N not used			
061	Mitchell	7/16/1969	A7L-106043-05/06	Mitchell
062	Scott	7/16/1969	A7L-106043-05/06	Scott
063	Lind	7/23/1969	A7L-106043-05/06	Lind
064	Young	7/23/1969	A7L-106043-05/06	Young
065	Medium	8/25/1969	A7L-106043-05/06	
066	Large	8/25/1969	A7L-106043-05/06	
067	Large	8/25/1969	A7L-106043-05/06	
068	Large	8/19/1969	A7L-106043-05/06	
069	Medium	8/29/1969	A7L-106043-05/06	
070	Large	8/26/1969	A7L-106043-05/06	
071	Large	8/27/1969	A7L-106043-05/06	

Source: Compiled from ILC documents.
Notes: a. "Suit pool" meant that the boots were set aside for use as needed to support all testing and training operations.
b. KSC = Kennedy Space Center.

Pressure Helmet

The bubble helmet, as it was also called, was a Lexan polycarbonate plastic material that could stand up to a lot of abuse while providing good visibility. The helmet contained a feed-port pass-through on the left side so that both liquid and paste-like food could be provided to the astronaut in the event that they could not remove their suits due to an emergency or an otherwise abnormal situation when they had to remain in the suits for an extended period. A padded piece of aluminum added to the back section of the bubble served two functions. It cushioned the head of the astronaut as he lay in the couch and experienced high G forces during launch and landings. It basically served as a pillow. It also helped direct air flow from the back of the neck ring, where the air duct blew air into the suit. This airflow would dissipate over the top and down the front of the bubble, taking in expelled carbon dioxide and providing fresh oxygen. This also cleared condensation from the inside of the helmet as the astronaut breathed out warm, moist air against it.

As I mentioned earlier in the story, a company named Texstars initiated the early development of the bubble helmet while working with Dr. Robert Jones and Jim O'Kane of NASA. This proved to be a big turning point for the development of the ILC suit because it radically changed the upper torso of the suit, making it much easier to accommodate the shoulder and arm openings. NASA eventually turned over the early bubble helmet work to ILC so that they could carry it through to the point that production units could be made. The techniques used to produce vacuum-forming polycarbonates were relatively new processes that were advancing rapidly with results that promised good optics and increased scratch resistance. Other coatings needed to be applied to absorb ultraviolet radiation and increase the optic capabilities. ILC gave a purchase order to the Missiles and Space Division of LTV Aerospace on June 10, 1966, to fabricate polycarbonate bubble helmets in four different phases per the terms of the purchase order. The results showed that at that time, a one-piece, formed polycarbonate bubble could not pass the optics test but a two-piece unit could. Thus, two sections were formed, a front and a back section that were bonded together. This polycarbonate bubble was then bonded to the metal neck ring that would attach to the mating suit-side connector, which had a small rubber O-ring that provided a tight seal. This new helmet weighted quite a bit less than the old-style ILC or Gemini helmets, and putting it on took just a few seconds versus as long as a minute with the old style.

In September and October 1966, NASA requested that a study be carried out to evaluate the ILC bubble helmet against a Litton dome helmet. At that time, NASA was very interested in evaluating the Litton hard suits for future missions that would likely include extended stays on the moon and they thought that the Litton dome helmets might have potential to take the place of the bubble helmet. In a controlled study, the ILC bubble helmet scored 394 points while the Litton dome helmet scored only 305. This was because the bubble helmet had a greater field of vision in many of the tests.

Eventually, Air-Lock Inc., the manufacturer of almost all the hardware for the Apollo space suit, got the contract to manufacture the pressure helmet for Apollo. They had excelled in the pressure-suit hardware business, which included connectors and bearings, since the 1940s. Today they continue to supply the hardware used on the International Space Station EVA* suits.

The Apollo helmet attached to the suit-side neck ring, where it was secured by locking pins that extended into a continuous groove machined around the base of the helmet hardware. Alignment marks were added to both the helmet and the suit hardware to ensure proper alignment. If the alignment wasn't correct, the oxygen vent in the back of the neck ring could be blocked and the astronauts would not receive the proper flow. Later, space shuttle–era helmets had a discontinuous groove around the base so the front pin on the neck ring of the suit would engage with a mating oval grove in the helmet that prevented the helmet from rotating and ensured that the vent ducts were aligned.

Richard Martin was a materials technician who worked closely with George Durney from 1964 to 1972. Although George was a tough guy to work for because of his demands and strong character, Richard recalls that the two of them got along quite well. When he had the chance, Richard was a free-dive spear fisherman who competed in national tournaments. One of Richard's projects at ILC involved developing a procedure for repairing polycarbonate pressure helmets that had developed scratches and light abrasions due to use, primarily during training. The helmets were quite expensive and NASA was happy to find ways to repair them instead of spending the money for replacements. NASA had been using a petroleum-based product and Velcro to mechanically polish the polycarbonate surfaces, but that ended up crazing the surface due to the chemical reactions. Richard used a process that involved hand-polishing the polycarbonate with a soft cloth and an abrasive paste that contained amorphous silica suspended in a soap solution combined with mineral oil. He used two different fine-grit

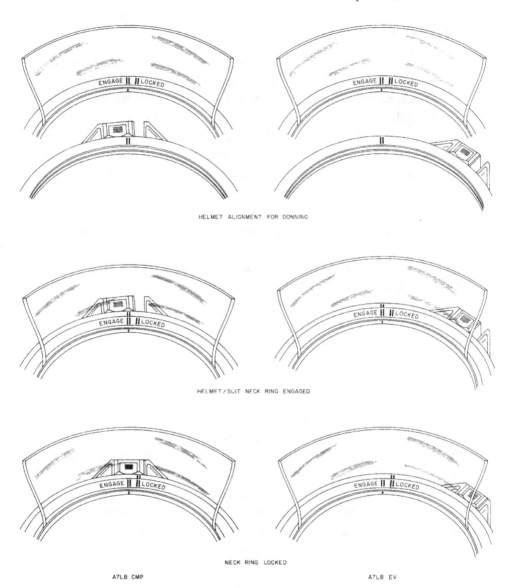

Figure 7.37. The alignment and locking operations of the pressure helmet and neck ring as viewed by the crew member inside the suit. Courtesy ILC Dover LP 2020.

sizes to polish the surface. He finished the process with a proprietary emulsive polish with an abrasive estimated to be 1 to 10 microns in size. Just days before the Apollo 10 mission, Richard was flown down to the Kennedy Space Center to polish the helmets of astronauts Young, Stafford, and Cernan using his technique.[40]

Extra-Vehicular Visor Assembly

LTV manufactured the extravehicular visor assembly, or lunar extravehicular visor assembly, as it was known more commonly during the lunar missions. Work on the extravehicular visor assembly started while ILC was developing the model A-6L suits. Thus, the first units developed for testing were given the A-6L part number that was later updated to A-7L numbers after some minor changes had been made. Only the model A-7L models flew on missions. The shell consisted of a molded, high-temperature polysulfone plastic that was bright orange, almost red. The shell was 0.125 inches thick. The assembly was designed to slip over the pressure helmet and be held in place by a clamping mechanism that fastened in the front just over the neck-ring hardware.

Apollo 9 astronauts Dave Scott and Rusty Schweickart used the first brightly colored extravehicular visor assemblies during their planned spacewalk that took place during Earth orbit on March 6, 1969. This was the first test of the entire extravehicular mobility unit, which consisted of the suit and the backpack. NASA conducted heat-load studies around this time and found that the uncovered polycarbonate shell contributed to higher than expected heat loads, so a Beta-cloth cover was designed for use on all remaining extravehicular activities. Heat loads were expected to be more significant on the lunar surface when taking into account the reflections off the lunar surface or the sides of craters when the astronauts were working around them. Underneath the Beta-cloth cover layers, the color of all the remaining Apollo lunar extravehicular visor assembly shells was still the bright orange-red polycarbonate plastic, but it was no longer visible.

Documentation shows that the first article configuration inspection for the lunar extravehicular visor assembly was conducted at the Missile Systems Division of LTV in Grand Prairie, Texas, on March 13, 1969. That happened to be the same day that the crew from Apollo 9, which had used a version of the extravehicular visor assembly for the first time a few days earlier, splashed down in the Pacific Ocean.

The lunar extravehicular visor assembly had one retractable, UV-stabilized, polycarbonate clear protective visor. This was to provide thermal control and protect astronauts against bump impacts and micrometeoroid impacts. On the outside of that was a retractable gold-coated polysulfone sun visor that protected astronauts against the light and ultraviolet rays and cut down the heat gain inside the helmet. Both visors could be raised and lowered as needed when a force of four pounds was applied to the

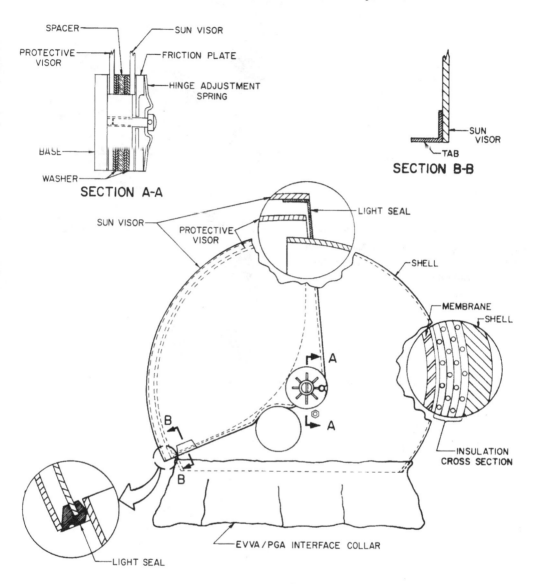

Figure 7.38. The early extravehicular visor assembly. Courtesy ILC Dover LP 2020.

tab. Finally, there were two side eyeshade assemblies made of fiberglass that could be lowered independently of the sun visor to restrict any sun penetration based on the sun angles an astronaut experienced. During the Apollo 12 crew debriefing, Pete Conrad suggested that NASA modify the visor to include a center top shield so that sun reflection could be eliminated based on the angle of the sun. This redesign was carried into the

Top: Figure 7.39. Dave Scott wearing the red polycarbonate extravehicular visor assembly used on the first Apollo extravehicular activity on the Apollo 9 mission in Earth orbit. (Photo courtesy NASA).

Middle: Figure 7.40. This configuration of the lunar extravehicular visor assembly was used on the Apollo 11 and 12 missions. It had a clear protective visor and a gold reflective visor and used only the two side shades. It had no center shade. Courtesy NASA.

Bottom: Figure 7.41. This NASA figure shows some of the details of the final configuration of the lunar extravehicular visor assembly , which protected against sun angles that could severely impair vision. This configuration of the assembly was carried on Apollo 13 through 17. Courtesy NASA.

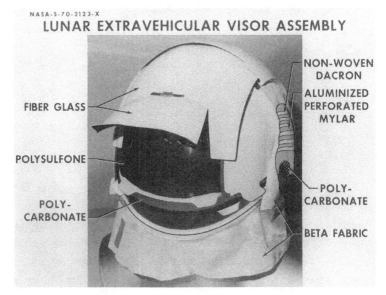

NASA-S-70-2123-X

LUNAR EXTRAVEHICULAR VISOR ASSEMBLY

FIBER GLASS

POLYSULFONE

POLY-CARBONATE

NON-WOVEN DACRON

ALUMINIZED PERFORATED MYLAR

POLY-CARBONATE

BETA FABRIC

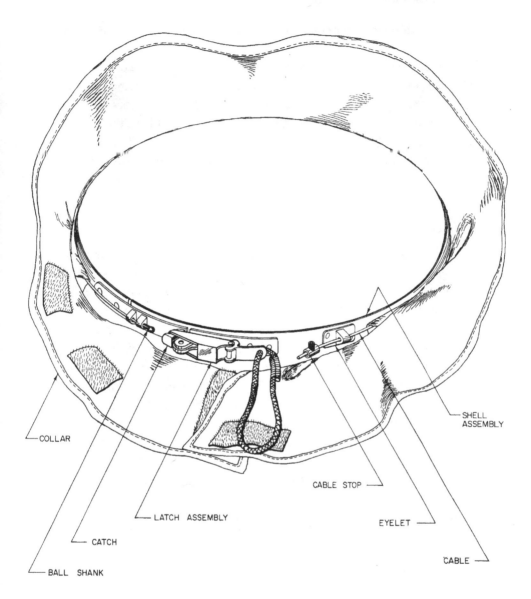

COLLAR

SHELL ASSEMBLY

CABLE STOP

LATCH ASSEMBLY

EYELET

CATCH

CABLE

BALL SHANK

Figure 7.42. Details of the latching mechanism located around the base of the lunar extravehicular visor assembly. It was slipped over the pressure helmet until its base was located over the hardware of the neck ring of the helmet. A slot in the end of the catch was inserted over the ball shank. The catch assembly was then pushed inward and a mechanical tension force locked it into position. It could be removed by pulling on the cord on the right side of the latch assembly to relieve the tension and separate the latch. Familiarization & Operations Manual for Model A7L, ILC document no. 8812700149B, June 6, 1969.

Apollo 13 mission, where of course it was not used. It was used on Apollo 14–17.

Sometime in 1971, Air-Lock Incorporated was given the contract to design and manufacture the Skylab extravehicular visor assembly. This unit was very similar to the Apollo units but had protective visors that could be replaced during orbit. The center eye shades did not have the outward-tilting brim; they had just a center shade that could be pulled down. The polycarbonate shell was molded in the color white, which was more than enough to radiate any solar heat loads and protect against micrometeoroid impacts while in Earth orbit.

Gloves

ILC began developing the pressure-suit gloves to mate with their suits in the 1950s. It was one of the most difficult parts of the suit to design and fabricate due to the nature of the work it had to support and the difficulty involved in designing something that had to fit the wearer so well and accommodate all the motions that the hand provides. That includes finger flex, hand flex/extension (fore and aft motion), adduction/abduction (flex from side to side), and wrist rotation. The gloves had to accommodate all these motions while pressurized at 3.75 pounds per square inch.

The gloves underwent many changes over the years leading up to the model A-7L suit. The basic design did not change radically from the earlier models. They were rubber dipped and supported between dips with a lightweight fabric to keep the rubber from stretching under pressure. A second separate skin of rubber-dipped material was added over the palm area that was called the fingerless glove. This provided added strength and guarded against abrasion. Metal cables were added to provide strength and take the loads that would develop when pushing fingers and finger crotches against the glove. Without this extra support, the gloves would rip from the metal disconnect between the glove and the lower-arm hardware. The glove's TMG underwent many changes but basically served the same purpose of providing resistance to cuts and abrasions and thermal protection between the astronaut and hot or cold objects they encountered.

Engineers began developing the model A-7L gloves that were the first to fly on a mission in June 1967.[41]

Pressure-Glove Molds and Dipping Process

All of the crews would fly with the latex-dipped intravehicular black gloves that had minimal layers of materials. They were custom made for a tight fit that gave the astronauts decent dexterity and tactile feel. This was important since they had to interface with relatively small switches and other controls within the command module and the lunar module.

The crew members assigned to perform extravehicular activity also carried one or two pairs each of EVA gloves that used the same basic latex-dipped intravehicular glove but also had a cover layer installed on top of it. This was made up of Chromel-R woven stainless steel fabric in the front palm and finger areas that were susceptible to high abrasion and the chance of cutting into the glove and causing leakage. The backs of the glove and the gauntlet were made from Beta cloth and the glove had many layers of aluminized mylar inside to reflect heat loads.

The process for manufacturing the Apollo gloves started out like many of the rubber-dipping processes for the convolutes throughout the suit. ILC designers reasoned that dipping a mold of a hand in a blend of neoprene and natural rubber, then sandwiching a thin, high-tensile, stretch-resistant fabric between dips would result in a flexible glove that would do the job. The problem was that the glove needed to hold pressure while at the same time taking a lot of abuse as astronauts operated spacecraft controls and used tools. It also had to flex and extend at the wrist (bend forward and backward) and flex at the fingers. There was also the issue of adduction/abduction, which is the side-to-side motion that happens when you swing a hammer, for instance. These motions had to be performed while the glove was pressurized. Added to that is wrist rotation that is needed to complete the full motion of the hand. That was taken care of by providing a rotating wrist bearing that also serves as the disconnect where the gloves are attached/detached.

I often explain to people that the whole reason to wear a space suit is to provide the necessary means to use your hands in some fashion. There have been no space missions where the astronauts simply ventured outside and looked at something and then returned to the safety of the spacecraft without extensively using their hands. That means that the gloves need to be the best fit possible and they need to hold up to brutal wear and tear. Another ILC employee liked to say that a space suit glove needs to be able to stop a bullet yet pick up a dime.

Early in the process, George and the engineers worked with the model

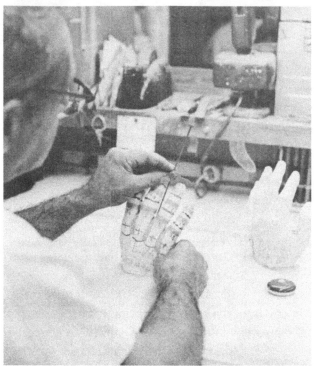

Figure 7.43. Model maker Julius Herrera works on the molds for gloves for astronaut Ron Evans. Courtesy ILC Dover LP 2020.

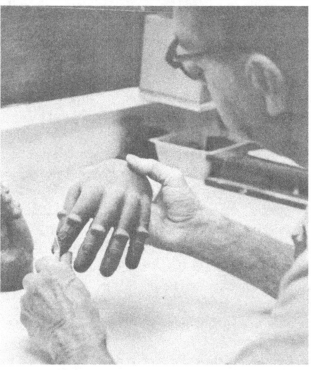

Figure 7.44. Model maker Harry Saxton puts the finishing touches on the dip form that was used to dip the custom-made latex glove bladder for each astronaut. Courtesy ILC Dover LP 2020.

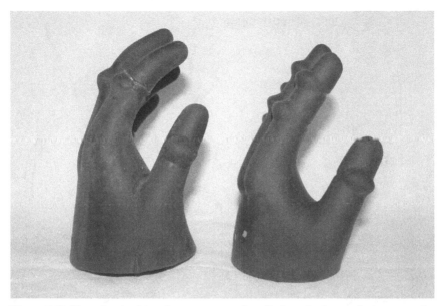

Figure 7.45. These hard rubber forms were used to dip into the latex solutions to make the pressure gloves. Model makers custom made them based on each astronaut's hand size. Glove engineer Dixie Rinehart redesigned the glove in a more natural curved-finger position (*left*), in contrast to the older, straight-fingered mold (*right*). Photo by Bill Ayrey

makers to design glove molds that took on the shape of a hand. However, the fingers in the earlier versions were straightened so the rubber could be easily stripped from the molds after curing. ILC brought on an engineer named of Dixie Rinehart in 1966 to work primarily on the gloves. Dixie came from North American Aviation with experience as a research analyst and had previous experience in the air force as a physiological training specialist. Dixie was well liked by all at ILC. Apollo veterans remembered him as the guy who favored turquoise jewelry and drove around Dover with a teepee on the roof of his station wagon. He also sported a bumper sticker that read "Custer had it coming." Several employees recall sitting around Sambo's Tavern after many long days of work listening to his stories and laughing at his jokes.

Early in 1969, Dixie and the model makers started working on new dip molds that provided a more exaggerated bend to the fingers. This followed comments from astronaut T. K. Mattingly in a memo dated July 22, 1968 that said "Develop a glove which requires less effort to maintain a grasp. Thoroughly investigate the feasibility of molding a bladder which retains a grasping configuration when pressurized (i.e., the claw)."[42] Stripping the

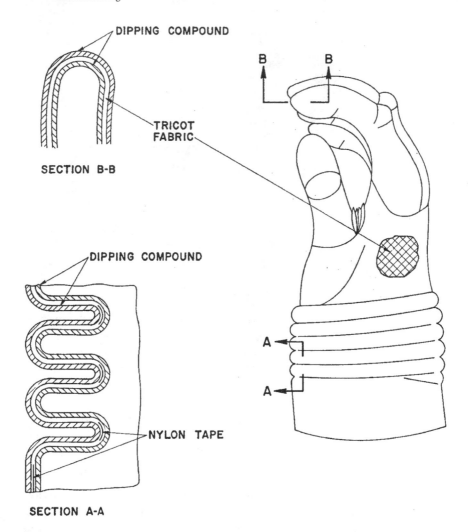

Figure 7.46. This figure and cutaway show the separate layers of the dipped glove bladder. Nylon tapes were used in the root of the convolute for structural support and the dipped rubber of the hand section was supported by nylon tricot fabric. Courtesy ILC Dover LP 2020.

rubber bladders from the molds with the curved fingers was not the problem that others had earlier assumed that it would be. These newer gloves were made available in time for the Apollo 10 mission. They were given the part numbers of A7L-201100-02 for the thermal cover layer and A7L-1030000-13/14 for the pressure-glove assembly.

Dixie called Julius Herrera, Harvey Kemp, Kenny Dennis, Harry Saxon, and Tom Townsend the Michelangelos of the model-making shop. These guys were the best at the artistry of making the various tools and molds for fabricating the suit parts.

These molds were dipped in a neoprene and natural rubber blend solution. The dipping technicians reinforced the layers between dips with a single layer of nylon tricot restraint fabric that provided strength and kept the rubber glove from expanding when pressurized. Additionally, nylon tapes were added to the inner curvatures of the rubber convolute to counter the stress brought about by the internal suit pressure and mechanical loads. Unfortunately, this rubber-dipping process was very difficult to perform and a lot of scrap was produced. According to Dixie, the reject rate would reach as high as 80 percent at times during Apollo. Inspectors would use magnifying glasses to inspect the entire surface of the glove bladders; there could be very few bubbles or other defects.

Thermal Micrometeoroid Garment

The TMG on the earlier model A-7L gloves were made from the very expensive Chromel-R fabric and covered the entire hand section except for the access flap for the palm-bar buckle on the back side of the hand. That was covered by Beta cloth and was secured with snaps and Velcro. The Beta-cloth gauntlet was noticeably longer; it was designed that way to cover the pressure relief valve on the lower left arm and the pressure gauge on the lower right arm.

As the pressure gloves were being redesigned, the TMG cover was also redesigned. In January 1967, astronaut Joseph Kerwin wrote one of the earlier memos related to the pressure glove after a manned qualification test in NASA's eight-foot vacuum chamber. He made it clear that the 32-layer extravehicular gloves did not perform well at all. He could not work the controls on either the remote control unit or on the primary life-support system because of the bulkiness of the gloves. When he switched over the intravehicular pressure glove, he had no problems working the switches and valves.[43] By April of 1967, NASA engineer Jim O'Kane was researching how the thirty-two layers of materials could be reduced to only seven layers and still pass all thermal testing.

The astronauts reported they could not pick up objects using the early-model gloves because of the way the Chromel-R material was terminated at the fingertips. In response, ILC added light-blue silicone fingertips. The tips were made from GE's silicone RTV-630 material, the same material used on the lunar boots. The tips were sewn onto the Chromel-R material. A shorter, less bulky Beta-cloth gauntlet was made to just cover the glove

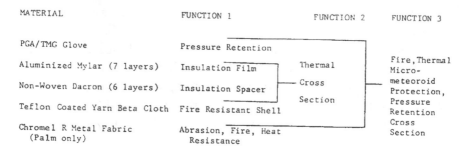

MATERIAL	FUNCTION 1	FUNCTION 2	FUNCTION 3
PGA/TMG Glove	Pressure Retention		Fire,Thermal Micro- meteoroid Protection, Pressure Retention Cross Section
Aluminized Mylar (7 layers)	Insulation Film	Thermal Cross Section	
Non-Woven Dacron (6 layers)	Insulation Spacer		
Teflon Coated Yarn Beta Cloth	Fire Resistant Shell		
Chromel R Metal Fabric (Palm only)	Abrasion, Fire, Heat Resistance		

Figure 7.47. Cross-section showing the details of the materials used in the Apollo extra-vehicular gloves from the inside (*top*) to the outside (*bottom*). "Familiarization & Operations Manual," Change 5, ILC document no. 8812700149B, October 1969.

disconnects but not the relief valve or the pressure gauge. The Beta-cloth flap on the back of the hand was removed and replaced with the Chromel-R fabric and secured with Velcro only.

Glove Assembly

The completed EVA pressure glove consisted of the latex pressure glove, the fingerless glove (assembled over the latex pressure glove), the palm bar restraint hardware, the thermal cover layer over the wrist area, and the wrist hardware, which was attached to the metal wrist disconnect.

Crew members had the option of wearing a comfort glove under all of this, but many thought it added too much bulk when combined with the other layers of the pressure glove and the TMG. A cover glove was designed to protect the glove that crew members used only occasionally, generally when they were training for their missions.

Small-diameter aircraft-grade cables attached to the metal disconnect hardware of the glove to give it added strength and permit the flex-extension and side-to-side motion. When they were attached to the disconnect hardware, they acted as a hinge for the wrist action in all directions. Because areas such as the palm of the glove would want to take a round tubular shape when pressurized following the basic laws of physics, the gloves had what was known as a palm bar that wrapped around the palm to the back of the hand and was bent inward on the palm side to conform to the natural cupped shape of the palm so that when the glove was pressurized, it was pulled in tight against the palm so the crew member could grasp objects, tools, and so forth. It also acted as a natural bending point when the palm was flexed.

The glove bladder, which was made of a blend of neoprene and rubber, was reinforced with a second layer of rubber bladder cloth that was called a fingerless glove. This covered the hand below the fingers and provided extra reinforcement, particularly in the finger crotch area. Early evaluations of the intravehicular gloves indicted that they abraded very easily. This was a concern, since crew members would be required to perform a multitude of hand operations inside the spacecraft if they had to remain inside the suits due to some unforeseen circumstance.

There was concern about protection should a fire break out in the spacecraft while on the launchpad, as happened with the Apollo 1 spacecraft. The latex rubber intravehicular gloves were exposed to potential fire damage because they did not have cover layers. At that time, NASA engineers were studying new materials that could offer improved protection from fire. One of them was a material known as Fluorel. This came in various elastomeric forms that could be molded or extruded but also could be made as solvents that could be brushed on.[44] ILC worked with NASA and arrived at a solution by painting on a 10-mil surface layer of Fluorel 1066 to the intravehicular gloves that raised the flammability temperature and provided a layer of protection against abrasion. Astronaut Dave Scott once raised a concern that the gloves were too stiff and did not provide the range of flex needed. He attributed it to the coating of Fluorel on the outer surface of the rubber glove. He suggested that ILC only add the Fluorel to the fingertips. In the concluding comments of an ILC memo written to address his concerns, it was recommended that Scott's gloves be fully coated to provide the fireproof coating and add the abrasion resistance that was necessary.[45]

ILC supplied the intravehicular gloves for all of the Apollo astronauts and built another identical pair that had the TMG cover layer permanently integrated onto it for the crew members who performed extravehicular activities. NASA had asked Dixie if he could design the EVA gloves so that the TMG could be removable if the crew returned to the lunar module from the surface and had problems repressurizing the module. The original plan was to repressurize the lunar module, then remove the EVA gloves and the primary life-support system and lunar boots. The astronauts would then connect their suits to the oxygen supply system of the lunar module, put on their intravehicular gloves (which did not have the TMG assembly), and then repressurize their suits. Then they would depressurize the lunar module, open the door of the lunar module, and throw the primary life-support system and the lunar boots onto the lunar surface. This was so they could offset the weight of the lunar rocks they were bringing back home with them. However, NASA was looking for an option where they could remove the relatively bulky glove TMG so they could operate the small switches on board the lunar module just using the thinner intravehicular glove in case they could not repressurize the lunar module and remove their EVA gloves. NASA always looked for contingency plans in the event of unanticipated problems. Dixie put a lot of effort into making the EVA glove so that the TMG was secured to the bladder with patches in twenty-one locations, including the sides of the fingers and at the base of the glove where Velcro was used to secure it. Yet the TMG could still be removed if necessary.

Dixie and the ILC team worked very hard to eliminate many of the earlier glove problems. As a result, they had mostly favorable comments from many of the astronauts, but they were by no means perfect. After the Apollo program concluded, many of the Apollo astronauts became more vocal about how difficult it was to work in the gloves, but that could be because as their memories faded somewhat, the most annoying part of the suit was the things they remembered the most. Many of the EVA crew members had problems with their fingernails turning black because they were held tight in the glove as they worked hard using various tools and performing other activities.

A 1990 NASA report titled "EVA Gloves" included interviews with Apollo astronauts Gene Cernan and Jim Irwin about their experience using the EVA gloves. Gene's comment was "I felt very much at home, and as far as we were in those days, we had as good a glove as we could have built at that point in time. It was not a limiting factor in the operation of the suit at all." Jim Irwin commented that he felt that the gloves provided all the dexterity he needed and gave him the ability to perform all the functions needed throughout the mission. He did complain that there was very poor air circulation to the hands, which meant that they were warm and wet. He felt tight in the fingertips and had to cut his fingernails as short as possible to make it bearable.[46]

I have spent many hours in the latest model gloves used on the space shuttle and now the International Space Station. I believe that if any of the Apollo crew members could try out this new glove compared to the old dipped-rubber model of Apollo, they would be truly amazed at the difference. The gloves are a prime example of how advancements in materials and design can significantly improve how humans function in outer space. Work on improving the performance of gloves is never ending because the users will always desire a glove that they forget they are wearing that at the same time provides a barrier between their flesh and a +280°F or -180°F

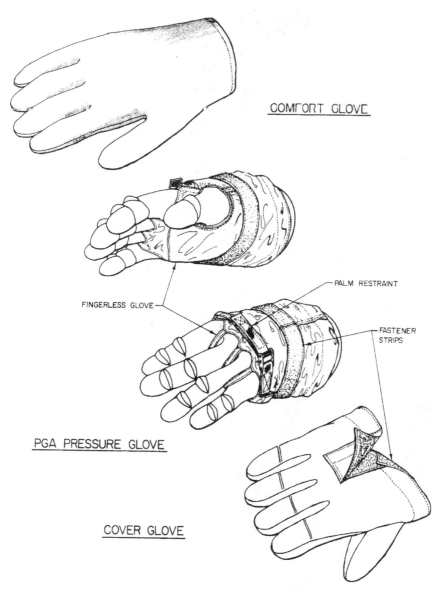

COMFORT GLOVE

FINGERLESS GLOVE

PALM RESTRAINT

FASTENER STRIPS

PGA PRESSURE GLOVE

COVER GLOVE

Figure 7.48. The top-level items that made up the EVA glove assembly. It consisted of the optional comfort glove and the pressure glove that had a fingerless glove integrated over it to provide support. The glove cover was used only to provide protection for the flight gloves when the astronauts were training before their mission. Courtesy ILC Dover LP 2020.

Figure 7.49. A pair of special protective cover gloves worn over the EVA gloves to provide protection from abrasion. Apollo 12 astronaut Alan Bean is in the suit for some training work, most likely for the backup role he played on Apollo 9. Courtesy NASA.

tool or handrail. They must also provide the correct gas pressure and protection against radiation and micrometeoroids.

Attachment Hardware

Occasionally I'm asked why the hardware that facilitates all the component connections comes in different colors. The glove and suit disconnects on the right are red and the left are blue. The oxygen and electrical connectors to the torso are red and blue. The neck ring started off as blue and ended up being red.

This coloring scheme started out during the Gemini program, when early ground testing revealed a problem that could occur if a pressure-relief valve was not inserted into the outlet side of the suit's gas manifold. One set of gas connectors allowed oxygen to flow into the suit through a hose from the gas source and the other set let it flow out of the relief valve to atmosphere. This relief valve was set to open and dump the excess pressure from the suit once it reached the prescribed pressure. If the relief valve was not installed, the suits connector was closed and did not allow the inside air to exit. As a result, the suit would overpressurize and burst at a very high pressure, potentially injuring the subject inside.

A technician from David Clark was working closely with Mr. James McBarron, the NASA engineer, when they discussed solutions to making sure the relief valve was inserted into the proper gas connector on the suit. As a result, the exhaust hardware was anodized in a red color, as was the mating relief valve. That way, everyone could be assured that the relief valve was in the proper hardware piece. Taking it a step further, it was decided that the inlet gas connectors would be colored in a blue anodize finish. Since the gas connector hardware was now going to be red on the right side and blue on the left, McBarron decided that to make it aesthetically appealing, they would keep everything on the right red and everything on the left blue, including the glove hardware. This was the order of the day and once ILC started the design of the early Apollo suit, other miscellaneous hardware pieces would be colored blue, including the electrical connectors, the water connector for the liquid cooling garment, and the helmet and the mating suit-side disconnect. That was the way it was until a change was made to the vent tube inside the suit where it interfaces with the back of the pressure helmet. ILC was having issues achieving acceptable oxygen flow into the suit by way of this vent into the helmet; the solution was to increase the size of the vent. This changed the hardware dimensions and made it necessary to change not only the vent (which was attached to the suit-side neck ring) but also to the vent pad opening on the back of the pressure helmet. Because so many suits and helmets had already been made with the smaller vent size, there was fear that the helmets with the smaller vents would be attached to the newer suits with the larger vents. Even the older suits could be modified with the larger vents, which would cause problems with the old helmets. As a result, the new helmets were made with the red anodized ring and the suit-side hardware with the larger vents came in the anodized red color. The blue helmets were not allowed to be used with the red neck rings, and so forth. This solved the problem and resulted in both blue and red helmet and suit neck rings.[47]

Liquid Cooling Garment

NASA's ideas about cooling garments were based on a 1962 design by Drek Burton, a human engineering expert with the Royal Aircraft Establishment in the United Kingdom.[48] In 1963, NASA hired a British doctor, John Billingham, to head its Environmental Physiology Branch in Houston. Billingham had a medical degree from Oxford University and had done physiology-related work as a medical officer with the Royal Air Force for seven years.[49] Before coming to NASA, Billingham had worked closely with Drek Burton, and the two had led efforts within the RAF to study the effect of liquid cooling for their pilots. The results showed promise.

The early Mercury and Gemini suits relied on gas to cool the oxygen provided by the on-board environmental control system, which pumped gas into and out of the suits. Since these early astronauts did not sweat to any measurable degree, the system could carry away a very acceptable 720 to 1200 Btus of heat per hour.[50] If much higher workloads were anticipated, it was estimated that gas flow rates as high as 40 to 80 cubic feet per minute would be needed, which is not practical or even possible in a space suit. The flow rate throughout the Apollo space suit was 12 cubic feet per minute maximum and only 6 cubic feet during the lunar extravehicular activities, when most of the energy was being expended. As the future missions were being laid out on the drawing boards, Billingham and others in NASA knew that they had to come up with a better solution for keeping the astronauts cool, since they planned much more work for them, such as venturing about on the lunar surface while under solar heat loads.

When astronaut Ed White took the first American spacewalk, he simply did a gentle float outside the spacecraft, expending very light to moderate energy. One year later, on Gemini 9, Gene Cernan had to cut short his EVA after two hours of wrestling the entire time with his positioning outside

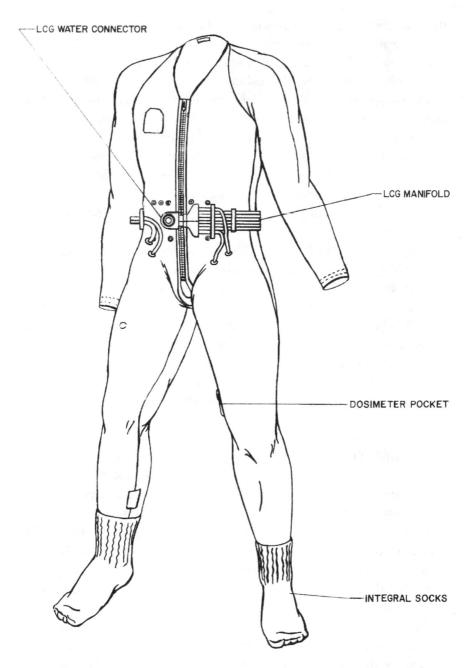

LCG WATER CONNECTOR

LCG MANIFOLD

DOSIMETER POCKET

INTEGRAL SOCKS

Figure 7.50. Illustration of the design of the liquid cooling garment. Courtesy ILC Dover LP 2020.

the spacecraft in zero gravity with a suit he described as having "all the flexibility of a rusty suit of armor."[51] A great deal was learned by his failure to complete the EVA, one of which was the fact that the spacecraft had no handholds. This meant that he had to work extra hard to position himself in the zero-gravity conditions while also fighting the umbilical line that supplied oxygen from the capsule. Cernan lost thirteen pounds of weigh due to sweating and his heart rate climbed to 180 beats per minute. His visor fogged to the point that he had zero visibility. When it was all over and Gene was back inside the capsule, he and everyone at NASA were aware of how close he had come to not making it back alive. By this time, work to solve the problem for future Apollo astronauts was well under way. The liquid cooling garment was the solution.

Initial testing at the Manned Space Center in 1964 under the guidance of Dr. Billingham used a British prototype garment that pumped ice water through small tubes at a rate of 3,400 Btus per hour. Testing revealed that a test subject enclosed in a pressure suit was working at a very high rate of 1,350 Btus per hour. The removal of excess Btus resulted in a very cooled-down subject.[52] Overall, liquid cooling was about 70 percent more effective than gas cooling. The liquid cooling system was also projected to save a lot more weight than a gas circulation system because the primary life-support system required less energy to pump water than it would to pump chilled oxygen.

Because Hamilton Standard and ILC had the contract for the suit work in 1964, when this research was being done, Hamilton worked with NASA to study the effects of liquid cooling. Their results confirmed the work that Billingham and Burton had started. Hamilton Standard obtained a patent on their garment design. NASA later obtained that patent so that ILC could use this basic design and refine it over the next few years. In addition to the very early work accomplished by NASA and Dr. Billingham, the development of the liquid cooling garment can be attributed to other organizations, including Garrett AiResearch, Honeywell, Webb Associates, Litton, and several others.[53]

During the early ILC-Hamilton teaming agreement, Hamilton had used B. Welson & Company, which was close to their Connecticut factory, to cut and sew the parts together for the development liquid cooling garment. The early Hamilton Standard evaluation units proved that the liquid cooling garment was effective but not ready for actual flight certification. That would require a garment that met strict quality and reliability requirements.

It would also need to be designed to survive the long-term wear and tear that would happen during extravehicular activities.

After ILC won the 1965 contest, it gave the initial contract for the liquid cooling garment to B. Welson since they were experienced with the basic design and manufacture of the garment. ILC staff and designers were overwhelmed with suit production work at the time and they did not need to be taking on another separate garment, regardless of how relatively simple it was. In a status report dated March 6, 1967, NASA noted that it was doing in-house development work to establish a way of providing head cooling while removing lower arm cooling and liner removal and studying their effects.[54] Ultimately, no head cooling was ever introduced as it was shown to be of little benefit.

The liquid cooling garments were made from spandex material that provided high elastic elongation, which was important because the vinyl tubes that provided the cooling water had to maintain contact with the crew member's body to pick up and remove body heat. The spandex used an open weave that made it possible to weave the tubing through the various strands that held it in place.

A liner was provided that was made from a chiffon fabric. This lightweight, thin material, which was sewn inside the liquid cooling garment, prevented the tubing from snagging the crew member as they were putting the garment on.

The vinyl tubing that provided the circuit for the cooling water had an inside diameter of one-eighth inch and was about 300 feet long. The ends of these small tubes were bonded into manifolds that consolidated all the tubing to form an inlet from and an outlet to the primary life-support system.

There were some design challenges during the period B. Welson had the contract because of the way they secured the vinyl tubing to the spandex material. Apollo engineer Ron Bessette recalled a very dedicated lady at B. Welson named Gerry who would use a sewing machine to make what was known as a drag-stitch, which would jump back and forth over sections of the small tubing at various intervals along the length of the tube to secure it in place to the spandex stretch fabric. This drag stitch would continue to the next area of the tubing in a continuous run of thread. The drag-stitch thread occasionally got caught in the zipper closure of the suit and there were other minor problems with the design of the garment, but the recollection of the ILC veterans is that overall, B. Welson was very good to work with and its employees always tried their best to work with ILC.

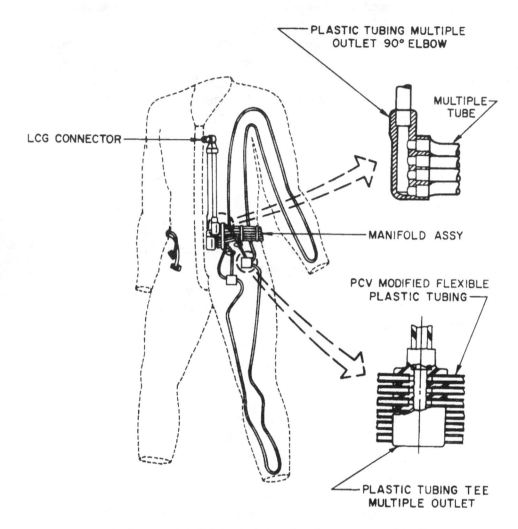

PLASTIC TUBING MULTIPLE
OUTLET 90° ELBOW

MULTIPLE
TUBE

LCG CONNECTOR

MANIFOLD ASSY

PCV MODIFIED FLEXIBLE
PLASTIC TUBING

PLASTIC TUBING TEE
MULTIPLE OUTLET

Figure 7.51. The drawing on the left shows the liquid cooling garment and the drawing on the right shows the details of how the tubes in that garment were joined together with the help of manifolds. The liquid cooling garment consisted of four spandex panels: left and right front and left and right back. Each section had pieces of ⅛" diameter tubing laced in. These spandex panels were sewn together and the ¼" tubing from each section was joined using the plastic tubing T outlet. Each of the four quarter-panels were joined together to the multiple-outlet 90° elbow made of plastic tubing. The liquid cooling garment had a total of eight T sections (water in and out times four sections) and two 90° elbows (water in and out to the four sections). Courtesy ILC Dover LP 2020.

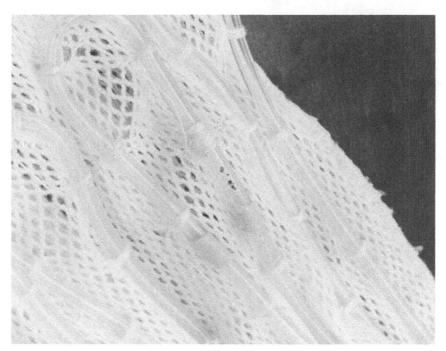

Figure 7.52. Detailed closeup photo of how the small ⅛" tubing was sewn onto the Spandex panels of the liquid cooling garment by B. Welson. Courtesy ILC Dover LP 2020.

During the time B. Welson manufactured the liquid cooling garment, the pre-Apollo 11 development models used clear vinyl "riser" tubes that connected the smaller tubing in the liquid cooling garment to the inlet and outlet tubes that fastened to the connector that passed the cooling water between the outside and inside of the suit. This was identified as part number A6L-400000-09.

At some point during this process, NASA discovered that a liquid cooling garment in one of their labs had a crimp in a vinyl riser tube that may have occurred when the liquid cooling garment was packaged for shipment. This caused alarm, since NASA did not want one of these tubes to become deformed from improper storage on board the spacecraft and thus shut off the flow of water into and out of the liquid cooling garment. As a result, the next version and the ones used on Apollo 11 through 14 used a wire-supported white silicone hose made by the Redar company. Redar also made the silicone oxygen hoses used in the spacecraft and on the primary life-support system. This updated liquid cooling garment became part number A6L-400000-10. The next version of suits that supported the Model A-7L and A-7LB suits had long sleeves with added cooling tubes to

Figure 7.53. The design of the pre-Apollo 11 liquid cooling garment, which had clear vinyl riser tubes attached to the pass-through connector made by Air-Lock Corporation. This permitted water to flow both into and out of the same connector and was called the multiple water connector. It was connected to the mating suit pass-through hardware located inside the pressure garment just after an astronaut had donned the suit. Courtesy ILC Dover LP 2020.

increase heat removal in the lower arms. Since the model A-7LB suits had the water connector pass-through lower on the torso, the riser tube was no longer needed, so the connector was directly attached to the cooling tubes coming in and out of the liquid cooling garment.

In the first quarter of 1970, ILC management dropped B. Welson as the subcontractor of the liquid cooling garment to bring further development and fabrication of the liquid cooling garment in house. ILC management admitted that this move made only a small difference in terms of profit.

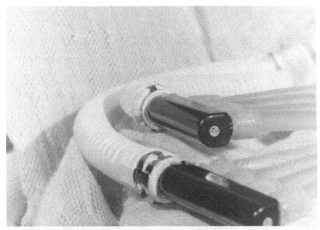

Figure 7.54. The manifolds manufactured by the Air-Lock Corporation and the wire-reinforced silicone riser tubes used on Apollo missions 11 through 14. These tubes eliminated the chance that the hoses would crimp and deform if they were not stored properly. Courtesy NASA.

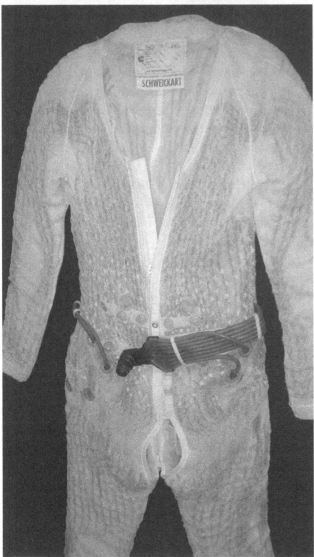

Figure 7.55. The liquid cooling garment used with the model A-7LB suit used a multiple water connector that was positioned lower on the torso since the suit pass-through was located lower on the space suit. Thus, a riser tube was not needed. Courtesy ILC Dover LP 2020.

The decision was driven by several primary reasons. One was that the logistics of coordinating activities with B. Welson was always a challenge. In addition, many saw it as an opportunity to prove ILC's skill in yet another area of soft-goods fabrication. ILC engineer Ron Bessette recalled that the owner of the B. Welson Company was probably not too sorry to lose the liquid cooling garment business because the quality requirements and amount of time he had to spend dealing with ILC and issues related to these garments was overwhelming and, in the end, he did not make much profit.

By the end of 1970, ILC was in the process of qualifying an updated version of their liquid cooling garment. Ron Bessette, the lead ILC engineer on the project, was working with Beverly Killen in the production area. Ron and Beverly came up with a better idea about how to secure the tubing into the spandex panels. Instead of using the drag stitch to sew it, they simply wove the tubing under the open-weave spandex thread bundles that held the tubing in place. The challenge was to locate the tubing in the positions in each garment that had been proven through testing to remove the most Btus. To do that, Ron cut a cardboard pattern for each quarter-panel of the liquid cooling garment and drew a sketch of the tubing that passed through each section. He then punched small dots in the pattern over the tubing lines at various intervals. The pattern was then placed over each section of the spandex fabric. A chalk-bag containing blue chalk dust was then patted over the holes that he made in the pattern. This left behind a light dusting of chalk on the spandex that provided a road map of sorts for the production personnel that showed them where the tube was to be woven through the spandex. The assembler would just play a game of connect the dots as they wove the tubing into the spandex panels. This eliminated the drag stitch and all the labor associated with mating the tubing to the spandex material.

At some point midway through the program, NASA engineer Jim O'Kane approached ILC and asked them to come up with a design for a redundant cooling loop for the liquid cooling garment system. Bessette recalls that although ILC attempted to satisfy O'Kane's request, the plumbing and connectors required to make it work were far more complicated than it was worth. ILC finally sent someone down to Houston to explain to O'Kane and others that this plan would not work.[55]

During qualification of the A-7LB suits and the A-6L liquid cooling garment, NASA approved the use of what was called the LCG adapter interconnect, which the Apollo 15 and subsequent crews could use to connect their liquid cooling garment to the cooling system on board the lunar module during rest periods when they were out of the suits. There is no evidence that any of the crew members on the lunar surface used it.

The specifications for the liquid cooling garment stated that water flowed through the liquid cooling garment at a rate of 4 pounds per minute (or about ½ gallon per minute). The primary life-support system provided water that was about 33°F as it left the backpack. A three-position adjustment was provided so the astronaut could set the temperature to roughly 45–50°F, 60–65°F, or 75–80°F.

The Model A-7L Intra-Vehicular and Extra-Vehicular Suits

ILC made what I consider to be four basic suit versions for the Apollo 7–17 missions, including Skylab and the Apollo-Soyuz Test Program. Some might say that there were only three versions. The first model A-7L consisted of the intravehicular activity and extravehicular activity models. The improved model A-7LB made some radical changes to the extravehicular suits, while the intravehicular suits used on Apollo 15–17 (including Skylab and the Apollo-Soyuz Test Program) were very similar to those used on Apollo 7 through 14. The intravehicular suits were also referred to as the command module pilot suits because they were intended for that crew member position where no extravehicular activity was performed. There were three exceptions when the command module pilots on board Apollo 15, 16 and 17 performed extravehicular activities on their return from the moon. Because of this added requirement, these intravehicular suits had to have the duel set of gas connecters added, this a minor change from the basic A7L intravehicular suits. Because of these changes to the mission requirements and the hardware, I consider them unique.

Space suits were very costly, so it was a benefit to NASA to save money where it could. One of the things NASA did was issue contract change authorizations that directed ILC to modify suits from the intravehicular version to the extravehicular version. This was one way to avoid buying a brand-new suit when refurbished ones would do for training purposes. Throughout the Apollo program, a total of 111 suits were modified at some point in time.[56] This was, amazingly, almost 50 percent of all the Apollo suits that ILC manufactured. One example of this was the model A-7L suit serial number 063. A contract change was issued to ILC in April 1969 to modify this suit from its original configuration as an intravehicular suit

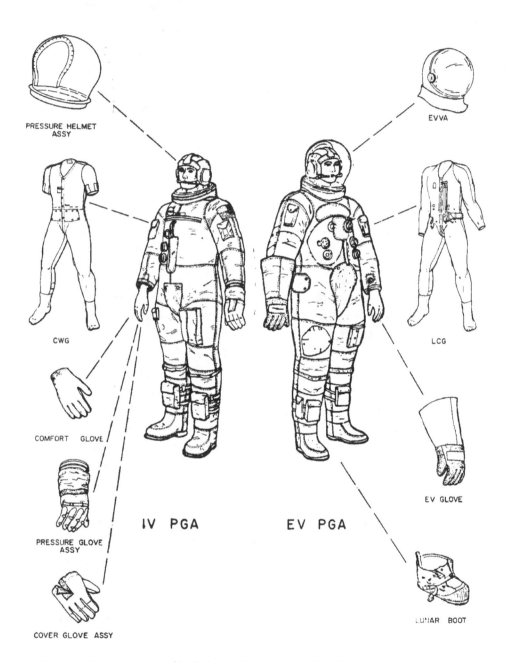

PRESSURE HELMET ASSY

EVVA

CWG

LCG

COMFORT GLOVE

PRESSURE GLOVE ASSY

EV GLOVE

IV PGA

EV PGA

COVER GLOVE ASSY

LUNAR BOOT

Figure 7.56. An overview of both the Model A-7L intravehicular and extravehicular suits for Apollo and the major pieces of support gear. These were the major components that ILC was responsible for providing. Courtesy ILC Dover LP 2020.

to an extravehicular suit for training use by astronaut Dave Scott for his Apollo 12 mission, where he was assigned as backup commander to Pete Conrad.[57]

A-7L Intravehicular Activity Suits

The basic model A-7L pressure-garment assembly for all crew members on Apollo 7 through 14 was essentially the same. ILC made two versions, one for use inside the spacecraft, referred to as the intravehicular suit, and one for use outside the spacecraft, the extravehicular suit. The two versions had several minor differences, including the connectors for the gas and the pass-through for the liquid cooling garment. Both the intravehicular and extravehicular suits used pull-down straps made of webbing located on the center top of the chest to assist crew members when they bent at the waist since there was no waist convolute, as there would be in the next generation A-7LB suits. These suits both used vertically installed zippers that ran from the upper back to the front of the crotch. The intravehicular suit had the following differences from the EVA suits:

Only one set of gas connectors on the chest (red and blue anodized finish)
No liquid cooling garment connector
No arm bearings (arms used the link-net restraint design)
No pressure gauge built into the suit
No pressure relief valve built into suit
Reduced outer TMG layers consisting of Teflon-coated Beta-yarn fabric, a center ply of Kapton, Beta marquisette laminate, and an inner layer of Nomex fabric (a second outside cover layer of the Teflon-coated Beta-yarn fabric was provided in areas of high abrasion)

A-7L Extravehicular Activity Suits

Although the Model A-7L EVA suits were very similar to intravehicular suits in terms of basic design and sizing, they had many more features that were necessary to protect the astronaut on the lunar surface and make his work easier. They included the following:

A dual set of gas connecters on the torso. This enabled the astronauts to connect to the oxygen supply on board the lunar module as they were donning and setting up the suits for their extravehicular activity and attach the gas connectors from the primary life-support system. Once suited, they would switch from the oxygen supply in the lunar module to the backpack after the lunar module was depressurized to the vacuum pressure as on the moon. They would then use the primary life-support system and exit to the moon's surface. The suit connectors had a spring-loaded valve that would close when the hoses were disconnected, preventing the oxygen from flowing out of the suit.

An oxygen flow diverter valve. This valve was located on the chest and was assembled into the vent plenum hardware located on one of the blue inlet fittings. When the crew members were connected to the environmental control unit inside the spacecraft, the flow rate into the suit was typically controlled at 12 cubic feet per minute. The diverter valve was positioned so that 50 percent of the oxygen (6.0 cubic feet) flowed to the helmet, providing fresh oxygen and clearing the helmet of any fogging. The other 50 percent, or 6.0 cubic feet, was directed to the vent system that directed flow to the hands and feet to provide cooling. However, when crew members were setting up to perform their extravehicular activities, the diverter valve had to be turned 90 degrees so that 100 percent of the oxygen flow was directed to the helmet. This was necessary because the primary life-support system supplied only 6.0 cubic feet per minute of airflow and it was necessary to direct all of this to the helmet to provide support for breathing and to defog the helmet. The liquid cooling garment provided quite enough cooling to compensate for the diminished airflow to the extremities.

One connector for the liquid cooling garment. This provided the chilled water from the primary life-support system through the connector and into the suit and the liquid cooling garment.

Arm bearings that ILC manufactured on the Apollo 11 through 14 model A-7L suits. These low-profile bearings provided that greater range of motion crews who would be using tools on the moon needed.[58]

Several more layers of insulation for the TMG (see Chapter 6).

Support accessories for the lunar surface that included EVA gloves, lunar boots, pockets for tools and sample collections, and the extravehicular visor assembly with the built-in gold visor and sun shields. The latter was assembled over the pressure helmet that used for both the intravehicular and extravehicular activity suits.

A pressure gauge on the left arm of suit.

A pressure relief valve on left arm for Apollo 9–14 suits. The suits for Apollo 7 and 8 had no relief valve; instead, they relied on the pressure control system aboard the command module. A solid plate was placed over the hardware for the relief valves on the Apollo 7 and 8 flight suits. A protective Beta-cloth cover was installed over the relief valves on all of the suits except for the suits for the Apollo 12 crew (the valve is visible in all photos).

Apollo 7–10

Apollo 7: First Flight of the Model A-7L suits

By late spring of 1968, ILC was in full production on the first suits, which would launch aboard Apollo 7 on October 11, 1968, whose crew consisted of Commander Wally Schirra, Pilot/Navigator Don Eisele, and Pilot/Engineer Walt Cunningham. ILC employed three mission managers—Walter Lee, Bill Cahall, and Joe Popano—who were responsible for getting all the flight gear together for each mission and ensuring that it was delivered to the Kennedy Space Center on time. The quality of the suits had to be well above average, but many other associated items also had to be perfect. One of them was the acceptance data pack. This booklet, which had to accompany each suit, had many pages detailing every piece of fabric, rubber, adhesive, nut, and bolt assembled into each suit. Any screw found in the suit could be traced back to the mine the metal came from to make that screw. Any piece of fabric in the suit could be traced back to the manufacturer who did the initial weaving of the fabric. That's how detailed this trace information was, but all of it would be needed if any component was found to be defective. Any defect in a space suit could lead to the death of an astronaut if it could not be corrected. Other lots or pieces of the defective material or the nut or bolt could have the same defect; thus, it was necessary to know what other suits had this defective part. The rigor NASA required of ILC to make certain that all the trace information was maintained throughout the manufacture of each suit was enormous. However, the result of all that work was that NASA complimented ILC on the quality of the Apollo 7 suits and the documentation that accompanied them. This followed years when NASA had justifiably criticized ILC because of its lack of experience with documentation and lack of experienced personnel to carry out the demanding processes. It looked like ILC was finally getting it right, and just in time.

Figure 7.57. Apollo 7 astronaut Walt Cunningham (*left*) introducing the first Apollo model A-7L extravehicular model space suit in front of national and international media at the ILC Industries plant on August 2, 1968. Test subject Kenneth Shane is modeling the suit made for ILC employee Tom Sylvester. Courtesy ILC Dover LP 2020.

While practicing months ahead of their scheduled Apollo 7 launch date, the three astronauts found it difficult to work side by side in the command module seats in their training suits. The problem centered on arm bearings that were too bulky and added to the width of the suit across the upper arms resulting in interference between the three crew members. ILC looked at all the options and following discussions and an agreement with NASA, they decided to remove the arm bearings for the first pre-lunar missions.[59] This decision was relatively simple to reach since limited arm mobility was required within the spacecraft and, except for the Apollo 9 mission, the suits would not be pressurized unless there was the unexpected loss of pressure. The plan was that the Apollo 9 crew would conduct the first Apollo-suited EVA while in Earth orbit, but even then, the arm mobility that would be required would be minimal compared to what would be expected for the lunar missions. This gave ILC time to work out the new bearing design in conjunction with an arm redesign that would reduce the size of the cross-section significantly. ILC had approximately ten months to come up with this fix. This is but one example of the challenges that the ILC suit engineers constantly faced.

The solution to the excessive arm width was to remove the arm bearing and replace it with nylon link-net material. This arm configuration was successfully used by all three crew members of the Apollo 7 through the Apollo 10 missions and then by the command module pilots on Apollo 11 through 14.

In addition to removing the bearing, ILC removed some insulation from the TMG layers under the arms of the two extravehicular suits. This area would not see excessive heat loads during an EVA, which was not even planned for the Apollo 7 mission. The insulation would be added back in for the Apollo 8 and subsequent missions, but I recall one of the Apollo veterans telling me that Neil Armstrong requested that some of the insulation be removed under the arms of his suit to provide more comfort in terms of fit and mobility.

Both the commander (Wally Schirra) and the lunar module pilot (Walt Cunningham) wore extravehicular suits, while the command module pilot (Don Eisele) wore the intravehicular suit.

On June 8, 1968, ILC employees Bill Dougherty and Bill Fry very proudly hand-delivered the first Apollo flight suits from the Dover plant to the Kennedy Space Center. They loaded the suits into a small charter plane at the nearby Cheswold, Delaware, airport and taxied down the runway for a short trip to Washington National Airport, where they transferred onto a commercial flight to Florida. The suits almost did not make it however, because shortly after takeoff, Dougherty noticed that a hatch had opened near the cases containing the suits. Fortunately, he was able to quickly close and lock the hatch before the precious cargo became bits of debris scattered across the farm fields of Kent County, Delaware. That would have been a very difficult call to make to NASA.

The Apollo 7 mission was long remembered for the many problems the crew experienced throughout their eleven-day mission. There were issues with motion sickness, and Commander Wally Schirra came down with a bad head cold. The mission rules were that the crew was to wear their suits and pressure helmets during reentry to Earth because of the unknowns associated with the ocean impact. NASA wanted the crew to have their helmets on for impact protection more than anything else. With a bad head cold and no way to clear his ears during the pressure changes that took place during reentry, Schirra rebelled and did not wear his helmet; the rest of the crew also refused to wear theirs. They never flew again in space.

Apollo 8: Historic First Flight to the Moon

The Apollo 8 crew had originally trained for a mission to test the new Saturn V rocket, command module, and service module in a high-earth orbit. The lunar module was still in development and would not be flown. There was no talk about a lunar mission until just five months before the launch date, when Assistant Director of Flight Crew Operations Deke Slayton summoned Frank Borman to his office. Slayton asked Borman if he'd be willing to take the mission to a lunar orbit; Borman didn't hesitate to say yes. There are two differing views about what drove NASA to choose Apollo 8 as the mission to the moon. Slayton told Borman that CIA reports indicated that the Russians were close to having a rocket to take their crews to the moon, while other researchers say that the facts indicate that NASA was simply staying on track with their accelerated plans to get the moon before the end of the decade in honor of Kennedy's challenge.[60] Regardless of the reasons, it was a gutsy call, and the mission proved to be a huge success on several fronts.

The crew wore the model A-7L suits. No EVA was planned, but both Commander Frank Borman and lunar module pilot Bill Anders wore an extravehicular suit model while command module pilot James Lovell wore the intravehicular suit model. During the production of the Apollo 8 suits, a change was made to use an intravehicular cover layer to all three suits instead of the extravehicular TMG. This reduced the weight of each of the suits by a few pounds since that removed many of the thermal layers under the Beta cloth.

Because of the problems Wally Schirra encountered on the Apollo 7 mission, a device was added to the pressure helmets so that crew members could perform the Valsalva maneuver by pressing their noses into the soft silicone material and blowing hard enough to equalize their sinus pressure.

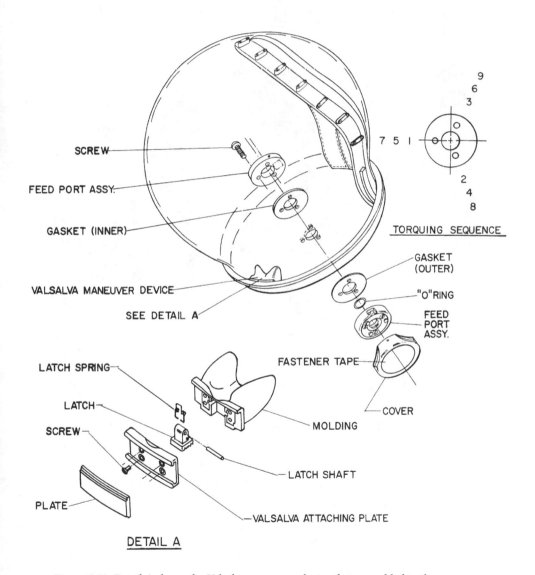

Figure 7.58. Detail A shows the Valsalva maneuver device that was added to the pressure helmet for the Apollo 8 crew. Courtesy ILC Dover LP 2020.

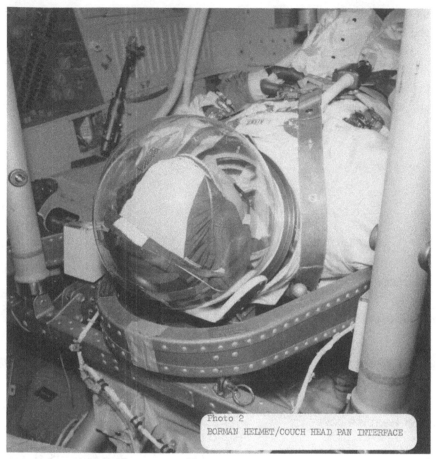

Figure 7.59. Testing the Borman helmet as it interfaced with the couch in the command module. The report from this test stated that the sides of the helmet contacted the wings of the headrest, but it did not touch the back support portion of the headrest. However, the helmet was still deemed as acceptable because it was adequately held in place. Courtesy NASA.

The dimensions of Mission Commander Frank Borman's head were somewhat larger than that of the general population. This primarily affected the area around the chin, where he had significant contact with the polycarbonate bubble. As a result, ILC worked with Airlock to design a larger helmet. They simply increased the bubble size to provide more room inside. Several tests had to be conducted on this new helmet to prove that the burst pressures, impact resistance, and functionality were adequate. No problems emerged with Borman's helmet, and it worked so well that the new design was used on all future missions, since more room was better than less in this case.

Apollo 9: The First Extra-Vehicular Activities

The Apollo 9 mission, whose crew consisted of Commander James McDivitt, Command Module Pilot Dave Scott, and Lunar Module Pilot Russell (Rusty) Schweickart, included several firsts. One of the primary objectives of this mission was to perform the first spacewalks, proving that the combination of the A-7L suits and the primary life-support system worked in the vacuum of space. It was also the first mission to fly a lunar module. This was planned to show that it was possible for astronauts to transfer between the lunar module and the command module in the event the two modules could not dock together on the return from the lunar surface. As it turned out, Russell Schweickart had a bout of space sickness that caused NASA to abbreviate the length of time and sequence of events. It was agreed that just a simple EVA outside the spacecraft hatches would be good enough. Even though the mission only orbited the Earth, it still represent most of what would be seen on the moon except for the lunar dust and walking in the reduced gravity. By the end of the Apollo 9 mission, the crew had spent approximately fifty hours in the suits with their helmets and gloves off. They had also spent about 94 minutes in pressurized suits for the EVA tasks.[61]

The extravehicular visor assembly used on Apollo 9 consisted of a polycarbonate outer shell painted a reddish color. One Apollo Experience Report states that the Apollo 9 extravehicular visor assembly had thermal insulation under the shell.[62] This setup later proved to be unacceptable in thermal vacuum chamber testing, so by the time it was used next on Apollo 11, the shell was covered on the outside with the Beta cloth with mylar film ply-ups to add the needed insulation.

In December 1968, NASA commended ILC employee Anthony Tomassetti for his development and delivery of a new ultraviolet stabilized polycarbonate sun visor material and the protective visor material for qualification prior to the Apollo 9 mission. New test data from NASA in October 1968 made it apparent that it was critical to use this newer ultraviolet material on the helmet and that time was running out before this mission to find the materials, form the visors, and qualify them. Tomassetti made it all happen within a month and flight units were delivered in November 1968.

The Apollo 9 EVA gloves were redesigned to include the blue silicone caps at the tips of the fingers of the TMGs. In addition, the gauntlet was shortened so that the pressure gauge on the lower arm of the assembly was visible. A protective cover was installed over the case of the gauge to protect it from any solar heat loading.[63] The new gloves were qualified for use

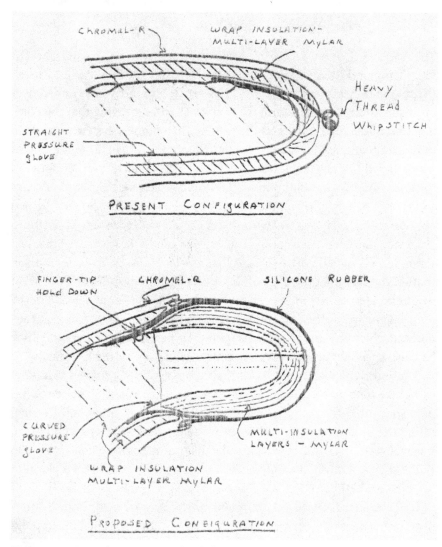

Figure 7.60. A rough engineering sketch from 1969 showing the change to the fingertips of the thermal micrometeoroid garment glove from Chromel-R only to the blue silicone caps that would become the hallmark of all future Apollo EVA and space shuttle gloves. Courtesy ILC Dover LP 2020.

by February 18, 1969, just two weeks before the Apollo 9 mission launch. Astronaut Dave Scott, the Apollo 9 command module pilot, whose suit had the earlier style glove, said in his post-mission briefing that he would not have been able to fly the command module with the earlier version of the EVA glove covers.[64]

The intravehicular gloves Dave Scott used on Apollo 9 (P/N A7L-103000-05/06) did not have the Fluorel protective coating that provided both resistance to abrasion and protection from fire because Scott had said that this coating made it harder to flex the gloves.[65] (All future gloves had the Fluorel coating.) During his EVA, Scott wore his intravehicular glove on his left hand and his extravehicular glove on the right as he floated in and out of the hatch of the command module to retrieve some thermal samples and take photographs. The thermal samples consisted of painted

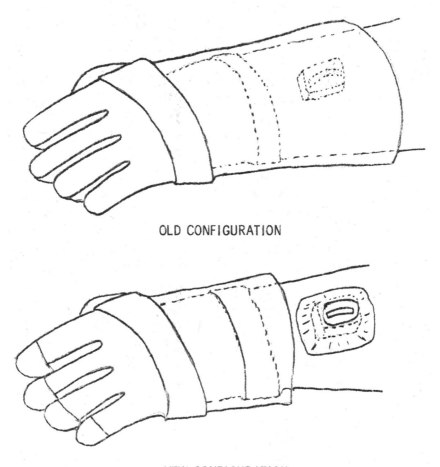

OLD CONFIGURATION

NEW CONFIGURATION

Figure 7.61. ILC sketch showing the change to the EVA glove for Apollo 9 that added the blue silicone fingertips and cut the length of the gauntlet to expose the pressure gauge to make it easier to read. A thermal cover was also added to the gauge to protect it from thermal solar loading, since the gauntlet would no longer protect it. Courtesy ILC Dover LP 2020.

metal pieces attached around the outer surfaces of the command module to study the heating effects experienced during launch. Scott reported that the intravehicular glove was warm to the touch but not uncomfortable.[66] He wore the one intravehicular glove in case the crew had to make an emergency reentry with an unpressurized command module. This was a possibility if they could not repressurize following the spacewalk. Scott felt he could not work the controls of the spacecraft with the EVA gloves on both hands if this happened.

The soles of the pressure-garment boots that were installed on all the suits up to this point tended to crack after moderate use in training. As a solution, ILC changed the material from a polyurethane material to more flame-resistant silicone and then finally to a non-flammable Fluorel rubber. It had similar properties to what we now refer to as Viton rubber. ILC designers also changed the orientation of the Velcro patch on the bottom of the sole to change the break points where the sole would bend. These round patches were there so the astronauts could secure themselves to Velcro on the floor of the lunar module during the landing phase on the moon and to the footrests in the command module.

During training, the TMG was experiencing heavy abrasion on the back where the primary life-support system was interfacing with it. This was overcome by adding a layer of the Chromel-R materials to the back. While this chromium steel fabric easily stood up to the abrasion, it was a very heavy and costly material. Several areas of the TMG evinced extensive wear due to abrasion. This was particularly the case in the upper arm, where the cable turnaround hardware is located. These hard spots on the suit likely rubbed against the hard areas within the spacecraft and cause the fragile Beta cloth to shred. ILC solved this problem by adding Teflon fabric in spots of abrasion wear; it held up remarkably well. After this finding, ILC added Teflon to the entire TMG outer layers to provide further abrasion resistance. Teflon was also much lighter in weight than the metal Chromel-R fabric.

Following the mission, several comments related to the pressure garments were recorded from the debriefs. The gloves were off the suit during the segments of the mission when the suits were being worn and the command and lunar modules were pressurized, and Russell Schweickart complained of abrasion to his forearms from the hardware where the gloves disconnected from the wrists. To resolve this on future missions, crew members wore the top section of a cotton sock on their wrist to provide abrasion resistance.

Apollo 10: Scouting Out the Lunar Landing Sites

When Apollo 10 launched on May 18, 1969, the astronauts were wearing the model A-7L suits. On board were Commander Eugene Cernan and Lunar Module Pilot Thomas Stafford, who both wore the extravehicular suit model, and Command Module Pilot John Young, who wore the intravehicular suit. Their mission was to fly to the moon and take the lunar module to within 50,000 feet (15,000 meters) of the surface to test out all the systems for the Apollo 11 mission, which was scheduled to follow just two months later. No extravehicular activities were planned and none were carried out. This was a good test of the entire system, including the use of the space suits while the lunar module was descending. Just to make sure that Commander Gene Cernan would not be tempted to land his lunar module on the surface and become the first to do so, NASA made sure that the crew had only enough fuel to make it from 50,000 feet above the surface to the awaiting command module.

The commander, command module pilot, and lunar module pilot wore their space suits for approximately 27, 22, and 25 hours respectively throughout the mission. Fifty percent of the time was spent with the gloves and helmet removed. The suits were never pressurized throughout the mission. The crew noted a total of twenty-two issues with the suits, but they were minor and were related primarily to wear-abrasion issues.[67]

The Lunar Missions: Apollo 11 to 14

The launch of Apollo 11 on July 16, 1969, began a series of the greatest human space missions the world has seen to date. Even though the Apollo 13 crew did not touch down on the lunar surface due to the explosion on board their spacecraft, everyone was glued to their TV or radio as the rescue of the stranded crew took place over the course of four very long days. When the crew safely splashed down in the Pacific Ocean, NASA and everyone else involved had proved the power of what humans can do technologically in dire circumstances. By the time Apollo 17 blasted off from the moon on December 14, 1972, the space suits had been used for a total of 82 hours, 27 minutes, and 23 seconds on the lunar surface. The suits performed marvelously throughout each mission. The minor issues that developed never cut any EVA short.

A tremendous amount of manned, pressurized hours had been accumulated in the A-7L suits prior to the Apollo 11 mission. But, as ILC's Apollo

Figure 7.62. Nelson Wyatt (*right*) was very proud to shake hands with Neil Armstrong when he came to ILC in May 1969 for the refitting of his S/N 035 training suit for the Apollo 11 mission. Nelson was responsible for shipping and receiving all the Apollo suits at ILC. Courtesy Nelson Wyatt.

Program Chief Engineer Homer Reihm noted, the Apollo 11 mission was the first true test of the Apollo suit. That's what made him the slightest bit nervous as he sat in Mission Control for the Apollo 11 landing on the moon at the request of NASA.

Apollo 11: First Lunar Suits

Making History

Most people only slightly familiar with the Apollo space program can recall the names of this famous crew. It consisted of Commander Neil Armstrong, Lunar Module Pilot Edwin "Buzz" Aldrin, and Command Module Pilot Mike Collins. While Mike might be lesser known since he was relegated to orbiting the moon while Neil and Buzz got all the attention and fame, he played the very important role of maintaining the command module that would provide the only way the crew could get home.

Neil Armstrong was assigned the primary suit serial number 056 and Buzz Aldrin was assigned primary suit serial number 077. These suits underwent tight scrutiny as they were being manufactured because everyone at ILC was aware that these would be the flight suits that would keep these two men alive in the otherwise deadly environment of the moon. Even though a backup suit was made for each of them (Armstrong's backup suit was serial number 057 and Aldrin's was serial number 076) it was not expected that they would fly with anything other than the suits designated as their primary suits. As the crews became more comfortable with their flight suits because of the fit checks that fine-tuned them and some training exercises that broke them in just a bit, they bonded with them much like a mother with her new baby. Mike Collins flew with his intravehicular model suit, serial number 033.

Neil was first fit checked in his serial number 056 suit at the ILC plant in Dover on December 5, 1968. He signed off on the fit check record at 5 P.M., affirming that he accepted the suit with all comments as noted. They included a note that there were pressure points on both shoulders. Mel Case, the ILC engineer who was responsible for the fit check, made a note in the "Corrective Action" section of the fit check form to "investigate possibility of re-patterning shoulder area on future suits." Neil's left arm was 2.3 inches longer than his right arm and although ILC had compensated for that difference in the build of the arms, Armstrong asked ILC to reduce the length of the right arm because it was too long. Attached to the fit-check form was a memo from ILC mission manager Walter Lee, who noted that manufacturing was to make adjustments to the arm length as noted on the fit check record. This became a moot point, since the arms were replaced in just a few months with a totally new arm design. Also in the memo was a note that stated that the vent ports in the torso would be replaced to accommodate the new helmet assembly that Air-Lock was scheduled to be delivered on January 2, 1969.[68]

New Arms for Improved Fit and Mobility

The most significant difference in the Apollo 11 extravehicular suits was the new arm design, which featured low-profile arm bearings. This was a marked improvement in the performance of the suit. Apollo 11 was the first mission where it was truly needed. ILC started building the primary suits for Neil Armstrong and Buzz Aldrin in October 1968. There was still a lot of processing of the suit through April 1969. At the same time, the new arm bearing was still being tested. The understanding was that ILC had a lot of work to do to get these bearings qualified for flight use and it was doubtful they could do so before the July 16 launch date for Apollo 11.[69] Both Armstrong and Aldrin gave the new arm bearings favorable reviews after some

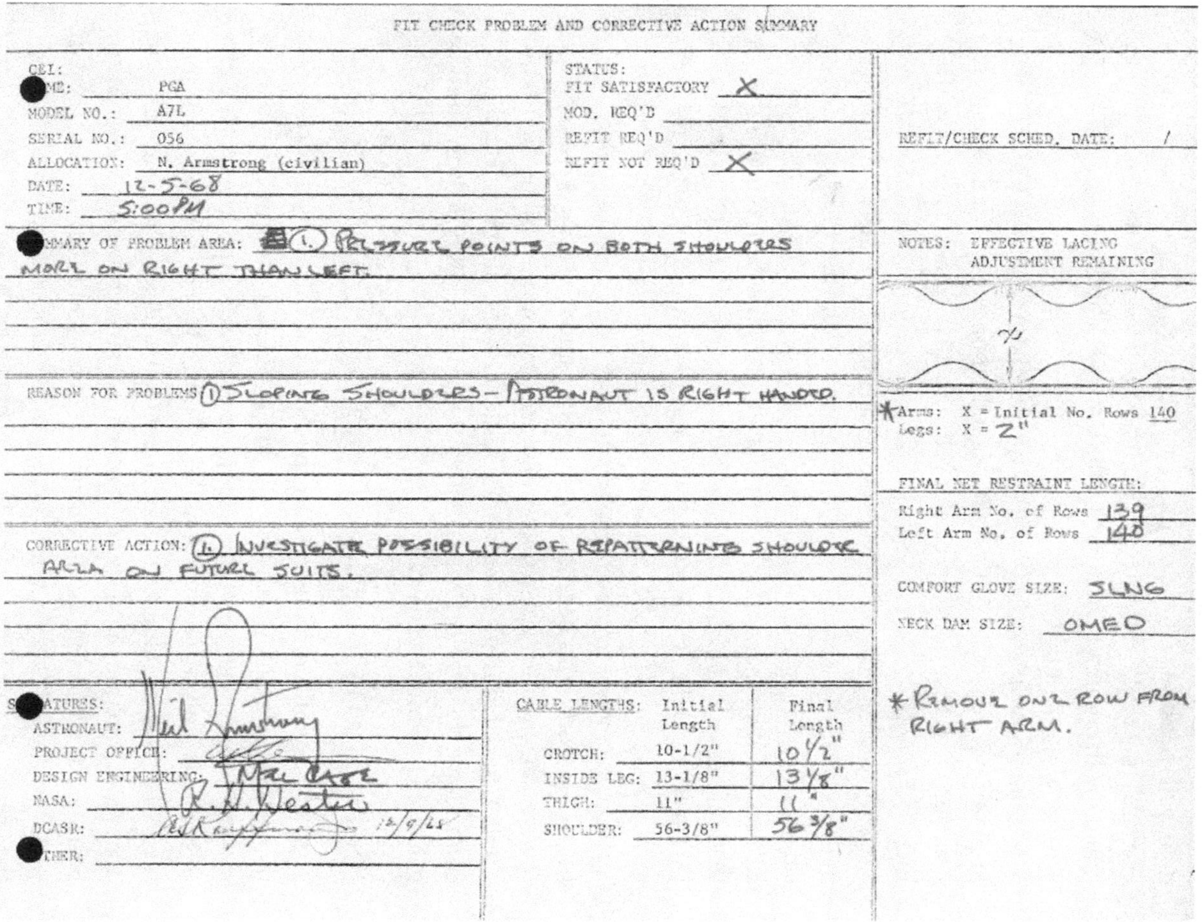

FIT CHECK PROBLEM AND CORRECTIVE ACTION SUMMARY

CEI:
MFG: PGA
MODEL NO.: A7L
SERIAL NO.: 056
ALLOCATION: N. Armstrong (civilian)
DATE: 12-5-68
TIME: 5:00 PM

STATUS:
FIT SATISFACTORY ___X___
MOD. REQ'D _____
REFIT REQ'D _____
REFIT NOT REQ'D ___X___

REFIT/CHECK SCHED. DATE: /

SUMMARY OF PROBLEM AREA: 1. PRESSURE POINTS ON BOTH SHOULDERS — MORE ON RIGHT THAN LEFT.

REASON FOR PROBLEMS: 1. SLOPING SHOULDERS — ASTRONAUT IS RIGHT HANDED.

CORRECTIVE ACTION: 1. INVESTIGATE POSSIBILITY OF REDISTRIBUTING SHOULDER AREA ON FUTURE SUITS.

NOTES: EFFECTIVE LACING ADJUSTMENT REMAINING

Arms: X = Initial No. Rows 140
Legs: X = 2"

FINAL NET RESTRAINT LENGTH:
Right Arm No. of Rows 139
Left Arm No. of Rows 140

COMFORT GLOVE SIZE: 5LNG
NECK DAM SIZE: OMED

* REMOVE ONE ROW FROM RIGHT ARM.

CABLE LENGTHS:	Initial Length	Final Length
CROTCH:	10-1/2"	10 1/2"
INSIDE LEG:	13-1/8"	13 1/8"
THIGH:	11"	11"
SHOULDER:	56-3/8"	56 3/8"

SIGNATURES:
ASTRONAUT
PROJECT OFFICE
DESIGN ENGINEERING
NASA
DCASR
OTHER

Figure 7.63. A copy of the fit-check form for the suit Armstrong flew in, serial number 056. This was the first fitting of a suit that would undergo several modifications between December 1968 and the July 1969 launch. Photo by Bill Ayrey.

evaluation testing. In a 1999 interview, NASA suit engineer Robert "Ed" Smylie recalled asking his fellow engineers what they could do to make the Apollo 11 crew more comfortable. The answer was to put the new arm bearings in the suit. Because there was no certainty that the new bearings would be qualified for flight use by launch day, Smylie and the team, with the concurrence of Neil Armstrong and Buzz Aldrin, decided to put the bearings in the suits designated as their primary suits and leave the backup suits with no arm bearings in case the bearings failed the qualification testing or if the testing was not completed before the launch date. As it turned out, the bearings were qualified in time and the crew used the primary suits with the new low-profile bearings, which significantly improved the arm mobility.[70]

Last-Minute Changes to the Thermal Micrometeoroid Garment

ILC seamstress and lead supervisor Eleanor Foraker later recalled that Neil Armstrong complained about the excess bulk between his arms and torso and asked the NASA suit engineers to investigate having ILC remove some insulation. It may not have been Neil alone that drove the change. He correctly felt that the heat loads in those areas would not be impaired if some insulation was taken out, but NASA would have obviously done some degree of analysis to validate that hypothesis. ILC did remove some of the insulation layers from under both arms of Neil's serial number 056 suit. Eleanor Foraker recalled that Armstrong's suit was returned so the excess insulation materials would be removed from the TMG a week before the Apollo 11 launch date. She and another worker closed the suit back up at 2:00 the following morning after working nonstop to accommodate the change. This was all in a day's work for the manufacturing folks of ILC.

The early A-7L suit design included a connector cover on the chest that provided protection for the various gas, water, and electrical connecters that plugged into the garment. More important in the design was the fact that the connector cover protected against heat leaks into the suit when solar loading might occur. The goal during the design process was to isolate any metallic objects on the suit that could conduct heat into the suit when intense solar radiation could strike the parts. The cover was attached to the TMG with Velcro, but it also included snaps because engineers thought the potential heat loads would weaken the ability of the Velcro to stay in place.

This connector cover design was in place and instructions for how to engage it were included in the EVA procedures checklist until the end of April 1969. In early March 1969, after a crew fit/crew function test to determine

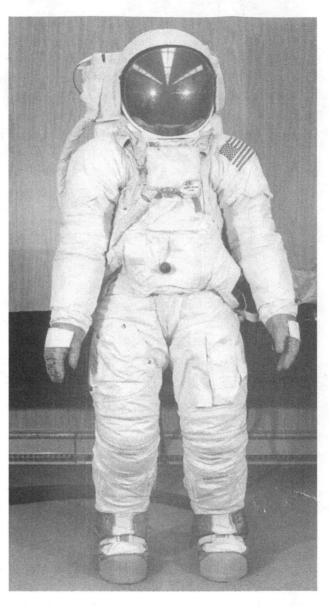

Figure 7.64. The early model A7-L lunar suit design with the connector cover installed under the remote control unit. Courtesy ILC Dover LP 2020.

how everything interfaced and operated as a system, Buzz Aldrin commented that "the present connector cover design is not consistent with the requirements to be able to don and fasten the snaps with the EV gloves on. Therefore, a look should be made to change the donning procedure, if possible, to completely don the connector cover before donning the EV gloves."[71] Aldrin figured that the problem could be solved if he and his crew attached the cover before they put on the EVA gloves.

On March 21, 1969, astronaut Don Mattingly sent correspondence to Deke Slayton, a former Mercury astronaut who was then senior manager of the astronaut office, to voice his concerns about the cover.[72] Mattingly identified the issues that astronauts on the lunar surface could encounter if one of the primary life-support system units failed to provide cooling water to the liquid cooling garment for one reason or another. There was a backup provision where a separate umbilical was to be carried along that could connect two space suits together so the functioning primary life-support system could provide cooling water to both suits. This was called the buddy secondary life-support system. Because this would require the lunar astronauts to disconnect and connect this emergency support system on the chests of both suits, the efforts would have been significant with the cover in place. His note to Slayton combined with Aldrin's concerns hit home.

By late April, testing was performed in NASA laboratories that simulated the lunar environment complete with the appropriate heat loads on the connectors with the cover removed. They found that, in a worst-case condition, the maximum heat gain was 75 Btu's per hour without the cover. The primary life-support system could dissipate 2,000 Btu's and the maximum expected with the astronauts working their hardest with the added solar loading would be about 1,400 Btu per hour, so the margin of safety was significant. As a result, the cover was eliminated soon after to everyone's satisfaction.[73] By 1970, the A7L TMG was upgraded to the point that testing showed it leaked only 180 Btus per hour at a temperature differential of 250°F, which demonstrated that it was very tight in terms of heat gain or loss and could evenly control the temperatures inside the suit.[74]

In addition to the connector cover, ILC also designed covers for the pressure relief valve located on the left arm and covers for the lunar module tether attachment clips located on both hips. The relief valve covers remained on the arms for the Apollo 11 mission however the covers for the lunar module tether straps were removed not long before the mission. The relief valve covers were removed for the Apollo 12 and later missions. Covers were also made for the pressure gauge on the right arm and these remained in place throughout the Apollo program.

With just one week before the launch of Apollo 11, the tension was high for everyone working on the program. A 29-year-old quality and reliability technician, Jim Rutherford, who worked for ILC at Cape Canaveral, tested the space suits in various stages of development and just before the missions. On the morning of July 9, 1969, Rutherford was assigned the job of performing a leak test of Neil's primary suit, serial number 056, following some activities where the suit was used. As it was under pressure, his focus was on the new arm bearing. He was flexing the arm assembly when he heard an air leak and discovered that the suit was failing the pressure test. This obviously was quite a concern. He immediately wrote the suit up on a discrepancy report and reported it to the ILC engineering staff on duty at the Cape. About an hour later, as he was taking a smoke break with several co-workers, the chief engineer for ILC approached Jim and threatened to fire him for failing the flight suits just a week before the launch. This reaction was likely due to the obvious stress of having to explain to NASA why the flight suits failed this close to the flight date, because otherwise the engineer might have taken a more reasonable approach to the matter. By the early part of that afternoon, both Neil Armstrong's suit and Buzz Aldrin's serial number 077 suit were loaded on board a chartered Lear jet that flew the suit directly to Dover so ILC could analyze the problem and fix it. By that evening, quality, manufacturing, test, and engineering personnel were on hand to process the suits back in and begin an around-the-clock effort to fix the problem(s). It was not long before they found that the arms had not been properly assembled. As Jim explained it, ILC personnel in Dover had failed to properly bond the bearing to the arm cone and had simply clamped it instead. The specific details of the problem are sketchy but nonetheless, Jim shed light on an interesting fact related to the Apollo 11 EVA suits. He took great pride in the fact that he found a problem that could have changed the course of history if the suits had not been fixed before the mission.[75]

Glove Updates

The gloves for both Neil Armstrong and Buzz Aldrin had the newly designed fingers that were more curved in the neutral position (as when the hand is at rest) so that they required less force to grasp tools and other objects. This update was first shown to NASA in September 1968 during the demonstration of the Omega III suit in Houston. NASA liked the upgrade and asked that ILC expedite the certification for the Apollo 11 mission.

ILC was given the task of making the gloves of the TMG detachable so that the thermal covers could be removed if crew members needed to keep their gloves on in the lunar module when they returned from the extravehicular activity. ILC glove engineer Dixie Reinhart worked on a system that would enable the crew to remove the TMGs while the suit remained pressurized. With the covers off, the crew was more capable of operating the switches and other controls in the lunar module during their ascent from the moon and when performing the rendezvous with the command module. Just before the launch of Apollo 11, Dixie was summoned to the Kennedy Space Center, where he trained Armstrong and Aldrin how to remove the TMG should they have to do so. Fortunately, neither crew nor any other needed to do so.

Helmets and Visors

The helmet pad in the back of the pressure helmet was changed to an aluminum base and the neck rings were changed from the blue anodized color to red following the change to the air vent. This created an enlarged opening that permitted greater air flow. A NASA document generated in early January 1969 identified the problem: "suit vent system pressure drop in LM-3 training PGA's was approximately 0.5 inches of water above specification at 6 acfm [actual cubic feet per minute]."[76] The solution was identified as follows: "The material in the PGA boot vent system has been modified with a more porous material, and the neck ring port has been enlarged, effectivity LM-5 and subsequent. Preliminary data shows a significant decrease in pressure drop." This confirmed that the change would be made on the LM-5 (Apollo 11) and subsequent suits.[77]

The lunar extravehicular visor assembly was like the one used for the Apollo 9 mission but it also had multilayer thermal insulation applied over the shell to provide the needed heat resistance. The polycarbonate sun visor was changed to a polysulfone material that could withstand the expected solar hear loads.

ILC Support at Mission Control during the EVA

NASA requested that representatives from ILC Industries be present at Mission Control in Houston while the Apollo 11 crew landed on the lunar surface, opened the hatch, and then closed it and repressurized the lunar module. That duty rested on the shoulders of George Durney and Homer Reihm. It made sense, since George had been involved with the development of the Apollo suit essentially since 1956 and Homer was the lead engineer who had worked side by side with NASA suit engineers. He and George Durney were sequestered in a back room behind Mission Control in Houston when Neil and Buzz climbed down the ladder to the moon's surface. They were both present in case a problem arose and NASA needed their help. Homer later recalled that he was several years younger than George and a bit more nervous. But according to Homer, even if George was nervous, which he almost undoubtedly was, he would have never shown it. Homer recalls that he became particularly nervous when Neil or Buzz would use their gloves to pound on any of the tools or objects that they were working with. Both George and Homer were aware of the strenuous testing the entire suit design had been subjected to in the ILC Lab, but the astronauts and suits were in uncharted territory. The world was watching this event, and God forbid Neil or Buzz should stumble upon some exercise in the lunar environment that would reveal some problem that no one considered before. All that both men wanted was to see Neil and Buzz wrap things up and climb back into the lunar module. Although George and Homer may have been a bit concerned about the outcome for obvious reasons, the men inside the suits had practiced enough before the mission to gain the confidence they needed. They knew that the suits would hold up to anything they could subject them to.

Meanwhile, back in Dover, all of the ILC employees were glued to their TV sets to watch the grainy images of Neil coming down the ladder, followed by Buzz Aldrin. What a thrill it had to be for all these folks, many of whom had touched the suits during production and testing, to see that. They were joined by the estimated 600 million people around the world who watched this historic event.

Anomalies after the Mission

When the Apollo 11 space suits and related gear returned to Earth, they were taken to the NASA suit laboratory in Houston and subjected to many tests. Included in the testing was Buzz Aldrin's liquid cooling garment. One of the tests involved overpressurizing the water lines because it had been determined that the vinyl tubes would become more stable when they were pre-stretched before being returned to the lower operating pressure. When the water lines of this garment were taken to 31.5 pounds per square inch for the pre-stretch and held for five minutes, one of the main feed lines to the garment burst. The company that made the silicone tube, Redar, did a failure investigation and determined that when the silicon hose was pressed over a metal fitting that had a flared end, a wire coil inside the silicon that happened to be of a smaller diameter than the flared fitting was positioned over the flared section and essentially cut into the silicone about one-third of the way around the inside diameter of the tubing. This caused a rupture to occur. The cut was likely made when the tubing was initially clamped to the fitting in manufacturing. It was just good fortune alone that the failure did not occur during the first mission on the moon. Because of the problem, Redar immediately made the end of the silicone tubing long enough to be pressed fully onto the fitting while the wire coil was an eighth of an inch from the flared fitting.

Crew Comments after the Mission

Both Neil Armstrong and Buzz Aldrin were generally very pleased with the suit's performance. Neil offered the following comments in his debrief:

> The operation of the suit, in general, was very pleasant. There was very little hindrance to mobility, with the exception of going down to the surface to pick things up with your hands which was a very difficult thing to do. As far as walking around and getting from one place to another, the suit offered very little impediment to that kind of progress. It was, in general, a pleasant operation. Thermal loads in the suit were not bad at all; I ran on minimum flow almost the entire time. Buzz found a higher flow to be desirable. This was consistent with our individual preflight experience. I didn't notice any temperature thermal differences in and out of the shadow. There were significant light differences and visibility changes but no thermal differences. The only temperature problem I had (and Buzz didn't have this problem) was with the gloves. I did not wear inner gloves. I chose to go without the inner liners in the gloves, and my hands were a little warm and very wet all the time. They got very damp and clammy inside the gloves. I found that this problem degraded my ability to handle objects and to get firm grips on things.[78]

Both astronauts made several comments about their inability to bend at the knees to get down to the surface. With the 1-G weight of the suits here on Earth, that activity was not too difficult but in the one-sixth gravity of the moon, there was less weight pushing the suit down to the surface, so it took more effort and both crew members felt unsteady when they stood back up. Armstrong suggested that having a pole or support of some kind would help.

Because of issues with arm length that Aldrin encountered during the initial fit check of his primary flight suit at ILC in Dover, it was suggested that a new elbow convolute be made that was 6 inches rather than 7 inches long. This was the best way to correct an arm that was too long. Making and then adding the new six-inch convolute was accomplished in parallel with the addition of the new arm bearings. Because of these last-minute changes, only the primary flight suits for Armstrong and Aldrin had the

NATIONAL AERONAUTICS AND SPACE ADMINISTRATION
MANNED SPACECRAFT CENTER
HOUSTON, TEXAS   77058

IN REPLY REFER TO: EC11IL022

JUL 2 4 1969

Dr. Nisson A. Finkelstein
ILC Industries, Inc.
P. O. Box 500
Dover, Delaware 19901

Dear Dr. Finkelstein:

It is with great pleasure that I offer my sincerest congratulations to you and the members of your organization for the
superb performance of the Space Suit during the Apollo 11 Mission.

This performance reflects great credit on all the members of
your organization and your subcontractor organizations who
worked long, hard, and often at personal sacrifice to assure that
the hardware was properly designed, tested, manufactured, and
delivered in a timely fashion. Each of these people have
contributed significantly to this historic achievement. I
consider it an honor and a pleasure to be associated with
such dedicated people and look forward to a continued association
for future Apollo missions.

Again, my congratulations for a job most well done.

Sincerely yours,

Robert E. Smylie
Chief, Crew Systems Division

RECEIVED
JUL 29 1969
N. A.
Finkelstein

Figure 7.65. A congratulatory letter from NASA to Finkelstein and members of ILC for
their work on the Apollo 11 suits. Courtesy ILC Dover LP 2020.

new arm bearings and only Aldrin's suit had the six-inch convolute. NA-
SA's conclusion was that there would have been significant problems if the
backup suits had to be used for Apollo 11 without these improvements.
Fortunately, they did not have to be used.[79]

Even with the new arms, Buzz Aldrin discovered during training that his
fingers were receding out of the tips of the gloves when he flexed his elbow.
This was attributed to the fact that he had larger biceps that interfered with

the bearings in the arms. When Aldrin bent his arm to adjust the remote control unit on his chest, it tended to pull his fingers inside the gloves by approximately ⅜ of an inch. This issue was first discovered during his fit check at ILC in Dover and the suggestion was made to install some pads beside the elbow area inside the suit. This was later tried at the Manned Space Center during simulator testing, but the padding added too much bulk inside the suit, so it was removed. After the mission, Aldrin made it clear that this was not a factor on the lunar surface, so no corrective action was required.[80]

One final note related to the arm was that the bearings did not have any stops built in that kept the lower arms positioned properly in relation to the upper arms. At no time during the mission did this become an issue, but the recommendation was made that two stops be provided so that the arms would not rotate more than 180 degrees. This change was added to the newer model A-7LB suits that were first seen on the Apollo 15 mission.[81]

Apollo 12: Venturing Farther from the Lander

Mission Details

The Apollo 12 mission was carried out by the all-navy crew of Capt. Pete Conrad as commander, Capt. Alan Bean as lunar module pilot, and Capt. Richard Gordon as the command module pilot. This crew wore the model A-7L suits. Pete Conrad later commanded the Skylab 2 missions, during which he spent twenty-eight days in space that included five hours of EVA time in an Apollo A-7LB suit. Alan Bean later commanded the Skylab 3 mission, during which he spent fifty-nine days in space and 2¾ hours of EVA time, also in a model A-7LB suit.

After his NASA career, Alan Bean went on to become a very well-respected artist who included the Apollo suits in most of his paintings. Featuring the space suits made sense, because they represent the successful human element of the Apollo missions. In the outline he included with his 1986 acrylic painting of Apollo 16 astronaut John Young in his suit, Capt. Bean wrote:

> Every human who walked on the moon did so in his own personal spaceship. We called them spacesuits and they performed brilliantly on all six lunar landings. I painted John Young all bundled up in his. John commented "I can't speak too highly for the pressure suit. Boy, that thing really takes a beating."

Figure 7.66. This special pouch was strapped onto the primary life support system of the Apollo 12 suit. It carried tools and provided a pocket for parts retrieved from Surveyor 3. Courtesy NASA.

Capt. Bean quoted Young as saying "Since it's the only thing between you and vacuum in plus or minus 250 degrees, it's a good piece of gear." Bean concluded by saying "Our space suits were an incredible American technical achievement. They had to reliably provide all of the functions of any spaceship, with one small exception—they contained no rocket engine, so we had to utilize our own two legs for propulsion."[82]

The Apollo 12 mission would include two extravehicular activities on the lunar surface, so it would be a test for the suits, since the Apollo 11 mission only had the one EVA for a total time of 4 hours and 22 minutes total for both suits. The Apollo 12 crew spent a total of 7 hours and 46 minutes on the lunar surface.

One of the primary mission objectives of the Apollo 12 crew was to land within a short distance of the Surveyor 3 spacecraft that had touched down on the moon a little more than two years earlier to study the moon's surface and perform a variety of other experiments. Commander Pete

Conrad made a pinpoint landing and came within 600 feet of the Surveyor. This gave the crew the opportunity to study the craft and see what damage might have occurred to the structure over those years. To carry the tools they needed to retrieve a few parts of the lander, ILC made a special pack to carry tools that was strapped on to the primary life-support system backpack.

Lunar Dust Problems

Conrad and Bean discovered that the greater accumulation of lunar dust attributed to the increased EVA time had begun to play a bigger role than it had during the Apollo 11 mission. They had difficulty connecting the gloves and the environmental control system hoses (which provided the oxygen flow in and out of the suits) on board the lunar module just before their ascent from the moon. Conrad was convinced that if there had been several more extravehicular activities, major problems might have occurred. He commented that the lunar boots had started to wear through the Beta-cloth cover layer and even into the Mylar layers of the garment legs. Conrad noted that after eight hours of EVA time, his flight suit had much more wear than his training suit had over hundreds of hours. He attributed this to lunar dust, which caused excessive abrasion. However, he also said, "Al and I had extreme confidence in the suits; therefore, we didn't give a second thought to working our heads off in the suits and banging them around— not in an unsafe manner but to do the job in the way we had practiced on Earth."[83] Between the first EVA and the second, both men were to rest for 12 hours and 30 minutes on board the lunar module. They decided to stay in their suits with the gloves and helmets off. This turned out to be a great decision because lunar dust may have had a negative impact on the pressure-sealing zipper, which would have allowed the dust to enter the suit. When they were in the suits, they could connect the suit's liquid cooling garments and the oxygen hose that provided airflow to the lunar module oxygen tanks, so they were quite comfortable, although Conrad did not hook up his cooling garment on board. His lower legs perspired heavily, but to alleviate that, he wisely connected the air hose that provided airflow into the suit (the blue connectors) to the red connector of his suit. This forced the dry oxygen inlet gas to reverse flow and distribute it to his lower legs, which dried the perspiration.

Command module pilot Dick Gordon once told me that when they rendezvoused the lunar module and command module upon the return from the lunar surface, he demanded that Conrad and Bean take their dirty suits off in the lunar module before entering his command module because he was not going to let them dirty up his clean capsule. When they had removed their suits, he opened the hatch between the two spacecraft and Pete and Al streaked into the capsule completely naked. Dick told me that it was the first historic streaking event in space.

Bad Suit Sizing

Just days prior to the mission, a leak was found in Conrad's left boot bladder and ILC needed to changed it out. The suit was put on a Lear jet and flown to Dover for immediate repair. Because the boot bladder had to be removed, the restraint had to be separated between the ankle and the lower leg section. This was secured together by lacing cords that acted as sizing adjustments for the leg length. The sizing cords on the lower leg had to be readjusted once the repair was made to fit Conrad properly. When the suit was returned to Cape Kennedy, Conrad was talked into getting inside the suit for the adjustment to the sizing cord wearing just long underwear instead of his liquid cooling garment because it had already been packed away for the mission. As a result, the leg sizing cords were taken in too much, as the added thickness of the liquid cooling garment had not been accounted for (according to Conrad in his post-flight debrief). This meant that the suit was tight against his shoulders and his legs. Conrad somehow managed to get four good hours of rest in the lunar module after his first EVA, but he woke up in pain because of the poor fit. Al Bean came to his rescue by pulling the left leg of the TMG up and exposing the lacing cords on the restraint layer. He then proceeded to let the length out so the fit was significantly improved. It took him an hour to accomplish the task in the cramped quarters.[84] During the crew debrief, Bean commented that:

> There were two sets (of cords) because everything was redundant.
> And they (the ladies at ILC) knotted the s*** out of those things.

Bean pressed Conrad on whether that the suit's tight fit might have been because the spine and hence the entire body grows over time in zero gravity. Conrad conceded that perhaps that had something to do with it. This explanation would make more sense.

The poor fit Pete Conrad experienced demonstrates how important suit sizing was and how many different variables had to be considered. Since the proper sizing is based significantly on human subjectivity and feedback, the result can be unpredictable. Although designers took approximately eighty measurements of each Apollo astronaut that were used for the manufacture of their suits, there was no guarantee that measurements alone would ensure a great fit. The actual dimensions were used to get the fit of the suit in the ballpark. The Apollo suits had sizing cords so final adjustments could be made based on feedback from the crew member when they were pressurized within the suit. During the fit checks of the suits in the ILC lab, the astronauts were kept in the suit for at least an hour or more doing many activities to find out what pressure points in the suit might bother them. Based on their input, the engineers would make the final adjustments and the astronaut would then sign off on a document formally accepting "their" suit. I have spent many hours inside the more modern space suits, and if they are not sized just right, the small, annoying pressure points that do not bother you initially become a big problem after an hour or so. In Conrad's case, it became unbearable after several hours.

In 1967, NASA unveiled two robots that were designed to test space suits. The idea was to take the human element out of the equation and replace it with actual data that provided the loads that were required to bend the arms or squeeze the fingers, for example. Without this calibrated test result, engineers had to rely on subjective input based on human feel, which is rather arbitrary, particularly when distinguishing between minor differences in suit designs. As good as the idea sounded, the robots never worked out. The various actuators within the suit were constantly leaking oil and given other issues, the project was scrapped. One of the robots is now on display at the National Air and Space Museum at the Steven F. Udvar-Hazy Center in Chantilly, Virginia.

Drink Bags

Conrad and Bean commented in their debrief that they became thirsty during their extravehicular activities and thought it would be a good idea to devise some sort of in-suit drink bag to sip from for future missions, particularly since they would be growing longer in duration. Walt Salyer Jr., an ILC engineer who worked at the Manned Spacecraft Center, immediately took on the task of defining and demonstrating the means of providing the drinking water to crew members while they performed their EVA. Walt tried many different techniques until he arrived at the best solution. In April 1970, Robert Smylie, chief of NASA's Crew Systems Division, wrote to Len Shepard to acknowledge Salyers's accomplishments:

> This effort included the fabrication, interfacing, preparation of procedures and manned testing of each system to evaluate the attributes

and short-comings of each. Through his contributions and outstanding performance, an in-suit drinking system design was achieved and tested. This device was intended to be utilized on Apollo 13 and will enable extended missions with increased EV activity to be planned and achieved.[85]

Whirlpool Corporation was contracted to produce an eight-ounce bag measuring approximately 5 inches by 3 inches that was fastened to the neck ring with Velcro. It contained a drink tube that activated under suction when the astronaut needed to draw water from it. Modifications were made to the drink-bag system throughout the Apollo missions until the bag was sized to hold approximately 32 ounces of water. This is just one story of how advancements were made to the suits from mission to mission based on the observations offered in the debriefs.

Blinded by the Light

For the Apollo 11 and Apollo 12 crews, the extravehicular visor assembly had two side shades that could be pulled down to help block the sun if it was just to the right or left of the forward-looking viewing area. Conrad noted that he liked the side blinders but that a top, center one was needed.

Figure 7.67. This image of Apollo 17 commander Gene Cernan, shows the bite-valve of his 32-ounce drink bag extending up from inside the suit on the right side of the neck ring. This valve permitted the crew member to position the valve inside their mouth and suck water from it as needed. Courtesy NASA.

During his EVA, as he looked forward in the direction of the sun, which was at a low angle in the sky, he had to hold his hand up in front of him in order to see without being blinded. Providing a center shade that could be pulled down halfway would be a big improvement. As a result, the extravehicular visor assemblies for all subsequent lunar crews had a center shade and two side shades. This design is still in use to this day.

After a Mission: Cutting Up a Suit

People at NASA Quality Control and many others were very interested in examining the space suits from Apollo 12. The Apollo 11 space suits were considered sacred objects and were already out on a tour of the forty-eight states of the US mainland. After that they went to the Smithsonian. The Apollo 12 suits, however, were fair game to be taken apart and inspected for damage. NASA Quality Control chose to dismantle Alan Bean's suit. The goal was to learn what could be discovered from a suit that had been subjected to an actual lunar mission as opposed to the suits that experienced the wear and tear brought about by random training and testing in labs that did not fully reproduce the conditions of space. There was interest in examining the TMG layers of the suit, since this took the more significant wear and had many layers inside that were covered up and could not be examined without completely disassembling the suit. There could be ramifications if significant issues were found with some of the Mylar layers in the suit, such as overheating due to thermal leaks. There was also great interest in seeing if any damage from micrometeoroids was evident.

Per a written test plan, the integrated TMG layer was removed from the torso of Bean's serial number 067 suit and sent to Building 8 at the Manned Spacecraft Center for thorough inspection.[86] Some activity took place at NASA's White Sands facility, where NASA engineers with the help of NASA quality personnel removed the left lower leg of the TMG for lunar dust propagation studies. The two lower arm assemblies were not dissembled.

The inspection of the TMG took two days to accomplish, since each layer had to be disassembled individually and thoroughly inspected. The first notes of the report stated that the overall condition of the TMG was much better than anticipated. No thermal overload problems that would indicate damage to the TMG were noted during the mission or during crew debriefs. The report noted that the Super Kapton material was delaminating and flaking and the perforated Mylar layer had sustained partial loss of its aluminum coating. After opening the cover layer to the liner seams and separating the two sections, the inspectors found Kapton flakes in the

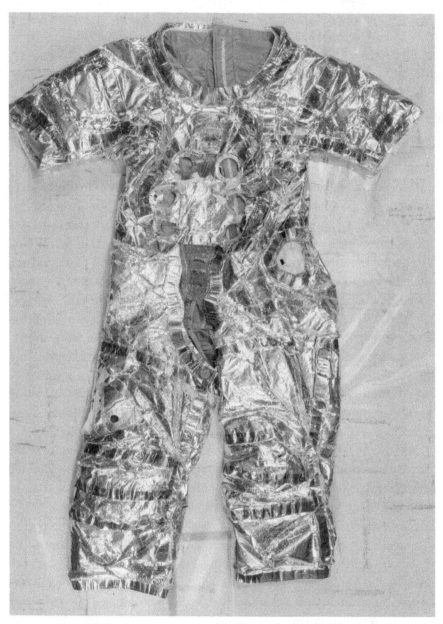

Figure 7.68. A NASA photo taken during the disassembly of Alan Bean's Apollo 12 suit after the mission. This image shows the layers of aluminized Mylar located under the outer Beta-cloth. Engineers were looking for damage caused by wear as well as any other obvious damage, including from micrometeoroids. Courtesy NASA.

right lower leg. They also found fabrication faults in the fourth layer of the perforated Mylar that were identified by stress failures. All of the other layers above and below were in good condition. The inspectors concluded that this layer had not been sized properly. They found no micrometeoroid penetrations during visual examination.

However, they found that the tape in the seams of each of the Dacron layers in the TMG had bled through to the next layers, potentially degrading the heat-leak performance of the insulation.

The result was that ILC had to consider why certain layers of the insulation were tighter than others, resulting in stress failures of the delicate films. Other means had to be considered for taping the seams so the bleed-through would not occur. Finally, the inspection team recommended that NASA look for a replacement material for the fragile Super Kapton layers and to prevent the aluminum flaking from the Mylar.

Apollo 13: The Addition of Red Stripes

Crew Identification

The crew of the fateful Apollo 13 mission consisted of Commander James Lovell, command module pilot Jack Swigert, and lunar module pilot Fred Haise. They wore the model A-7L suits. Lovell was the first to sport the new red stripes on his suit that identified him as the commander and made it easy to tell him apart from Fred Haise when they were on the lunar surface.

As Eric Jones recounts in the *Apollo Lunar Surface Journals*, British researcher Keith Wilson told the story of how the red stripe came about. According to Wilson, NASA had hired a new public affairs officer at the Manned Space Center named Brian Duff. Immediately following the return of the Apollo 11 mission, he was with many other top NASA officials who were anxious to get the best photos of Neil Armstrong on the lunar surface. Everyone realized, however, that they could not tell Neil or Buzz apart in the photos. Something would have to be done for the future missions so that the crew members could be properly identified. Brian Duff was given partial credit for coming up with the idea of the red stripe located on the arms, legs, and the extravehicular visor assembly of the commander's suit. There was no time to institute the red stripes for Pete Conrad on Apollo 12, but by the time Apollo 13 launched, the stripes were in place. They were used for all subsequent missions, including the missions to the International Space Station (see Figure 7.72).[87]

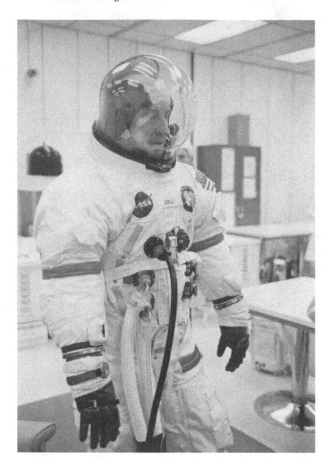

Figure 7.69. Apollo 13 commander James Lovell with the new red identification stripes. Courtesy NASA.

The Apollo 13 crew was the first to fly with the drink bags Pete Conrad had suggested during the Apollo 12 mission debrief.

Apollo 14: Last Mission for the Model A-7L EVA Suits

Rubber Reversion

Apollo 14 was flown with Commander Alan Shepard, the first American to fly in space in the Mercury program, lunar module pilot Edgar Mitchell, and command module pilot Stuart Roosa. This was the last mission that used the model A-7L suits.

One of the most significant issues related to the Apollo 14 mission suits was that the rubber convolutes showed clear signs of failure due to their age, which was only around fourteen months or slightly older. Everyone thought the properties of the rubber were enough to last for a few years at least, but until this mission there really was no track record to prove otherwise. Given the way the rubber compound was formulated, a few years was far too much to expect, as became evident when one of the boot convolutes in an Apollo 14 suit failed during a test just a month before the mission. Fortunately, the immediate problems were solved by installing new rubber convolutes in all of the flight and backup suits just a couple of weeks before the flight.

Wrist Disconnect Hardware

Earlier crew feedback about the gloves and the associated hardware indicated concerns that the diameters of the wrist and the interfacing glove disconnects were too small. The inner diameter was 3.5 inches, which was a relatively tight squeeze for larger hands. The first larger-sized disconnects were installed on the serial number 088 suit John Swigert used for Apollo 13. Edgar Mitchell also had a pair on the serial number 073 suit he used on Apollo 14. It wasn't until the Apollo 15 mission that all of the model A-7LB suits had the larger disconnects. When ILC was developing the Omega suit in late 1968, they included the larger-diameter disconnect as the standard size. A test conducted on September 29, 1969, using the larger wrist hardware concluded that the increased size permitted less pressure drop through the vent system which increased the airflow across the back of the hand at all times throughout the test. The smaller-diameter glove hardware did not provide this continual airflow, resulting in more discomfort and humidity inside the glove.[88]

The Buddy Secondary Life Support System

The crews for Apollo 14 through 17 would be venturing farther away from their lunar module as they explored the many geological wonders around them. They had the lunar rover, which could carry them much farther than crew members had traveled on previous missions. When the Apollo 17 astronauts left the lunar surface, they had traveled close to 7.6 kilometers (4¾ miles) from the landing site. The mission planners were understandably concerned that the backpack on one of the suit would fail. In a worst-case scenario, this failure would occur when the crew was the farthest away from the safety of the lunar module. The Apollo 14 crew ventured 1.5 kilometers (0.9 miles) from the landing site, so getting back to the safety of the lunar module was imperative if systems in the primary life-support system failed.

The two main areas of concern were the primary oxygen supply and the water cooling system. The primary life-support system had what was

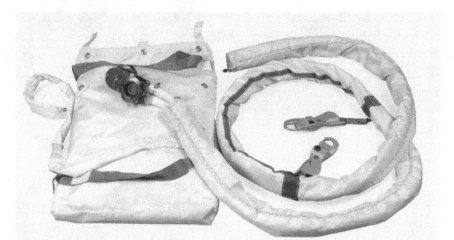

Figure 7.70. The buddy secondary life support system assembly and pouch. Missing is the standard water connector on the other end of the hose (upper center of photo) that would have been exchanged with the connector on the failed suit. The blue multiple water connector pictured on top of the pouch would have connected to the functional suit after the water connector was removed from the primary life support system. One end (facing away from the photo) would be plugged back into the suit and the primary life support system connector would then be plugged into the fitting shown facing the camera. The water would flow into the first suit and travel down the hose to the malfunctioning suit. Courtesy RR Auction.

known as the oxygen purge system, which contained 5,880 pounds per square inch of high-pressure oxygen in two bottles that would have provided for a safe return to the lunar module in just about all cases should the primary system fail. The other concern was providing cooling water to a crew member if that system in the primary life-support system failed. The potential for overheating was a great concern, particularly if astronauts were forced to walk back to the lunar module. As a result, the NASA engineers came up with a rather simple system that enabled one astronaut to share cooling water from their good suit with the failed suit.

This new system, which was carried on the Apollo 15–17 missions, came to be known as the buddy secondary life-support system. If a failure in the primary life-support system water cooling system happened, this hose assembly could be removed from its Beta-cloth pouch that was stored behind the seat of the rover and connected to the two suits. That was accomplished by removing the liquid cooling hose from the damaged primary life-support system at the suit interface and installing one end of the buddy secondary life-support system, which had the typical multiple connector fitting.[89] The crewmate with the functioning primary life-support system would also remove their liquid cooling hose and attach the other end of the buddy secondary life-support system hose, which had two connectors. I have seen several incorrect statements that the buddy secondary life-support system supplied backup oxygen, but that is not the case. It only supplied cooling water.

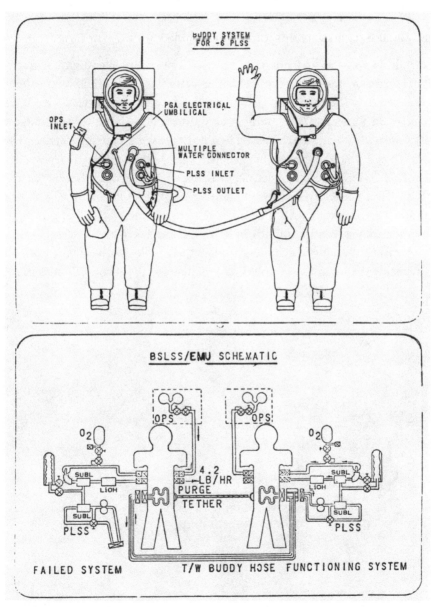

Figure 7.71. These sketches illustrate how the buddy secondary life support system worked. Courtesy NASA.

The system included an 8.5-foot cable with tether hooks on each end that secured to attachment rings on the waist of each suit. This provided enough strain relief so the water lines would never be under tension. Even though the water flow to both suits would have been cut in half, it was adequate to relieve the heat loading and permit safe return to the lunar module.

Modular Equipment Transporter System (METS)

Apollo 14 was the first mission that used the modular equipment transporter system for lunar extravehicular activities. The METS was a two-wheeled cart that enabled crew members to carry small tools, cameras and extra film cartridges, sample containers, and the buddy secondary life-support system. It also served as a portable work bench if needed. The METS had a handle that enabled an astronaut to pull it along like a wagon. In the mission debriefing, astronaut Mitchell commented that he felt the METS was generally worthwhile and saved them a lot of time. Mitchell made a comment about the stability of the cart and noted that pulling it along changed his gait and threw off his balance just a bit.

By April 1971, just three months after the Apollo 14 mission, it was still unclear if the lunar rover would be available for the Apollo 15 mission. It

Figure 7.72. Apollo 14 Commander Al Shepard pulls the METS during training in July 1970. Note the commander's stripes on the arms and legs of the suit. Courtesy NASA.

Figure 7.73. Apollo 15 astronaut James Irwin experimenting with a webbing system connected to the bottom torso bracket of the suit and clipped to the handle of the METS positioned between his legs. This system was tested to see if it would take a pull-load of 250 pounds, which it was able to do. Courtesy NASA.

was quite possible that the METS would be used for Apollo 15 as well. Plans were in the works to make it easier to pull the METS based on the debrief of astronaut Mitchell. On April 19, 1971, a memo circulated through NASA noting that Apollo 15 lunar module pilot James Irwin was looking at the possibility of pulling the METS along with a strap that was fastened to the lower torso bracket that supported the primary life-support system straps. He reasoned that with a webbing strap secured to this bracket and tucked between his legs, he could clip the strap to the handle of the METS that would trail behind him, leaving his hands free and providing more stability during his moon walks. ILC's Houston engineer Walt Sawyer explored whether this was possible. Ultimately, however, the lunar rover was used on Apollo 15, so the METS were no longer needed.

Glove Restraint Cable Problem

The one incident of note that occurred on the moon was recorded during the second Apollo 14 EVA conducted by Edgar Mitchell. Shortly following his suit up in the lunar module and before starting his EVA, he radioed to Houston that he felt as though a restraint cable on the wrist of his right glove may have broken because he felt that his hand was being pulled forward and to the left slightly. Following some exchanges between Mitchell and ground control, it was determined that he could press on with the EVA but that he should monitor how it was behaving. (No further comments were made but it was noted that further investigation needed to take place when they returned to Earth.)

By Apollo 14, scientists were no longer concerned that astronauts would bring microbial contamination back from the moon, so Mitchell's gloves were released rather quickly to the quality group in Houston to investigate the problem he had experienced. They immediately pulled back the TMG cover of the glove far enough to look at the restraint cable system. To their surprise, there was no broken cable. Even to this day I see references to a broken cable, but that was not the case. The quality folks were perplexed when they didn't find a broken cable, so they decided to send the gloves and the suit back to ILC in Dover and have Mitchell get into the suit and attempt to replicate the problems he had experienced in front of the ILC designers. Soon after, a dirty suit, gloves, and supporting hardware arrived in Dover, all covered in lunar dust. With Mitchell pressurized in the suit, the engineers narrowed the problem down to the fact that the restraint cables were attached to the wrist disconnect hardware non-symmetrically, causing the glove to be pulled forward and to the right when pressurized. The result of that investigation was that all future gloves would need to be inspected to make certain that when they were pressurized, the restraint cables attached to the base of the glove disconnect hardware were positioned exactly in line with the hinge point of the wrist so that it did not pull harder in any direction. No similar problems were experienced after this.

As an interesting side note, when NASA sent Mitchell's suit to ILC Industries in Dover for the evaluation, the suit container was opened in the

Figure 7.74. Astronaut Edgar Mitchell demonstrating the issues he experienced with his suit when he was on the moon. ILC engineer Richard Pulling looks on. Courtesy ILC Dover LP 2020.

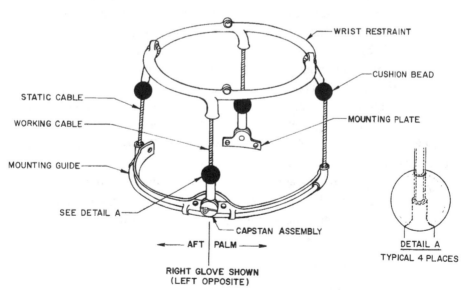

Figure 7.75. Details of the glove restraint that allows the astronaut to flex the hand fore and aft (flex/extension) as well as side to side (abduction/adduction). On Mitchell's glove, the mounting plate was fastened to the glove disconnect hardware too far forward (toward palm side). This threw the balance of the glove off when it was pressurized. Courtesy ILC Dover LP 2020.

receiving department so it could be inspected for obvious shipping damage and to confirm that the contents were what the shipping label said they were. It was not long before one of the ILC employees took a piece of scotch tape and gathered up some of the precious moon dust on the outer layer and then stuck the tape on a piece of ILC letterhead paper. Within minutes, a few others followed, and they had themselves the best souvenir a person could ever imagine. NASA had earlier made it clear through the media that it was not advisable to take any moon rocks or dust since it was NASA property. Because of these threats, this sample collecting was done very discretely; many in the plant knew nothing about it. Soon after, one of the employees allowed their child to take the sample into school for show and tell. In the classroom was another child whose father was in the upper management at ILC. That child went home that evening and asked his father why they did not have a moon-dust sample. All hell broke loose at ILC the following day and most—but probably not all—of the samples were turned in.

Leg Restraint Cable Pulley Redesign

Around the time that the suits were being fabricated for the Apollo 14 mission, ILC was in the middle of testing the new A-7LB model suit that would be used on Apollo 15 and later missions. During this qualification testing in the ILC Lab, the test subject cycling the suit noticed a rather loud "pop" sound and felt something give in the upper leg. Testing stopped and the cover layers were pulled back to expose the problem. It turned out that the pulley that guided the restraint cables had broken where it attached to the restraint webbing sewn to the leg of the suit. This created a rather significant failure analysis on the part of ILC and NASA. The issue, which turned out to be metal fatigue, drove the redesign of the pulley block. After the block was beefed up, tensile testing revealed that the cable did not break until the load reached 2,000 pounds but the pulley was fine at this load. That redesigned pulley replaced the old style in the Apollo 14 suits and all subsequent suits.

8

The Model A-7LB Space Suit

The Next Generation, 1971–1975

To tell the story of how the next generation Apollo suit came to be, we should step back in time to September 12, 1968, and a meeting held at NASA's Crew Systems Division in Houston, Texas. The ILC team had requested the meeting so Homer Reihm, George Durney, John McMullen, and Mel Case could deliver their Omega suit presentation.[1] Reihm later recalled that he came up with the name Omega because as the last letter in the Greek alphabet, it literally means "Great O" or "final." Reihm said that Omega would represent the "great" and "final version" of the suit.[2]

It is unclear exactly who the NASA audience was that day, but an internal NASA memo dated two days earlier announcing the presentation was written by Charles Lutz and addressed to R. Johnson, R. Bond, K. Mattingly, A. Shepard, J. W. McBarron, and the Apollo Space Suit Section. Most on the list were probably at the meeting since this was a significant presentation.[3]

After Reihm's introduction, Durney spent the next ninety minutes presenting the technical details of the Omega suit they were proposing to NASA. The original logbook for the Omega suit showed that Mel Case was the test subject on this date.

The basic ideas and earliest concepts for this suit had been hatched by Case, Durney, and Reihm in 1967. The actual work on this advanced suit started in early 1968, when these three men kicked off a sustained engineering project funded by NASA. One of the definitions of sustained engineering is "the development of necessary design changes to resolve operational issues." The A7L suit was good but it had shortcomings and the ILC engineers wanted to address and resolve the operational issues to the best of their ability with this new suit design.

The first order of business was to figure out how to put a convolute in the waist area so that the wearer could flex without having to fiddle around

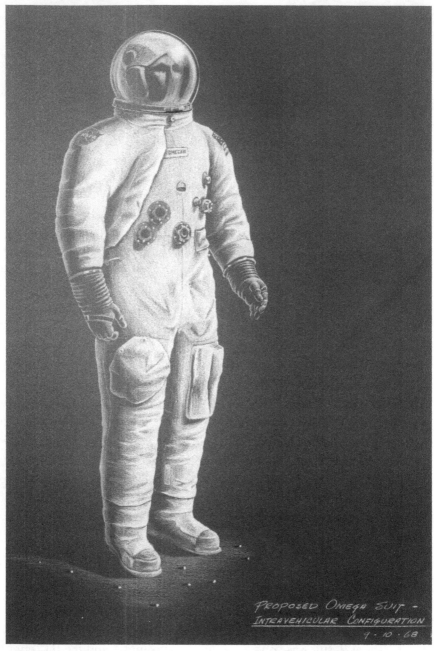

Figure 8.1. The Omega III suit as first depicted in a 1968 sketch. Courtesy ILC Dover LP 2020.

Figure 8.2. Mel Case working with one of the ILC suit subject pressurized inside the model A-7LB suit, which was a great improvement over the A-7L. Case was credited as the principal designer of the A-7LB suit, for which he received the Silver Snoopy award (a prestigious award NASA has given to only a few). Courtesy ILC Dover LP 2020.

with pulling and releasing the clunky pull-down torso strap. The problem with the A-6L and A-7L suits was that the zipper closures ran vertically from the top of the back of the torso to the front of the crotch, cutting through the waist area. It was nearly impossible to make anything flexible where a zipper ran through it. The upcoming lunar missions would certainly require more flexibility so that the astronauts could carry out all of the tasks required of them.

Case reasoned that relocating the zippers away from the waist would free up that area so a series of four wire cables could be installed horizontally across the hips to provide break points to assist with waist flexibility. Mel set out on a course to relocate the zipper onto the chest area, since much of the very early suit work that ILC did was based on George's efforts to put the closure in the front. Two mockup suits were built to test this frontal entry concept, but both proved that this idea did not work. As with the

Figure 8.3. Note the zipper placement on the torso of the model A-7LB suit. One pressure-sealing zipper was located under two restraint zippers—one on the front and the other from the hip to the back. The pull-tab hardware intersected at the hip and a mechanical lock was designed to secure them together. Courtesy NASA.

very early ILC suits, the only option was to have a removable neck ring so that the wearer could get in through the neck opening once it was opened for donning. The neck ring was then slipped over the head and attached to the torso opening using a clamp to seal it tight. The conclusion was that the frontal entry was basically unreliable and difficult to close. In addition, it interfered with neck mobility. It was just a bad idea all around.

Since the closure had to stop above the waist area so the convolute could be included for waist mobility, little room was left between the bottom and the top of the closure. The top was located just below the helmet neck ring. The idea of locating a zipper that spiraled around the torso from top to bottom had its merits. The main issue with the spiral zipper idea was the fact that the old pressure-sealing zipper made by B. F. Goodrich could not take any bending from side to side because that would cause a crease in the closure surfaces that would allow the gas inside the suits to leak out. Fore and aft flexing was OK, but the angles this zipper placement would create eliminated the Goodrich zipper from consideration.

Following many failed attempts, Mel was fortunate enough to find a pressure-sealing zipper that was much more flexible and could maintain a tight seal when bent and flexed, the OEB Slide Fastener made by the Talon Corporation. "OEB" stood for omni environmental barrier. This was a key element of selling the Omega suit idea to NASA. The ILC technical description document that George Durney used as a backup for his presentation stated:

> The OEB pressure-sealing slide fastener has one unique feature. When relaxed, both tapes lie on top of each other and form a very flexible band. Bladder patterns were developed that would allow this closure to assume this position and go down and around the torso in a relaxed position without bunching or attempting to change the length relationship to the tapes or tooth spacing. This technique removes all stress from the closure, permits easier operation and extends closure life.

Translated, this meant that the engineers could install the one-piece pressure-sealing closure so that it would start at the right side of the upper chest and spiral under the right arm, go horizontally across the back, and terminate on the left side of the front torso just above the waist. This placement could take the side-to-side loads without leaking, as the Goodrich zippers did. The more common-looking restraint zipper was located on top of the pressure closer and consisted of two independent sections. The diagonal chest zipper ran from the right-side torso and the second section ran horizontally across the back, where they both terminated on the left side of the torso area.

With trial and error, engineers found the best placement for the zippers so they were as straight as possible while at the same time getting the hardware placement correct so the astronaut could see and access all the air, water, and electrical connections.

By all accounts, the NASA contingent was pleased with the demonstration of the Omega prototype suit on September 12. When it was over, Reihm recalls that one of the NASA management people pulled him aside and told him that this new "marketing tool" demonstration suit looked great but that the end product had better provide great reliability. Reihm

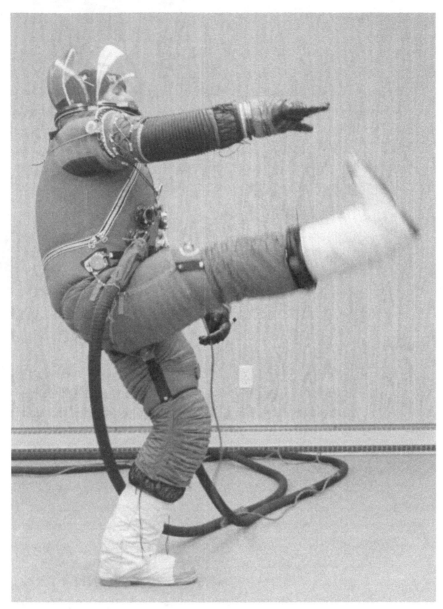

Figure 8.4. A side view of the Apollo Omega suit showing the two separate restraint zippers with the one-piece pressure zipper located beneath them. This freed up the area of the torso below it for wire cables across the front that provide break points that permitted flex between the hips and the torso. This demonstration shows hip-leg flexibility that was not possible in the A-6L or A-7L suits. Courtesy ILC Dover LP 2020.

was aware of that challenge and the ILC team was under constant pressure to make sure they delivered quality and reliability along with the product itself.[4]

ILC allowed NASA to hold onto the Omega III suit after the demonstration. Between September 12 and October 8, eleven astronauts including John Young, Charlie Duke, Gordon Cooper, Buzz Aldrin, Gene Cernan, and Fred Haise tried on the suit for evaluations. An entry was made in the logbook under astronaut Cooper's test run that stated "subject very pleased with the mobility of the suit."[5]

The message ILC delivered was that the Omega suit was a significant improvement over the model A-7L suit.

> The new pressure-sealing zipper closure would be able to withstand 500 cycles of opening and closing without failure.
> Improved neck mobility allowed crew members to flex the helmet forward when driving the lunar rover while in the seated position.
> Abrasion patches were added inside the suit to eliminate wear in certain areas.

During the presentation, Reihm and Durney also proposed that intravehicular and extravehicular suits be combined into one design. From the waist down, the thermal micrometeoroid garment (TMG) would have the cross-section of the lunar integrated TMG while the upper torso would have the lightweight intravehicular cover layer so that the bulk inside the capsule would be decreased and the mobility of the upper torso would be increased. The EVA crew member would put on another TMG cover layer "jacket" for lunar or EV operations. Taking this approach would enable NASA to change which crew members would do which mission functions up to the last minute if necessary without any impact on basic configuration or delivery schedule. This approach would also keep suit costs lower because the design would be standardized for all crew members.

Reihm and Durney also offered up what they called the omni-extravehicular visor assembly. As proposed, this would offer total protection from solar radiation by combining a sun visor and coated helmet. The proposal said, "The visor is locked into position with the thermal protective shell to permit omni-lateral rotation of both the visor and shell to provide protection from low angle solar exposure from either side without a reduction in field if vision."[6]

Other proposals included a new pressure helmet design that removed the feed port and relocated it to the upper torso section of the suit just

Figure 8.5. The proposed thermal micrometeoroid garment jacket for the Omega suit. The new suit as initially proposed would have been identical for all crew members, but the astronauts who did extravehicular activity would slip on the extra protective jacket layer that opened in the back since the primary life support system would cover that area. Courtesy ILC Dover LP 2020.

below the neck ring. The helmet provided a spherical shape that permitted more fore-and-aft mobility; Reihm and Durney claimed that it also offered better optical qualities. They also demonstrated two glove designs for intravehicular activity that included curved fingers, which reduced the torque required to grasp and hold small objects. One of the glove designs provided for the demonstration had a larger wrist disconnect that was 3.9 inches in diameter in place of the current 3.5-inch hardware. At that time, many astronauts were complaining of tightness and discomfort in the wrist.

Lastly, Reihm and Durney proposed a new liquid cooling garment that was not radically different from the A6L model. It would include a new water connector that was located at the waist, eliminating the riser tube that was needed to connect with the pass-through on the upper torso of the A7L model chest. Other improvements would be made to address issues related to comfort and wear.

In the end, however, NASA management rejected both the spherical pressure helmet and the torso-section feed port. NASA management also rejected the common suit design for all crew members, including the slip-on TMG for EVA use. However, the larger glove disconnect and the clenched finger design were included in future A-7LB suits.

On December 11, 1968, ILC submitted Engineering Change Proposal number 397-33 for the "New Model EMU Qualification Unit." If NASA gave the go-ahead by January 6, 1969, ILC would set a target date for completion

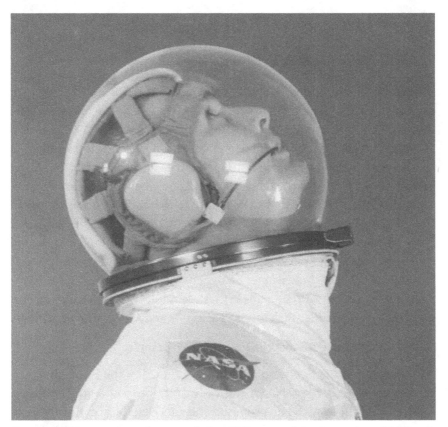

Figure 8.6. The new spherical pressure helmet ILC proposed. Design engineer Mel Case is in the suit. Courtesy ILC Dover LP 2020.

of thirty-eight weeks from that date. The estimated total program cost was $636,000 ($4.6 million in 2019 dollars).[7]

On February 12, 1969, NASA issued a purchase requisition to ILC in the amount of $700,000 ($4.9 million in 2019 dollars) for the design, development, and qualification of an A-9L pressure garment. They required delivery of the prototype by May 13, 1969, and qualification of the garments by September 12, 1969. [8] NASA's identification of this next-generation suit as the A-9L replaced the Omega name. According to NASA's Jim McBarron, the A-8L suit would have been the A-7L intravehicular suit, modified to be an A-7L EVA suit for the Apollo 15 through 17 missions, where the command module pilot was expected to do an EVA in space.[9] As it turned out, the command module pilot suits would continue to be referred to as the A-7L model through Apollo 14. For Apollo 15, 16, and 17, they were called the A-7LB model. The A-8L designation was never used.

Competing against Hard Suits

Litton and Garrett-AiResearch Hard Suits

Turbulence characterized 1969 as NASA worked to figure out the best path forward for suit development to support future missions that were in a state of flux. While ILC continued to work on the new Omega, or A-9L suit, as it was also known, NASA was still pressing forward with the development of two separate hard suits, one by Litton Industries and one by Garrett-AiResearch Corp. NASA awarded those contracts on August 1968, just a month before ILC presented the new Omega suit. In March 1969, it was proposed at a NASA Management Council Meeting that NASA use the A-7L suit through Apollo 14, that it use the new A-9L suit for Apollo 15 and 16, and that is use the Litton or Garrett-AiResearch hard suits to support Apollo 17 through 20.[10] NASA Management Councils were made up of upper management within NASA that made critical decisions that impacted the cost and delivery of product such as the space suits and supporting hardware.

NASA Temporarily Adopts the Hard Suit Philosophy

At a meeting of the Special Management Council at the Kennedy Space Center on May 16, 1969, NASA decided not to proceed with the model A-9L suit ILC Industries had proposed and canceled the purchase requisition it had issued to ILC in February. They decided that they would use the A-7L suits through Apollo 17, then switch to hard suits for Apollo 18

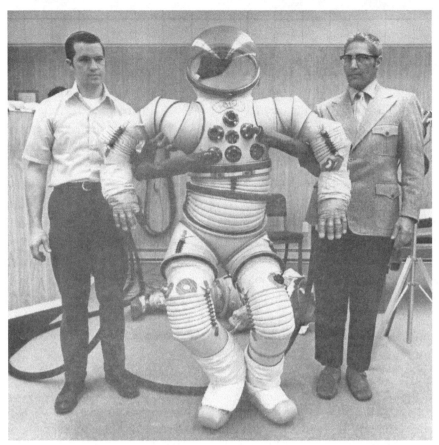

Figure 8.7. The advanced extravehicular suit that Garrett-AiResearch proposed as one of two suits NASA evaluated for the later Apollo lunar missions. This photo shows the suit while pressurized in the ILC laboratory around the time ILC was working on the advanced A-7LB suit. Courtesy ILC Dover LP 2020.

through 20. Although the reasons why NASA made this decision are not clear from the memos, it is likely that the costs of funding three different suit development programs were just too much and that NASA felt that it would select one of the two hard suits.[11]

The NASA council, which was somewhat confident about the future of lunar missions, decided to proceed with the pre-qualification development and testing of prototype hard-shell suits at a cost of $1.6 million ($10.4 million in 2019 dollars) during a Phase I contract award. The period of the contract would be approximately nine months, but it could be shortened if the government deemed it advisable. Phase I would provide funding to Litton and Garrett-AiResearch to develop their best hard suits with delivery by May 1970. In Phase II, the two contractors would compete against each other so NASA could determine who had the best suit. At this point, in mid-1969, NASA projected a budget that would carry Apollo beyond Apollo 17 to at least Apollo 20, where hard suits would be needed for longer-duration lunar extravehicular activities.[12]

In 1968, as NASA increasingly talked of using hard suits, ILC management agreed to fund a program to look at the possibility of making a hard suit for the later Apollo missions, since they did not want to be left behind if that was the direction NASA chose to take. They had confidence that their Omega suit would win over the hard-suit approach, but they wanted to be prepared for any NASA decision. Thus, ILC did some limited design and development work using aluminum joints in sections of the suit. Work did not progress very far, however, because of the federal government's cuts to the NASA budget that resulted in a quick end to the hard-suit program. In addition to the budget issues, there were many challenges with a hard suit, such as difficulty storing them in the already cramped spacecraft. ILC recognized many of these challenges early on and focused their energy on making the best soft suit they could, which ended up being the A-7LB model.

Punt, Pass, and Kick in the New A-7LB Suit

While all the turmoil was taking place with regard to the future of the suit development, ILC wanted to make sure everyone at NASA was aware of the progress they were making on the new Omega suit, so someone came up with an idea. One day in May 1969, they loaded several compressed air tanks onto the back of the company pickup truck and took them across the street to the Dover High School football field along with fifty feet of air hose and the modified A-7L suit—now the Omega suit—that had been custom-made for ILC test subject Tom Sylvester. They also brought along a football and a video camera.

Tom got in the suit and they hooked up the air supply so he could be pressurized to 3.7 pounds per square inch. With the cameras rolling, Tom did calisthenics to show the mobility of all the limbs of the suit. He then took the football and, with the help of a co-worker, proceeded to perform the best job he could in a demonstration of punting, passing, kicking, and even rolling around on the ground. It had to have been quite a sight to see for anyone who happened to drive by on that day. Although the film footage alone did not convince NASA to purchase the new model suits, it certainly could not have hurt.

The A-7LB Suit Gets the Final Go-Ahead

Only four months passed between the time NASA decided to pursue hard suits and the time it decided to continue with ILC's advanced suit program. In September and October 1969, managers in NASA's Apollo program agreed to discontinue Phase II efforts to develop hard suits. On October 20, 1969, NASA gave the contract to ILC for the model A-7LB suit. NASA had decided to look at continued use of the A-7L suit through Apollo 20 and the possibility of developing the A-7LB suit for use on Apollo 16 through 20.

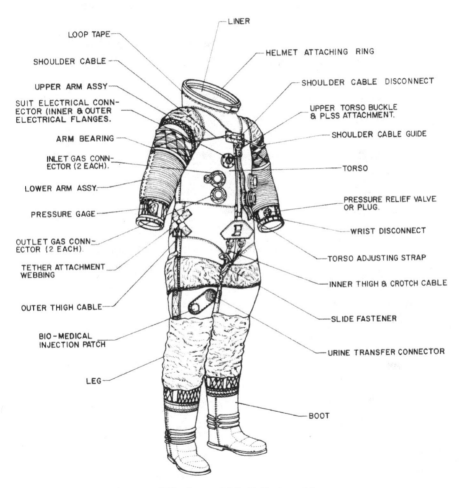

LINER
LOOP TAPE
SHOULDER CABLE
UPPER ARM ASSY
SUIT ELECTRICAL CONN-
ECTOR (INNER & OUTER
ELECTRICAL FLANGES.
ARM BEARING
INLET GAS CONN-
ECTOR (2 EACH).
LOWER ARM ASSY.
PRESSURE GAGE
OUTLET GAS CONN-
ECTOR (2 EACH).
TETHER ATTACHMENT
WEBBING
OUTER THIGH CABLE
BIO-MEDICAL
INJECTION PATCH
LEG

HELMET ATTACHING RING
SHOULDER CABLE DISCONNECT
UPPER TORSO BUCKLE
& PLSS ATTACHMENT.
SHOULDER CABLE GUIDE
TORSO
PRESSURE RELIEF VALVE
OR PLUG.
WRIST DISCONNECT
TORSO ADJUSTING STRAP
INNER THIGH & CROTCH CABLE
SLIDE FASTENER
URINE TRANSFER CONNECTOR
BOOT

Figure 5-7 Torso Limb Suit Assembly

Figure 8.8. This sketch from the model A-7LB maintenance manual shows the command module pilot suit used on Apollo 15 through 17 minus the outer Beta-cloth cover layers. Note the two sets of gas connectors and the arm bearings. Courtesy ILC Dover LP 2020.

In January 1970, NASA announced that it had cancelled Apollo 20 so they could support the Skylab missions. At that same time, NASA announced in a letter to contractors that they would proceed with the modifications to the A-7L suit and drop Phase II of the contract for developing hard suits.[13]

Why was it called the model A-7LB? As it was once explained to me, astronaut Jim McDivitt had input with management at NASA and voiced his concerns that if the advanced suit model was changed to reflect a completely different model, such as A-9L for example, there could be concerns that the contract would have to be bid again or that other contractors would protest that it had not been offered for bid. Thus, it was easier to simply make a slight modification to the contract and call it an update. The new model A-7LB suits would start with serial number 301. An ILC memorandum dated September 3, 1970, suggests that the serial numbers for the new model A-7LB suits should start with 301 since the TMG that was being manufactured for the first suits was starting with the serial number 301. This would have clearly established this new model as a different variety based on the new series of numbers.[14] Thus, with the start of the A-7LB version, ILC chose to start the serial numbers at 301.

In March 1970, the Skylab Program Office announced that they would go with the A-7LB suit for their later missions. Two days later, Litton filed a protest in the hope of getting some chance to make their hard suit for NASA. But by the end of the month, Litton had withdrawn their protest after NASA agreed to let them conduct command module interface tests. This was nothing more than the small offering it appeared to be. Both NASA and Litton (and Garrett-AiResearch) knew there was little chance NASA would use hard suits for any future Apollo missions. Despite that reality, in April 1970, Litton still offered NASA an unsolicited proposal to furnish their suits for Apollo 17–19 and Skylab.

On June 11, 1970, Robert Gilruth of NASA outlined the procurement plan at that time. He stated that ILC would make thirty more suits to satisfy Apollo 18 and 19 and fifty-four suits for the Skylab missions. Gilruth wrote, "Phase I of the advanced extravehicular suit (AES) effort is now complete and in the view of no present need for the AES, it is recommended that Phase II of the AES program not be initiated."[15]

In September 1970, NASA announced that the Apollo 18 and 19 missions would be eliminated. Even as the Apollo 11 crew was on their way to the moon and making history, prospects were already dimming for NASA lunar missions to advance beyond Apollo 17.[16] The fate of the hard suits was pretty much sealed; NASA knew that they would be throwing more money at something that would likely never be used. Besides, they knew what they had with the ILC Apollo suits, and given the advancement in design expected in the next generation of ILC suits, they decided to shift their financial resources back to ILC.

Homer Reihm recalled that while it was obvious that many at NASA wanted to pursue the hard-suit concept, NASA's Charles Lutz was not a proponent of that approach and felt that ILC's suits should continue to be developed further to eliminate any issues with the suits rather than spend the money on a project that presented more challenges than opportunities.[17]

Because of all the work that would have to be done to qualify the new A-7LB suit, the earliest missions they would be used on was estimated at the time to be Apollo 16. Len Shepard received a telegraph on April 24, 1970, from Robert Smylie, chief of the Crew Systems Division in Houston, that reviewed the present plans for the Apollo 16 and 17 missions that would use the A-7LB suit for all crew members. As it turned out, the qualification testing for the A-7LB suits started on September 21, 1970, and was completed on June 25, 1971, just one month before the Apollo 15 mission. Thus, the Apollo 15 crew used the A-7LB suits. Smylie later stated that after a thorough review of that plan, NASA felt it would be more cost effective to use the A-7L extravehicular suits minus the water connectors and lunar boots for the command module pilots. NASA asked ILC to provide pricing updates based on supplying twenty A-7LB suits for training and flight and ten A-7L suits for training and flight.[18] This laid out the plan for the final missions, where the proven A-7L suit was used for the intravehicular crew members and the A-7LB for the EVA crew members.

The A-7LB Suit Design

The Model A-7LB Intra-Vehicular Suits

The model A-7LB command module pilot suits came in the following configuration:

I. Torso Limb Suit Assembly
  A. Torso: Model A-7L extravehicular model less the liquid cooling garment connector.
  B. Arms
     A-7LB configuration with standardized dipped convolutes in all A-7LB suits and improved abrasion resistance by the addition of a nylon cloth scuff layer bonded to the inner surface of all convolutes.

A-7L configuration arm bearing with 360 degree rotation.

A-7L configuration cable turnaround hardware at the shoulders—no pulleys.

Wrist disconnects of the larger 3.9-inch diameter configuration.

Elimination of the lunar module tether mount brackets and lower primary life-support attachment bracket.

C. Legs:

A-7LB configuration with standardized convolutes.

Laminated boot bladders for improved structural integrity and increased abrasion resistance.

D. Medical Injection Patch: Located on right leg

E. Pressure Gauge: 2.5 to 6.0 psid scale, located on right arm.

F: Pressure Relief Valve:

Located on left arm

Redesigned to incorporate manual override cap.

Opening pressure: 5.0 to 5.75 psid

Closing pressure: 5.0 psid.

Flow rate: 12.2 pounds O^2/hr. minimum at 5.85 pounds per square inch absolute.

II: Helmet: A-7LB configuration with Teflon-covered vent pad for improved abrasion resistance.

III: EV Gloves: A-7LB configuration with large-diameter disconnects.

IV: Thermal Micrometeoroid Garment: A-7L extravehicular configuration with an all-Teflon outer layer for improved abrasion resistance (as opposed to having Teflon covering only specific areas that had been identified as being susceptible to abrasion.)

On the Apollo 15, 16, and 17 missions, the command module pilot exited the capsule and took a spacewalk while the crew was traveling back to Earth from the moon. The spacewalk was necessary so that the crew could retrieve two film canisters from cameras located at the aft end of the service module before reentry. These two cameras were used to photograph the lunar surface from above during a crew's many orbits and provided details that scientists would study for years to come. The film canisters were relocated to the command module because that was the only place for the safe return of anything to Earth; everything else that was not inside the capsule would be discarded or burned up upon reentry. The command module pilot performed this spacewalk in the wide expanse of outer space. Finally the command module pilots would get their turn to experience an EVA, even

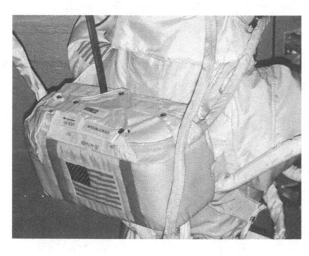

Figure 8.9. Apollo 17 command module pilot Ron Evans wearing commander Gene Cernan's oxygen purge system unit on his back as he practices the EVA he performed in transit from the moon to the earth. Courtesy NASA.

if it was not on the lunar surface. These intravehicular suits were designed based on the requirement that they perform a one-hour EVA in deep space.

On the Apollo 15, 16, and 17 missions, a crew member carried one of the two oxygen purge systems that sat atop the primary life-support system back from the lunar surface so it could be used to support the command module pilot's EVA (see Figure 8.9). These later intravehicular suits had two full sets of gas connectors. With one set, the suit would use an umbilical from the environmental control unit located inside the command module to provide the oxygen to the suit as the commander performed his deep-space EVA. This approximately 25-foot-long umbilical included a steel cable tether secured to the oxygen panel inside the command module and to a harness secured around the suit. The oxygen purge system was strapped to his back behind the helmet to provide backup oxygen to the other gas connector on the suit in the event that he lost his source from the command module. On the suit outlets, he had a pressure control valve that was designed to regulate the gas flow out of the suit and permit the suit to stay pressurized at the required 3.75 pounds per square inch. On the other outlet connector was a purge valve that also acted as a relief valve should an overpressure condition happen, particularly if the oxygen purge system was activated. During his EVA, in addition to using an oxygen purge system that had been used on the lunar surface, the command module pilot used one of the extravehicular visor assemblies his lunar-walking crewmates had used.

The Model A-7LB intravehicular suit followed the basic architecture of the model A-7L suit in that it had the vertical zipper that ran down the

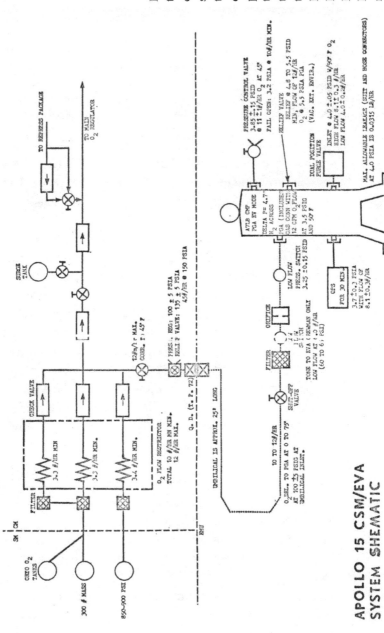

Figure 8.10. Details of how the Apollo 15, 16, and 17 command module pilot suits were connected to the oxygen supply systems during the EVAs that took place in deep space in transit from the moon to the earth. Diagram attached to memo from H. D. Reihm, ILC, to J. McBarron and T. Riggan, NASA, "A7LB CMP PGA Maximum Allowable Leakage," April 6, 1971, ILC memo.

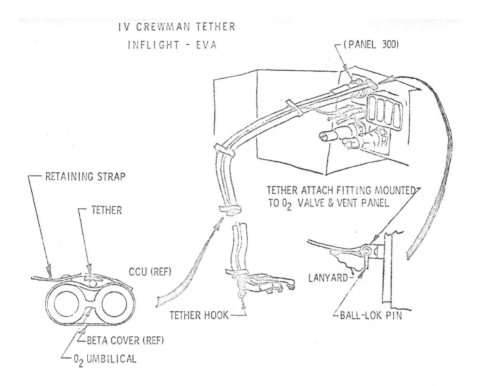

IV CREWMAN TETHER
INFLIGHT - EVA

(PANEL 300)

RETAINING STRAP

TETHER

CCU (REF)

TETHER ATTACH FITTING MOUNTED
TO O_2 VALVE & VENT PANEL

LANYARD

TETHER HOOK

BALL-LOK PIN

BETA COVER (REF)

O_2 UMBILICAL

Above: Figure 8.11. Details of the tether arrangement the three space-walking command module pilots used on Apollo 15, 16, and 17. Diagram attached to H. D. Reihm, ILC, to J. McBarron and T. Riggan, NASA, "A7LB CMP PGA Maximum Allowable Leakage," April 6, 1971, ILC memo.

Left: Figure 8.12. Apollo 17 command module pilot Ron Evans during the fit check of his A-7L suit at ILC. This was the first fitting of the day, during which the thermal micrometeoroid garment was uninstalled. Note the arms, which contained bearings for better range of motion. Also, you can see the markings for the liquid cooling garment pass-through on the left side of the chest that was absent in intravehicular suits but was available in case the suit was ever modified to an EVA suit for training. Photo courtesy NASA.

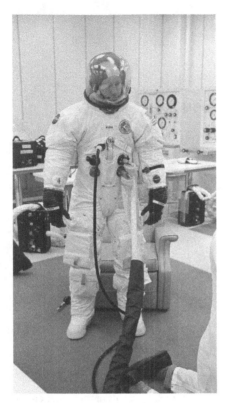

Above: Figure 8.13. Apollo 15 command module pilot Al Worden during his suit up in the A-7LB, which was designed to be used for extravehicular activity between the moon and the earth. Note that there are two sets of gas connectors but the liquid cooling garment connector is missing. Courtesy NASA.

Right: Figure 8.14. Sketch of the model A-7LB space suit. Courtesy ILC Dover LP 2020.

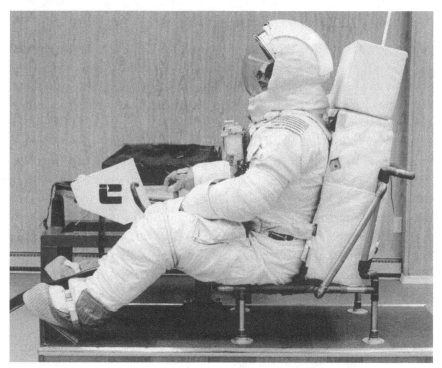

Figure 8.15. This photo, taken on August 7, 1970, shows astronaut John Young in the model A-7LB suit serial number 303 while he evaluates the position within the pressurized suit as he sits in ILC's mock-up of the lunar rover. You can see that his eyes are positioned very low in the helmet. This suit was used for development testing, so it was not specifically sized for Young. As ILC refined the suit design and eventually made Young's suit based on his dimensions, this issue improved and his visibility when seated was fine. Courtesy ILC Dover LP 2020.

back of the torso and into the crotch area. This meant that there was no convolute in the waist for bending. Instead, it had the webbing lanyard and the block-and-tackle system to make it flex at the waist when pressurized. It did have two sets of gas connectors, as all the extravehicular suits did. The three intravehicular suits used for Apollo 15, 16, and 17 did not use the liquid cooling garment because the spacewalk was relatively short and the astronauts did not expend that much energy and thus they avoided any significant heat buildup.

Because of the mobility required to perform the deep-space EVA where the hands and arms would get the bulk of the workout, these suits were provided with the low-profile arm bearings.

The Model A-7LB Extra-Vehicular Activity Suit

Although each model A-7LB suit was customized to a specific crew member, several adjustments could be made to improve the sizing and thus the comfort and mobility of these extravehicular suits. The height of the neck, the angle of the neck, the width of the shoulders, the height of the elbow convolute, the length of the arms, the height of the crotch, the angles of the crotch and the limbs, and the length of the legs all could be adjusted.[19]

Apollo 15 was the first crew to use the new model A-7LB suit. This mission used the new lunar rover that would take the crews much farther from their touchdown locations. This would be a true test of the durability and dependability of the space suit, since it might not be possible for a crew member to simply hop a short distance to the steps of the lunar module should the suit or the backpack have any issues. Table 8.1 shows the total distance and the maximum distance each crew traveled from the lunar

Table 8.1. Total and maximum distance each crew traveled from the lunar module

Mission	Total distance	Maximum distance
Apollo 11	~3,300 feet	200 feet
Apollo 12	~7,600 feet	1,350 feet
Apollo 14	~13,000 feet	4,770 feet
Apollo 15	17.3 miles	16,470 feet
Apollo 16	16.7 miles	15,092 feet
Apollo 17	22.2 miles	25,029 feet

Figure 8.16. The model A-7LB suit without the thermal micrometeoroid garment cover layer being evaluated in the mock-up of the lunar rover seat. Note that this neck ring is angled farther forward than the one in Figure 8.15 for better forward visibility. The helmet in this photo was developed for the Skylab program. Left to right: ILC resident government representative John Leshko, ILC systems engineering manager John McMullen, and ILC engineer Steve Rubin. Courtesy ILC Dover LP 2020.

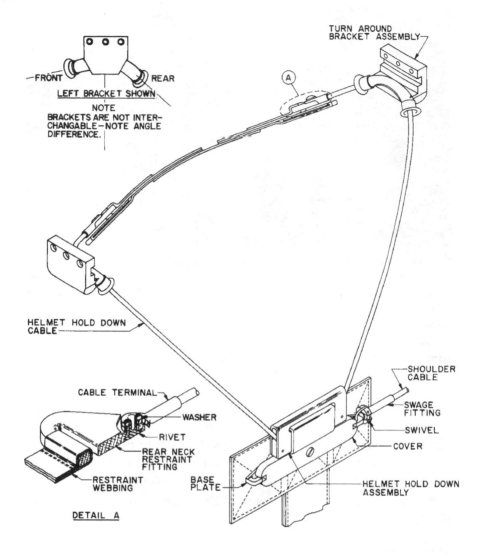

TURN AROUND
BRACKET ASSEMBLY

Ⓐ

FRONT   REAR
LEFT BRACKET SHOWN

NOTE
BRACKETS ARE NOT INTER-
CHANGABLE—NOTE ANGLE
DIFFERENCE.

HELMET HOLD DOWN
CABLE

SHOULDER
CABLE

CABLE TERMINAL

WASHER

SWAGE
FITTING

RIVET

SWIVEL

REAR NECK
RESTRAINT
FITTING

COVER

RESTRAINT
WEBBING

BASE
PLATE

HELMET HOLD DOWN
ASSEMBLY

DETAIL A

HELMET HOLD DOWN INSTALLATION

Figure 8.17. Details of the model A-7LB neck ring/helmet hold-down cable. The base plate and patch assembly were secured to the upper torso. The turn-around brackets, which were secured to both sides of the neck ring, permitted the cable to slide through it. Once the suit was pressurized, the astronaut could move the helmet forward or backward, A convolute in the neck area would flex and permit the cable to slide through the turn-around brackets. Friction between the cable and the brackets would secure the helmet in position. Courtesy ILC Dover LP 2020.

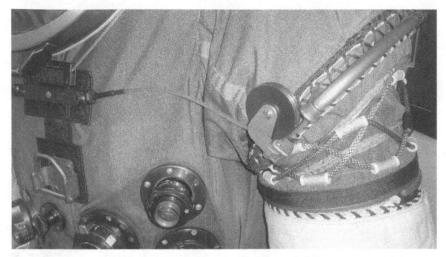

Figure 8.18. Details of the model A-7LB shoulder cable and pully arrangement. The one-piece steel cable wrapped around the entire torso and over both shoulders. It passed through two aluminum tubes, one over each shoulder, and carried structural loads but also permitted the shoulder to flex both up and down and forward and backward. As the cable moved in and out of the tubes as the shoulder assembly moved (particularly forward and backward), the pully would reduce the bend in the cable where it exited the tunnels. Earlier suits had cable failures in the test labs due to sharp bends and abrasion against the tube openings. Courtesy ILC Dover LP 2020.

module. The Apollo 15 through 17 missions pushed the suits to their limits. The crews had to have total confidence in the suits or else they would be very reluctant to stray even a few feet beyond the safety of their module.

ILC made several key improvements to the basic model A-7L suits in addition to adding waist convolutes. The sections below include some of the details for each update.

Torso-Limb Assembly

The neoprene bladder layer had a nylon cloth scuff layer that was bonded to the entire surface of the neoprene bladder, as opposed to a basic neoprene-coated nylon ripstop that was added only in critical wear areas. The new Talon OEB pressure-closure zipper was installed in the spiral fashion so the waist section could accommodate the convolute necessary to allow the bending motion that the A-7L model did not provide. Two sections of the restraint zipper were installed over the pressure zipper. Two sections were added because the restraint zipper was too rigid to allow for the bending that the pressure zipper could withstand.

Main Closure Assembly

Zipper-lock mechanisms were installed to ensure that the lock would not come undone accidentally during extravehicular activity.

Convolutes

A nylon cloth scuff layer was bonded to the inside surfaces of all convolutes to help prevent abrasion that could cause leakage should abrasion become significant.

A new rubber neck convolute was designed for the A-7LB EVA suits so that crew members sitting on the lunar rover had a better horizontal field of view. Early trials by astronauts at ILC showed that when the suit was bent at the waist, the hardware on the neck ring would ride up over the chin because of the angle of the torso relative to the neck ring.

The combination of the rubber convolute and a sliding restraint cable arrangement on both the torso and the helmet meant that the astronauts could angle their helmets forward. This provided much improved visibility. Without this convolute, astronauts would not have been able to see the remote control unit on their chests and would not have had the best visibility as they drove along in the rover. In addition, Mel Case and the team added steel restraint cables on both sides of the torso and the neck ring to add the structural loads to the neck ring. Mel designed it so that when the helmet was attached to the neck ring, it could be manually pulled forward or backward to give the best angle for visibility. Although this proved to be a big improvement, not everyone was on board with this change at first. Astronaut John Young voiced his opinion in a memo dated July 29, 1970, that the design was not comfortable and that the plan should be to go back to the fixed design of the A-7L version.[20] Among his many concerns were that for lunar surface operations, the neck is adjusted full forward and thus there was no need to move it back. He also was concerned about some fit and comfort issues he had because of the vent tubes in the back near his neck. Adding to the controversy was the fact that one of the restraint cables broke during the design cycling in the ILC Laboratory. Young expressed other issues, but Maxime Faget, the director of engineering and development at NASA, provided a rebuttal in a memo dated August 20, 1970, that addressed each of Young's concerns. The bottom line was that ILC resolved the problems as the design was tweaked. In addition, the major issues were related to the proper sizing for each astronaut and at that time, ILC had not been able to adequately size the new A-7LB suit for everyone since

Figure 8.19. The front hip section of the model A-7LB suit with the convolute. The convoluted section was formed by four cables that were sewn horizontally in place across the hips and acted as a natural break point when the astronaut bent forward. The four vertical cables carried the structural loads from the legs to the front and rear torso. The heavy-duty webbings attached to the torso restraint fabric (attached to the two D rings) helped distribute the loads. Courtesy ILC Dover LP 2020.

the design was still a work in progress. Once the suit was sized for John Young and Charlie Duke a few months later, their feedback was much more positive.

The new waist convolute provided fore and aft flexibility. This was not a rubber convolute; it was a cable system that was integrated across the lower waist area of the suit's nylon restraint layer. These cables provided a natural break area in the suit that permitted flex while retaining the shape in that area of the suit when it was pressurized.

Lower Torso Cable Restraint System

The extra waist mobility that the Model A-7LB suit provided gave rise to a new problem: the restraint cables wore into the restraint layer. There was a concern with wear that became evident in the rear crotch of the lower torso. To fix the problem, ILC engineers came up with a unique device that contained a Teflon plastic slider that moved within a track as the suit flexed at the waist. The restraint lines from the front of the crotch were tied into this Teflon device that isolated the cables from the restraint layer. Qualification testing of the A-7LB suit showed that this Teflon slider eliminated the abrasion that was found in the design verification testing suit identified as DVT-001.

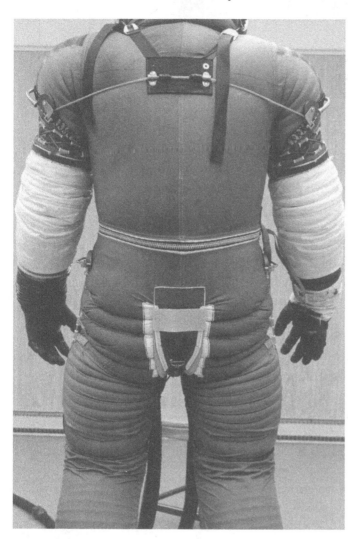

Figure 8.20. The device designed to prevent abrasion to the rear and crotch of the model A-7LB suits, known as the beaver tail. Courtesy ILC Dover LP 2020.

Pressure Garment Assembly Boot

A nylon cloth scuff layer was bonded inside to form a sock that addressed concerns about abrasion and provided an increased level of structural integrity.

Medical Injection Assembly

The medical injection patch assembly was moved from the right leg to the left to accommodate the relief valve located inside the pocket on the upper right thigh.

Hardware

Gas connectors were relocated to accommodate the new zipper closure location.

A single diverter valve was mounted and integrated into the gas connector housing.

The multiple water connector for the liquid cooling garment was relocated because of the new zipper placement.

More alignment marks were added to the liquid cooling garment multiple water connector for proper positioning and lock/unlock indication.

The electrical connector was relocated to accommodate the new zipper location.

The pressure gauge was relocated from the right arm to the left.

The pressure relief valve was redesigned to incorporate a manual override. It was relocated from the left arm to the right thigh under the flap for the urine collection transfer valve.

The neck ring was attached to the restraint and bladder soft goods of the torso with a flange mount rather than a compression band. The latch for the neck ring was relocated off center by approximately three inches for ease of view and access.

A pulley system was added to the shoulder cable restraint system to lower the resistance (torque) loads.

The wrist disconnects were the larger 3.9-inch inside diameter rather than the previous 3.5-inch diameter.

The arm bearings provided two integral stops so that the lower arm would always be aligned properly with the upper arm assembly.

Helmet

The helmet vent pad was made of Teflon to provide improved abrasion protection.

Thermal Micrometeoroid Garment

The entire cover layer of the TMG would be made of Teflon cloth for improved abrasion resistance. The TMG for the A-7L model used Teflon only in the highest wear areas.

Apollo 15 through 17: Using the Advanced Suits

Apollo 15: First Use of the Model A-7LB Suits

The Apollo 15 crew consisted of Commander Dave Scott, lunar module pilot James Irwin, and command module pilot Al Worden. Since this was the first mission using the new model A-7LB extravehicular suits, there was a flurry of activity related to fixing last-minute problems and concerns the crew had with the new design. A "NASA-MSC Fit Check Status Report" dated January 22, 1971, outlined many problems with the new suits and the corrective actions that were needed. There were issues such as tightness and pinch points due to sizing that needed to be adjusted and visibility problems when crew members were seated in the lunar rover vehicle mockup. As the astronauts were getting familiar with the new suit design, the ILC engineers were also adjusting to some of the differences that had an impact on fit and function. Dave Scott made several comments regarding the fit of his backup suit (serial number 311). He was not pleased with the initial sizing and complained, as he had on previous occasions, about the stiffness and sizing of the gloves. He felt that they were too tight and that the Fluorel fire and abrasion resistant coating added to the problems.[21]

The flight suit and backup suit for Al Worden (serial numbers 094 and 081) were built before the new A-7LB design, so ILC was contracted to upgrade these suits with the new arms that included the bearings. In addition, new legs were installed that included laminated boot bladders that consisted of abrasion patches over high wear areas of the boot bladder. The contract also permitted newer thigh convolutes to be installed that had improved lamination. NASA's requirements stated that the rubber used in the suits had to have a minimum of twelve months of shelf life on the date of launch. In an ILC memorandum that addressed the configuration of the intravehicular suits for this mission (one flight suit and one backup suit) the rubber in the serial number 094 suit Al Worden was to fly would have had only seven months of shelf life remaining and ILC suggested that they replace the entire torso. It is unclear if ILC was given the go-ahead to replace the torso section of this suit or if NASA approved the use of it with bladder layers made from the older rubber.[22]

Apollo 16: Model A-7LB Suits

This crew consisted of mission Commander John Young, lunar module pilot Charles Duke, and command module pilot Thomas Mattingly. Modifications were made to John Young's primary suit (serial number 322) during the month's slip in the launch date. Young had earlier demonstrated a new procedure, eventually known as "Young's rocks," that he devised to gather lunar rock samples with his hands rather than with tongs. His technique consisted of jumping up to allow the weight of the suit with him inside to spread the legs apart on impact with the ground. This applied extreme loads to the crotch and knee restraint cables in the suit. This caused quite a concern among the NASA and ILC engineers because just around this time as the suit design was being qualified in manned testing, a crotch restraint cable had pulled out of the termination hardware that was used to attach the cable to another piece of hardware. In addition, a knee convolute cable had failed during tests. Because the Apollo 16 launch was just four months away, all involved were very concerned about these issues. ILC engineer Al Gross recalled a certain level of frustration in the group because the testing that NASA was requiring went well beyond what would have been carried out with the suit in the 1/6th gravity of the moon. Young's rock exercise was excessive and he never used it during his extravehicular activities, but neither John Young nor NASA wanted to assume anything. (For more information, see the section of Appendix A that details the steel restraint cables.)

Command module pilot Al Worden performed the first spacewalk to retrieve film canisters from the service module on the previous mission. The command module pilots for Apollo 16 and 17 did similar EVAs. Mattingly wore the model A-7L suit ILC had built for him for his Apollo 13 mission. He had been fit checked for that suit on August 18, 1969, but NASA had scrubbed him from Apollo 13 due to fears that he may have contracted measles.

Apollo 17 Mission: Model A-7LB Suits

The Apollo 17 astronauts—Commander Eugene Cernan, lunar module pilot and geology scientist Harrison Schmitt, and command module pilot Ronald Evans—were the last crew to visit the moon for who knows how long. Cernan and Schmitt's were on the lunar surface for three days and three hours. They spent a total of 22 hours walking and riding about on the surface during three separate extravehicular activities. The first EVA lasted for 7 hours and 12 minutes, the second for 7 hours and 37 minutes, and the third for 7 hours and 15 minutes. This helped reinforce the belief that the suits were in fact designed, tested, and manufactured to stand up the abuse that could be dished out in the most hostile of conditions. Both Eugene Cernan and Harrison Schmitt had full confidence in the performance of their suits and did not experience any unanticipated issues during the mission. When looking back at all the comments of the Apollo 17 crew over the years, the one issue that rises to the surface is related to hand discomfort experienced while using the pressurized gloves. Although these gloves were sized for the crew, their hands took a real beating over the twenty-two hours they spent in their EVA suits. They were constantly using their hands to drive the lunar rover or to work with tools to perform other tasks. These rubber gloves wrapped with chromium steel fabric were the best on offer at the time, but it was clear that future space missions requiring long-duration extravehicular activities would need a much more comfortable glove. Eugene Cernan commented years after his mission that he felt at home in the gloves and that they were as good as could be expected given the technology of the time.

In August 1972, an ILC Work Request Form was initiated that outlined the work required to develop and fabricate "wrist disconnect dust covers" that would be placed over the glove disconnects on the suit side. These would consist of a rubber-neoprene mixture made in a mold. The idea was to install these covers over the glove disconnects in order to keep the lunar dust out of the bearing and attachment hardware. That was based on feedback from the earlier crews who complained about how the lunar dust affected the operation of the hardware once it penetrated even the smallest of openings.[23] An ILC memorandum dated November 7, 1972, included a test report that noted that the crews preferred the tape idea over the rubber dust cover. However, the tape was not acceptable because it would not stick properly and because one of the crewmen would have to apply the tape with his gloves on, which was too difficult.[24]

Suit Use on the Later Apollo Missions

At the start of the Apollo missions, the crew was required to wear their pressure suits for various phases of the mission including launch, rendezvous, and reentry. Recall the problems that the Apollo 7 crew had with their ability to wear their helmets on reentry due to head colds and associated sinus problems; they needed to squeeze their noses to equalize sinus pressure

TABLE 2-1
(12/6)

SUIT WEARING SCHEDULE

ACTIVITY	PRESSURIZED (HARD SUIT)	SUITED (SOFT SUIT)	PARTIAL SUIT WITH-OUT HELMET & GLOVES	SHIRTSLEEVES (ICG)
LAUNCH		ALL		
EARTH ORBIT THRU S-IVB EVASIVE MNVR			ALL	
TLC & TEC EXCEPT TEC EVA	*			ALL
PGA TEST			ALL	
LM ACTIVATION			ALL	
UNDOCKING		CDR & LMP	CMP*	
UNDOCK +5 MIN THRU CIRC			ALL	
POI thru TD		CDR & LMP	CMP	
LUNAR STAY EXCEPT EVA				ALL
LUNAR SURFACE EVA'S & EQUIP JETT	CDR & LMP			CMP
LIFT-OFF PREP			ALL	
LIFT-OFF THRU DOCKING		CDR & LMP	CMP	
DOCKING TO LM JETT			ALL	
LM JETT		ALL		
POST LM JETT THRU TEI				ALL
TEC EVA	ALL			
ENTRY				ALL

*CMP DON HELMET & GLOVES FOR DOCKING LATCHES RELEASE.

Figure 8.21. Apollo 17 suit-wearing schedule showing usage and configuration throughout the mission. Courtesy NASA.

as the capsule gained internal pressure on descent into the earth's atmosphere. Following these early missions, the NASA doctors and engineers looked more closely at the problems and eventually agreed that it would no longer be necessary for crew members to have their suits on for reentry.

What was the schedule for wearing the suits on later missions aside from the scheduled extravehicular activities? Figure 7.21, which is copied from the Apollo 17 Mission Plan, provides an example of suit usage and configurations throughout the mission. The only time the suits were pressurized was during extravehicular activities on the moon (commander and lunar module pilot) and during the trans-Earth coast phase of the mission, when the command module pilot opened the hatch and performed his EVA to gather film cartridges from the service bay. The entire crew had to be in their pressurized suits when the hatch was opened.

During the various phases of the mission when potential problems could arise, such as launch, docking, and undocking, crew members put on their suits, including their helmets and gloves. During less critical times, the suits needed to be worn but the helmet and gloves did not need to be on. The reasoning was that there would likely be time to put on the helmet and the gloves if problems developed.

Early in 1969, NASA deleted the requirement that the suit to be self-donning. That meant that crew members would have to be able to help each other put their suits on at times throughout the mission. For that reason, the other crew members helped the command module pilot put on his suit before the crew split up when the commander and the lunar module pilot undocked to descend to the lunar surface. Part of the reason for that was that if the command module pilot had to return to Earth by himself due to some unforeseen catastrophe, for instance if the lunar module could not return for rendezvous, he would have his suit on since no one would be there to help. The rule was that if a crew member was going to be inside the suit for an extended time, he had to have the fecal containment unit on to contain any potential waste.

Apollo Lunar Missions: What Did We Learn?

The Apollo lunar missions taught us lessons about space suits that I hope will be put to use in the not-too-distant future, whether we return to the moon or go the distance to Mars. NASA has established requirements for future suits, taking the data gathered from Apollo into account, and they continue to refine the requirements based on new technology as it advances.

There was much we didn't know about exploring the moon when Neil Armstrong set foot on it, but we had learned a great deal by the conclusion of the Apollo 17 mission. Given the state of technology at the time, Apollo was a tremendous success. The space suit was one of many technologies that was brand new at the time. While the suits received minor upgrades between Apollo 7 and Apollo 17, making any significant improvements between each Apollo mission was impossible. We were just learning to crawl at the time and the schedules didn't allow much time between missions. The kind of improvements I speak of come with years of experience and advancements in technology that include new materials and manufacturing methods. The combination of these improvements and the lessons we have learned should yield good results in future suit design.

During Apollo, we learned rather quickly that digging in the lunar soil contributed to problems related to the dirt sticking to the suits and clogging up the metal bearings and disconnects. The dirt acts like sandpaper and could easily contribute to abrasion. When crew members returned to the lunar and command modules, they brought that dirt back with them, which caused obvious problems as it floated around the cabins.

By the mid-1970s, as NASA was beginning work on the space shuttle program, new requirements for the next generation of EVA suits were being established. Among the requirements were that all future suits should avoid the use of rubber, zippers, or steel cables that could cut like a knife into fabrics. These cables included questionable terminations that secured them.

The rubber used during Apollo had a short shelf life and there was no guarantee that the physical properties would be acceptable from batch to batch. Although the zippers did the job, they could cause the death of an astronaut if a significant failure occurred. Several failures were recorded, primarily on zippers in test suits that were cycled far more than they would have been on a mission. The steel cables would come apart at the termination ends when subjects were putting high loads on the suits in the test laboratories. The steel cables also cut into the relatively soft fabrics during extended testing in the labs. As bad as that all sounded, the suits were designed to carry out each mission with room to spare for safe operation. No significant problems were ever encountered with a suit that supported a flight. At the end of most missions, the suits were sent to the Smithsonian, where they were retired.

To remedy the issues experienced during Apollo, the rubber was replaced by modern coated fabrics and the zippers were replaced with strong metal disconnect joints that separated major components, such as the upper and lower torso sections. This provided a way for crew members to get into and out of the suit and ensured a secure joint once fastened. The steel cables were replaced by Dacron webbings that could be sewn to form restraint lines that took very high loads and provided a breaking strength that was repeatable and always well beyond the maximum working loads of the suit.

The space shuttle suits were designed for an entirely different mission during which crew members work in microgravity. In contrast, a spacewalk was just a space float that required little mobility in the lower torso. The

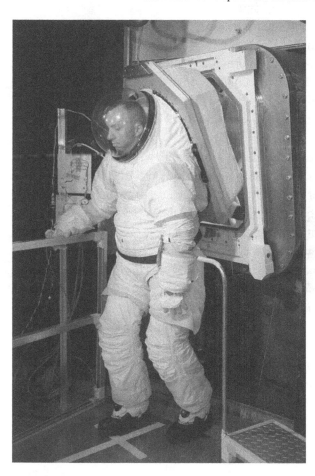

Figure 8.22. The ILC prototype model Z-1 suit with an adapter on the backpack that enables the crew member to back up and lock into the airlock interface. Once they are locked in, they can open the backpack door and exit into the airlock, leaving the dirty suit outside. Procedures were formulated to retrieve the suits for servicing, cleaning, and repair as needed. Courtesy NASA.

weight of the space shuttle suits was not as significant an issue as it was for the Apollo suits. (However, in future reduced-gravity environments it will play a critical role.) The shuttle suits were the first to use flat-pattern convolutes that replaced the rubber pieces. They also used composite materials for the first time to make the rigid upper torso section. The hardware used to join the various suit sections was more advanced and allowed for the replacement of parts so that suits could be reused and provide modularity to accommodate other astronauts yet provide a reasonably comfortable fit.

Future suit designers will look to the past to provide the suits of the future. Dirty suits coated in lunar or Martian dust may be left outside the spacecraft or habitats; crew members may get in and out of their suits by way of backpacks that will open and close like a door. The backpacks would be mated to a doorway leading into and out of the pressurized airlock environment of their living quarters. Additionally, future space suit designs that can operate at higher pressures, say in the range of 8.0 pounds per square inch absolute, can use more of a gas mix (oxygen and nitrogen), which has the advantage of reducing the time needed to pre-breathe pure oxygen to get the nitrogen out of the body to avoid the bends. Future suit designs will also require improved joints that flex and bend easily when pressurized at the higher pressures.

The bearings and disconnects on future suits will have to be protected against dust intrusion. Hard composite sections will likely be used to provide structural support and provide a means of securing hardware sections and allow for the unique sizes of crew members.

Glove development has come a long way since Apollo. Gloves now provide greater mobility and crew members need to exert less torque to flex the many joints of the hands. This development will continue, perhaps using technology such as motorized devices that help open and close the hand, for example. There is no end to the use of new systems that employ miniaturized systems using computer feedback that may one day aid the astronauts who walk on Mars.

The Apollo program cannot be remembered simply as the one that President Kennedy established to win a space race. We learned so much from the Apollo missions, and the future of space exploration will depend on how wisely we use the information we gathered during those years.

III

POST-LUNAR MISSIONS

9

Skylab, the Apollo-Soyuz Test Program, and Other Development Suits

The Manned Orbital Lab Suit

In 1966, ILC engineers Mel Case and Homer Reihm devoted a relatively small amount of their time to developing a space suit for a contract to support the Manned Orbital Laboratory, a program the air force was looking to carry out from 1970 through 1975.

The ILC group was walking a tightrope at the time, since they were under pressure from NASA to turn out the Apollo suits and under great strain just to keep up with the numerous and ever-changing contract demands. Diverting any amount of support away from Apollo to focus on the air force initiative was a stretch. Reihm recalled that NASA was not too interested in hearing that ILC was planning to provide a suit for the contest. His recollection was that NASA urged contacts within the air force to look to other contractors for the suit. This makes sense, since other contractors had the resources and the suits to meet the air force requirements. Hamilton Standard entered a suit they had developed for the contest based on their earlier Apollo work. The air force ended up selecting that suit, but then the Manned Orbital Laboratory program was cancelled in 1969 because technology had advanced enough that unmanned satellites could carry out missions. The ILC manned orbital lab suit consisted of the basic A-6L suit with the link-net arms and a new neck cone assembly that provided for the attachment of an oval pressure helmet that provided the full-angled visibility that Air Force specifications likely required. The shoulders used a cable-and-pulley arrangement that appeared to be in the early stages of development. No system like this was used in any further suits ILC manufactured.

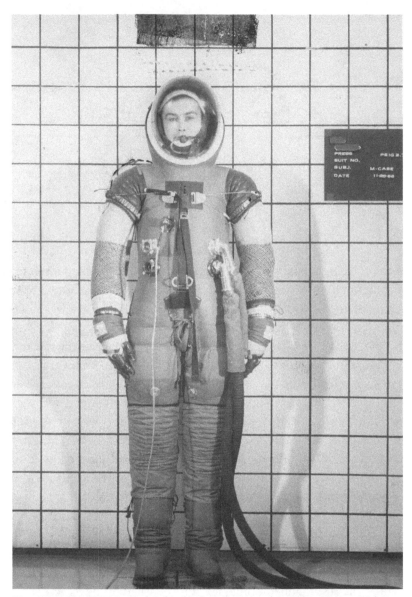

Figure 9.1. The ILC Manned Orbital Laboratory suit that ILC engineers Mel Case and Homer Reihm were working on in 1966. It is basically the Model A-6L Apollo suit with modifications to the helmet and the arm-cable restraint system. Mel Case is in the suit. Courtesy ILC Dover LP 2020.

Skylab

Work on the Skylab missions began as far back as 1959 with Wernher von Braun, who had envisioned an orbiting space station since his youth in Germany. In 1945, von Braun settled in the United States after being moved here after World War II under a military program known as Operation Paperclip. That controversial program allowed the United States to bring to this country some of the smartest minds that had served the German war machine. Von Braun professed throughout his life that he only ever wanted to use rockets for peaceful purposes and that the Germans had forced him to develop V2 rockets for their army. After settling into this country and developing the rockets that would eventually take our astronauts to the moon, he became a hero of sorts. To further his goal of the peaceful promotion of rockets to conquer outer space, he teamed with anyone who would listen to him, including Walt Disney, science writer Willy Ley, and artist Chesley Bonestell. (Bonestell's artwork appeared in stories von Braun published in *Collier's* magazine that romanticized space travel and space stations.) In 1959, when he was the head of the Development Operations Division of the Army Ballistic Missile Agency, von Braun proposed a mission to the moon in which he mentioned parking the first stage of the rocket in orbit around the earth for use as a space station. He called the program Project Horizon.[1]

The Skylab program began to take shape somewhere around 1963, when a group formed within NASA called the Apollo Applications Program. Their forward-thinking mission was to look at various ways that NASA could continue to use spacecraft and equipment developed for the lunar missions in other space programs. They correctly realized that it was in the interest of NASA and the nation to have programs beyond lunar missions to sustain their own future and the advancement of science. It also ensured job security for the many NASA employees and contractors. This group looked at ways of using the lunar modules, command modules and Saturn rockets for other space applications. Finally, on January 26, 1967, NASA associate administrator George Muller announced in a press briefing that by 1968 or early 1969, NASA would have a space station orbiting 275 miles above the earth. He reasoned that NASA would have a surplus of Saturn rockets that could be converted for use as a space station. He also reasoned, incorrectly, as it turned out, that the budget would cover the costs of both the lunar missions and this space station.[2]

Sadly, the very next day, the Apollo 1 fire took the lives of the crew. This cut short any grand visions that NASA had of a space station by 1968 or 1969. Congress ultimately slashed the 1968 budget for the Apollo Applications Program due in large part to the war in Vietnam. Finally, in August 1969, McDonnell Douglas won a contract to develop a space station using two excess Saturn S-IVB stages.[3]

In August 1970, Dale Myers, the associate administrator for NASA's Manned Space Flight program, sent a letter and an outline to Manned Spacecraft Center director Robert Gilruth that provided justification for going directly to ILC for the suits that would be used on Skylab. NASA had to use care to ensure that they addressed all possibilities of looking at other contractors capable of providing the space suits for each of the different missions. Because Skylab would be different than the Apollo lunar missions, did that mean that other suits would have been better candidates for Skylab? Companies such as Hamilton Standard and Litton would have certainly liked to have NASA use their latest-model suits for future missions, but that was not to be. NASA did not make it easy for ILC, however. The contract award followed a long period of negotiations between NASA and ILC to get the price down as low as possible. Homer Reihm recalled basically living in Houston between October 1969 and April 1970 and meeting with NASA program people to go over very detailed costs and personnel numbers. The goal was to reduce the ILC workforce that NASA was paying for to the bare minimum.[4] John McMullan recalled that Hamilton Standard had proposed to NASA that they use the Hamilton Standard manned orbital laboratory suit for Skylab that had been designed a few years earlier for the canceled air force program. ILC worked hard to ensure their suit was competitive in all areas, including cost.[5]

In the document attached to the opening letter, Myers made it clear that NASA had chosen to stay with the ILC Apollo suits because they had proven to be very reliable throughout the years.

Based on the above considerations, it is concluded that ILC is the only source capable of producing and providing production quantities of flight qualified suits at reasonable costs with the proven performance and demonstrated confidence required. It is not considered to be in the best interest of the government at this time to enter into a competitive solicitation for manufacture and support of space suits for Skylab, when the overwhelming circumstances in support of retaining the Apollo type equipment for Skylab are as stated. To enter into such a competition would promote duplication of efforts in facilities, tests, development, support and thereby unnecessarily cause additional expenditure of government funds.[6]

Included in the justification letter was a request that ILC provide a total of fifty-four suits for Skylab. These would be the model A-7LB EVA suits. The total number was later amended to thirty-seven suits. Because the design changed slightly from the lunar A-7LB suits, the serial numbers changed to the 600 series; the Skylab suits were given the serial numbers 601 to 637. The fit-check records for the Skylab program show that four model A-7LB suits of the series 300 (307–310) were used for training early in the Skylab program before ILC qualified the new design.

The Vought Missiles and Space Company (a subsidiary of LTV, the makers of the Apollo extravehicular visor assemblies) wrote a letter to ILC on January 29, 1971 that detailed the cost of twelve new Skylab extravehicular visor assemblies.[7] The net cost was $206,144 ($1.3 million in 2019 dollars). The changes from the lunar extravehicular visor assemblies included:

Elimination of the center eyeshade
UV-stabilized polycarbonate visors
White external surface of high emittance and low absorptivity
Dark-colored non-reflective interior surfaces of the shell
Increased downward visibility
Positive downward positioning of the protective visor
Both visors replaceable in flight

In the end, the Skylab extravehicular visor assembly helmets were designed and manufactured by Airlock, Inc. Airlock had proven themselves to be the best at making the majority of the Apollo hardware. The helmet shell of the Skylab extravehicular visor assembly was made of a white plastic composite that was not insulated, as was the case for the helmets for the lunar missions. No contract information could be located regarding the costs that Airlock charged per helmet but it can be assumed it was in line with or slightly less than the Vought Missiles and Space Company estimates since reducing the costs of this program was important to NASA because of government funding cuts.

The thermal layers of the Skylab suits were reduced from those of the Apollo lunar suits since the mission requirements were less severe. This had the benefits of weight reduction and increased mobility.

Because the Skylab astronauts were expected to perform extravehicular activities where they would have to do a lot of interface work with the outside of the space station in the zero-gravity environment, a foot restraint system was developed that permitted the crew to attach their feet to a secured foot-restraint plate attached to the station to hold themselves in place. To secure the boots to the plate, the pressure boot was modified so that it would lock under what was called a Z-plate.

Mission Highlights

There were three manned Skylab missions to the first space station the United States placed in orbit. Skylab II launched on May 25, 1973, with Commander Pete Conrad, James Kerwin, and Paul Weitz on board. The unmanned Skylab I Station had launched eleven days earlier, but the vehicle sustained significant damage to the micrometeoroid shield and solar arrays before reaching orbit. This situation set the stage for significant troubleshooting to fix the problem since the Skylab Station was reaching high internal temperatures because the meteorite shield was torn away. ILC Industries engineers Bob Wise and Ron Bessette and seamstress Eleanor Foraker were sent to Houston to assist NASA and the ILC Houston office with the development of a fabric shielding device for the station to provide solar protection. Within days, and following many hours of nonstop work, the ILC team came up with a solar shield that consisted of an aluminized Mylar laminated to an orange nylon ripstop fabric. This was referred to as a parasol. The team also had to develop outer protective gloves to guard against cuts to the pressure gloves as crew members worked around the damaged metal sections of the station.

When the Skylab II crew arrived at the station, they had to perform a stand-up EVA outside the command module to try to deploy the jammed solar panel with the aid of a special ten-foot pole that had a hook on the end. That failed to work. Soon after, they attempted to dock with the station, but the docking probe failed to work. This required Conrad to perform an EVA to disassemble and repair the latch. That did the trick and the crew was finally able to dock with the station. Once inside, they deployed the new parasol from inside through a small airlock. Once they had deployed this solar shield, the internal temperature of the station dropped from 125°F (52°C) down to a more manageable 80°F (27°F). Two weeks after they arrived, Conrad and Kerwin donned their suits and performed an EVA that finally freed the stuck solar panel.

The Skylab 3 crew consisted of Commander Alan Bean, Owen Garriott, and Jack Lousma. They performed three extravehicular activities, one to

Figure 9.2. The A-7LB model Skylab suit. The most obvious differences between this and the Apollo model included the new Skylab extravehicular visor assembly, the astronaut life support assembly (worn on the chest), and the secondary oxygen pack (worn on the right leg). Garrett AiResearch made the life support assembly and the oxygen pack. Courtesy ILC Industries.

make a final repair to the solar shield that involved installing a twin-pole sunshade over the existing parasol. These three EVAs totaled almost fourteen hours.

The Apollo-Soyuz Test Program

In October 1970, NASA administrator Thomas Paine proposed to the president of the Soviet Academy of Sciences that both countries engage in a space mission with the goal of linking an American and Soviet spacecraft together in orbit so that the crews could show that cooperation in space between the two Cold War enemies and rivals in the space race was possible. With the end of the lunar missions fast approaching, NASA had time and surplus Apollo spacecraft available to support the US portion of such a mission. The Soviets were interested not only for some political gain (as it would show that they could cooperate with their former rival) but also because it would show everyone that their technology was equal to that of the United States. In early 1972, the two nations reached a formal agreement, and on May 24, 1972, President Richard Nixon and the Soviet Council of Ministers chairman Aleksey Kosygin signed a formal document called "Agreement Concerning Cooperation in the Exploration and use of Outer Space for Peaceful Purposes." The proposed date of the mission was July 1975. We called it the Apollo-Soyuz Test Program, but the Soviets referred to it as Apollo-Soyuz Experimental Flight.[8]

The three-year window between signing the agreement and the launch date gave NASA time to lay out the requirements for the space suits that would support such a mission. These suits were truly just pressure suits that would be used if there were problems with the Apollo command module during various phases of the mission, including docking and undocking with the Soyuz capsule. The suit would of course provide flame protection, but no thought was ever given to an EVA on this mission.

Manufacturing these suits was easy, since the same requirements for the Apollo mission launch and docking phases would apply. NASA had no intentions of looking elsewhere for another suit for this final mission and awarded the contract to ILC. Crew Systems Division chief Robert E. Smylie made these points in a memo to NASA's chief of program procurement in September 1972. Smylie noted that although Hamilton Standard had a suit that they had made for the Air Force Manned Orbital Lab, they no longer had the capability to make a suit in a timely manner. In addition, Smylie said, if any other suit was selected such as anything from the David

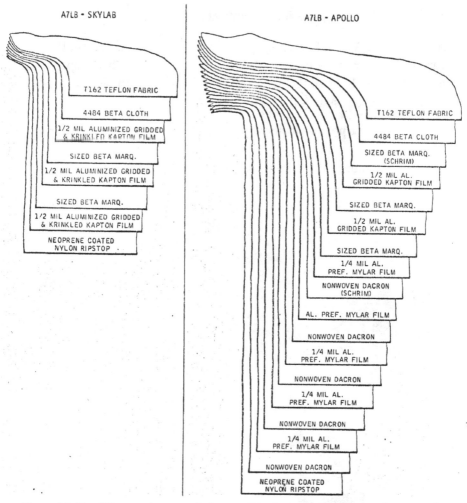

A7LB - SKYLAB

T162 TEFLON FABRIC
4484 BETA CLOTH
1/2 MIL ALUMINIZED GRIDDED & KRINKLED KAPTON FILM
SIZED BETA MARQ.
1/2 MIL ALUMINIZED GRIDDED & KRINKLED KAPTON FILM
SIZED BETA MARQ.
1/2 MIL ALUMINIZED GRIDDED & KRINKLED KAPTON FILM
NEOPRENE COATED NYLON RIPSTOP

A7LB - APOLLO

T162 TEFLON FABRIC
4484 BETA CLOTH
SIZED BETA MARQ. (SCHRIM)
1/2 MIL AL. GRIDDED KAPTON FILM
SIZED BETA MARQ.
1/2 MIL AL. GRIDDED KAPTON FILM
SIZED BETA MARQ.
1/4 MIL AL. PREF. MYLAR FILM
NONWOVEN DACRON (SCHRIM)
AL. PREF. MYLAR FILM
NONWOVEN DACRON
1/4 MIL AL. PREF. MYLAR FILM
NONWOVEN DACRON
1/4 MIL AL. PREF. MYLAR FILM
NONWOVEN DACRON
1/4 MIL AL. PREF. MYLAR FILM
NONWOVEN DACRON
NEOPRENE COATED NYLON RIPSTOP

Figure 9.3. The differences between the layers of the thermal micrometeoroid garment for the Apollo lunar surface A-7LB EV version suit and the Skylab A-7LB EV suit. Source: Figure 1 from Head, Crew Systems Resident Office, NASA, to Resident Manager, Skylab Program Office, "Crew Systems Division (CSD) Hardware Changes for the Skylab CSM 116," October 31, 1972, NASA memo.

Clark Company, it would have to be fully qualified, which would add to the costs.[9]

The contract called for a total of nine suits. These would have the serial numbers 801 through 809. The Apollo-Soyuz Test Program suits were basic model A-7L intravehicular suits with arm bearings. They did not include a pass-through for a liquid cooling garment since the suit did not require that feature.

The Intra-Vehicular Space Suit Assembly (ISSA)

As the end of Apollo/Skylab/Apollo-Soyuz Test Program neared, ILC began looking for every opportunity to keep their name in the space suit business. In early 1970, ILC set aside funding to develop what was called the ISSA suit after setting the following design goals:

Weight = 10.2 pounds including boots, gloves, helmet, and two arm bearings.
Stowage Volume = 0.5 Cubic feet or 0.22 cubic feet when vacuum packed.
Operating Pressure = 3.75 pounds per square inch (same as Apollo)

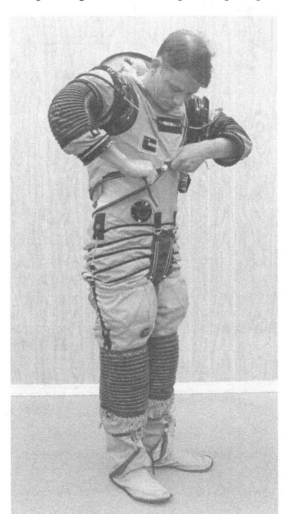

Figure 9.4. Apollo glove engineer Dixie Rinehart models the intravehicular space suit assembly that ILC made with corporate funding in 1970. It resembled the Apollo A-7LB suit in many ways but was much lighter and occupied less volume. Courtesy ILC Dover LP 2020.

NASA approved a contract for the first prototype based on this ILC development suit in October 1970; the suits were to be ready for delivery in May 1971 but never supported any NASA missions.

The Emergency Intra-Vehicular Suit

The emergency intravehicular suit of 1973 was the first flat-pattern ILC suit made. Jack Rayfield, a former NASA engineer who had begun working at ILC, began pushing others at ILC to develop a flat-pattern suit. The suit, which was designed to eliminate rubber convolutes, was made of materials sewn together at the various joint sections (arms, knees, ankles, etc.). Excess material was provided to assure the flexibility needed when the suit was under pressure. The suit also used a metal disconnect ring in the waist area to separate the upper torso from the lower torso, which made the suit much easier to get in and out than the zippered Apollo suits. At this point, NASA was very interested in getting away from the rubber components, steel cables, and zippers, and ILC was interested in showing NASA that it

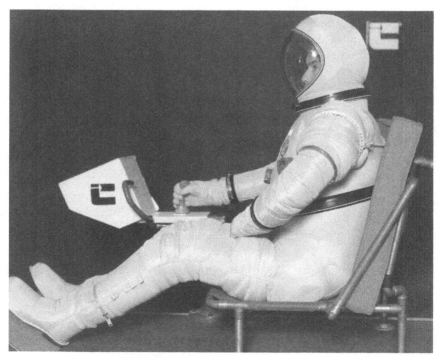

Figure 9.5. A side-view of the emergency intravehicular suit made of DuPont Kevlar that ILC made for NASA in 1973. Courtesy ILC Dover LP 2020.

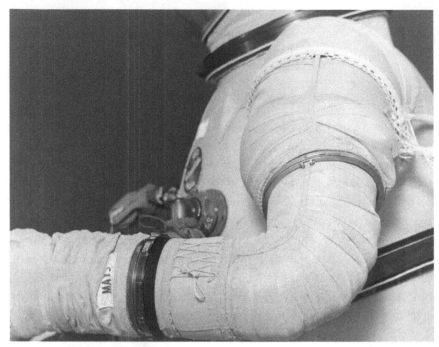

Figure 9.6. A closeup photo of the arm of the emergency intravehicular suit showing one of the first examples of what was called a flat-pattern joint that took the place of the rubber convolute. It also eliminated the steel cables and replaced them with webbing on the inseam and the outer seam of the arm that acted as a hinge. Courtesy ILC Dover LP 2020.

was all possible. This design foreshadows some of the features of ILC's later space shuttle suit.

The emergency intravehicular suit was made around the time that DuPont released the first Kevlar fabrics for commercial use, which were thought to be the savior for applications that required high tensile strength (such as the restraint layers for a space suit). ILC used the Kevlar materials a few years later when they developed the first space shuttle suits but soon discovered during testing that repeated flexing of Kevlar caused a weakening in the crease area and a significant tensile strength reduction. ILC engineers looked toward using Dacron materials, which were lightweight, had high tensile strength, and could stand up to the flexing, in future suit development.

10

End of a Historic Era

Selling Space Suits to the USSR

Leading up to the Apollo-Soyuz Test Program, there was a growing sense that maybe the Soviets were on the verge of becoming more friend than foe. We had beat them to the moon, so the space race was essentially over. Now both nations were cooperating toward the rendezvous mission between the Apollo and the Soyuz spacecraft. This thawing of frozen relations could not have been felt more than at ILC, given a prospect that emerged in early 1973. News broke that the Soviets were trying to negotiate with ILC to buy ten Apollo model A-7LB suits at a cost of $185,000 each ($1.1 million in 2019 dollars). Evening news reader Walter Cronkite announced that fact on his CBS news broadcast on March 15, 1973.[1]

The Soviets approached Intercontinental Computer Exchange, a company in Washington, DC, to act as the deal broker. The USSR Academy of Science and Technology approved the deal and released the news of this offer. They made it clear that their suit technology worked just fine, but they wanted to dwell on other technologies such as spacecraft design and simply wanted the suits for conducting comparative studies in support of human factors technology. The only things standing in the way of completing the deal were the State Department, the Pentagon, and NASA. All the news releases announced that the deal would happen within thirty to sixty days. That is where all the news releases end. Obviously, the US government had legitimate concerns about space suit technology falling into the hands of the Soviets regardless of the state of affairs between the two nations, not to mention the obvious issue of a rather complex technology paid for by US taxpayers would be going to the Soviets. It was a nice try on both sides. ILC could have really used the infusion of money at this point since the Apollo suit business was essentially at an end.[2]

ILC Parking Lot Sale

After the Apollo-Soyuz Test Program, the only chance for the company to remain in existence was winning the contract for suits for the space shuttle program that was still a few years away. In 1974, ILC brought in a new president, Fred Suffert, who was known as a hatchet man because of previous assignments he had had in industry. Fred began clearing house in terms of personnel and physical resources. Things were desperate, to the point that ILC decided to hold a yard sale, or as they preferred to call it, a parking lot sale. The Associated Press ran stories about the sale across the country.

In the early morning hours of Saturday, March 15, 1975, John McMullen, the man who had overseen the Apollo program as the systems engineering manager, looked out over a sea of discarded Apollo suit items that he and hundreds of others had so meticulously worked on over the previous nine years. Anything left over from Apollo that the government did not own was there, carefully laid out on the asphalt parking lot of ILC for the big sale. The early hard suit that ILC had developed was there. A complete Apollo thermal micrometeoroid garment was there. When the gates opened at 8:00 A.M., the crowd rushed in. Someone reportedly bought the early metal development suit because they thought the appendages of the suit would make a great flue-pipe for their woodstove. Several of the polycarbonate bubble helmets were snatched up by a lady who reasoned that they would be really nice to put over her tomato plants so the frost wouldn't kill them. A local science teacher bought the complete Apollo cover layer of a thermal micrometeoroid garment for $100 ($475 in 2019 dollars). When a reporter asked him if he thought that was a bit expensive, he responded, "I figure it will be worth a lot more money someday."[3] The more recent online auctions of space collectibles have proven that he made a good investment.

Several years after the sale, McMullen and his wife attended the Paris Airshow and were wondering around looking at the goods on display. Aside from aircraft, there was also a Russian space suit on display that caught John's attention. As he got closer and looked at the gloves, he realized the fingertips were remarkably like those of the ILC Apollo gloves. He quickly snapped a few photos. He recalled a box full of the blue-gray silicone fingertips being sold at the parking lot sale. Meanwhile, John's wife, who was sitting off to the side, noticed that two observers—thought to be the Russian owners—got stern looks on their faces and headed in John's direction. John's wife hurried over and grabbed his arm and off they went. When John got back to Dover and returned to ILC, he took the developed photos into work, where a couple of the engineers confirmed that the fingertips were almost certainly ILC manufactured and had possibly been bought on March 15, 1975, from the parking lot of ILC Industries.[4]

The Space Shuttle Era: A New Beginning

By July 1975, the doors were shut on Pear Street in Dover and anything left had been moved to the Frederica, Delaware, building that had made all the cover layers for the Apollo suits. John McMullen was one of the last employees to shut the doors on the Dover facility; he tells the story of how NASA required ILC to return much of the government-furnished property and that they also had a representative on site who was required to witness the destruction of many remaining Apollo-related documents and miscellaneous space suit items. Since ILC was essentially downsizing to the other building, there was no room or desire at the time to retain all the old records. John recalls dumping large volumes of documents and other Apollo items in dumpsters as the government representative looked on. Sadly, a lot of the suit history was lost that day.[5]

The twenty-five remaining employees consisted of a core group that could pull things together if ever anyone could. Homer Reihm became ILC's president. He had proven his leadership abilities far beyond question as the Apollo engineering manager and he had won the respect of all those involved in the program, including NASA. Now it was his turn to lead the company—or what was left of it. Into 1976 and 1977, ILC formulated new ideas for the space shuttle EVA suit using concepts developed during the closing years of Apollo. In bidding on the shuttle contract, the few remaining ILC employees knew that they could not provide a space suit without help. Recently, doors of communication and respect had opened between ILC and Hamilton Standard. The team at ILC had proven they were the best in the space suit business and wanted to continue that tradition and Hamilton knew that they could continue to produce the primary life-support system that NASA wanted for Shuttle, so the two organizations joined together, knowing they would make a formidable team.

By agreement, ILC would design and produce the soft goods for the space suit and Hamilton Standard would supply the primary life-support system and other hardware components such as the hard shell of the upper torso. Hamilton was the prime contractor and oversaw the shuttle suit program. Air-Lock, the other team member, was responsible for the hardware, as they had been for the Apollo space suits. NASA awarded the contract for the shuttle suits to the Hamilton/ILC/Air-Lock team in 1976 based on that team's SX-1 space shuttle suit. By 1980 it had evolved through several development changes and had become known as the space shuttle extravehicular mobility unit.

Homer Reihm understood that ILC had to develop a more diverse product mix and make sure there was other business under the roof. ILC brought in large aerostat contracts that George Durney managed. [6] They also landed a large contract for a protective suit for the army and took on small jobs to make sun, wind, and dust goggles for the military. Some days, Reihm and other executives could be found on the production line assembling these goggles so they could meet delivery schedules and pay the bills to keep the lights on and make payroll. Things continued to improve substantially, and over the years ILC has invented and produced valuable products. Three sets of rovers—Sojourner, Spirit, and Opportunity—explored the Martian surface thanks to the soft landings provided by impact bags developed by the next generation of talented ILC engineers such as Ralph Weis and Chuck Sandy. ILC is now the leader in the innovative design and manufacture of disposable bulk packaging and processing systems for top pharmaceutical companies around the world. With the threat of global warming and the implications that follow, notably seen in the aftermath of Hurricane Sandy, they are engineering and producing flood protection systems to be used in New York City that will likely catch on in other parts of the world.

ILC Dover continues to be the world leader in space suit design and manufacture. They provide the talent necessary to sustain and improve the suits made for EVA work outside the International Space Station and they continue to look at new designs for future space missions. The list goes on, but it's a good bet that the future will continue to be bright for ILC. All of this traces back to the fact that Len Shepard and George Durney never gave up on their dream of making the first space suits for our nation. And behind them was company founder, Abram Nathaniel Spanel, who discovered some of the brightest visionaries alive at the time and supported their creativity to the extent that their minds would take them, which of course was to the moon and back.

CONCLUSION

Preserving Our Treasures

The Smithsonian National Air and Space Museum

Preserving History

All good things must come to an end. When the Apollo program was over, the items that were returned to Earth became treasures to be shared with the American taxpayers who funded this great adventure and with people from around the world. Of course, many of the items that supported the Apollo missions never returned to Earth. Many items still sit on the moon today, including the lunar rover and various components of space suits such as lunar boots and the primary life-support systems.[1] Other items burned upon reentry to the earth's atmosphere after their onetime use. The first stages of the Saturn rockets settled to the bottom of the Atlantic Ocean after performing flawlessly. The suits are the only life-support equipment that was taken to the moon's surface and brought back to Earth. That we have relatively few items back from lunar missions makes the preservation of equipment and moon rocks all the more important.

Fortunately, America has some of the best museums in the world. The Smithsonian stands out on top. Nineteen museums, a zoo, and several research centers make up the Smithsonian Institution and each is accountable for researching, preserving, restoring, and displaying everything from the great arts to dinosaur remains to Indian and African American cultures. The Smithsonian's National Air and Space Museum, located on the Mall in Washington, DC, began as the National Air Museum in 1946. At the end of World War II, the nation had accumulated many aircraft from that war and from WWI and from the infancy of aviation itself. Initially, many items were stored and displayed at various facilities such as the Arts and Industries Building in Washington. Some items, such as the early rockets, were exhibited outdoors. As the surplus accumulated, a central museum focused on the care and display of these artifacts became necessary. The growing surplus of treasures of the space program emphasized that need. It also drove the name change. In 1966, the National Air Museum became the National Air and Space Museum. Funding for the present building was approved in 1971 and the museum opened its doors on July 1, 1976. The first director was none other than Apollo 11 command module pilot Michael Collins. As the inventory continued to grow, a companion museum named the Steven F. Udvar-Hazy Center was opened in 2003 next to Dulles International Airport in Chantilly, Virginia.[2] Each year, millions of visitors make the trek to both facilities to see some of the greatest treasures from early aviation to the space programs, including Mercury, Gemini, Apollo, and the space shuttle. Seven million people from around the world visited the National Air and Space Museum in 2017 alone—almost a third of the visitors to all Smithsonian museums combined and 3.2 million more than the next closest Smithsonian Museum.[3] That says something about the collections the National Air and Space Museum houses.

From Mercury to Apollo, NASA was well aware of the significance of all the items that supported their manned space missions. NASA ordered that all flown suits go through a post-mission inspection and test followed by basic cleaning to remove most of the moon dust on the lunar suits. They were then packaged and offered to the National Air and Space Museum. Based on the historical significance, the museum had the option of accepting them or turning them down. Most of the items the museum chose to accept were forwarded to its Paul E. Garber Preservation, Restoration and Storage Facility, located a short distance from the center of Washington, DC. Until the new museum opened its doors in 1976, some of the suits were stored in their relatively small, cramped storage containers for several years.[4]

NASA sent many other items, including training and backup suits, gloves, and helmet hardware (whose historic value was considered to be less significant) to the Cosmosphere Science Center and Space Museum, in Hutchinson, Kansas, which is affiliated with the Smithsonian. Other artifacts can be found at smaller museums and NASA visitor centers scattered across the country.

Currently, forty-three flown Apollo space suits are in the care of the National Air and Space Museum, including all twelve that were worn on the surface of the moon. Few would argue that the most significant suits in the collection include the twelve Apollo lunar suits. The top prize in the collection goes to Neil Armstrong's model A-7L suit, serial number 056, which was the garment that protected the first human to touch another world beyond our earth. Following close behind would be the second suit on the moon, Buzz Aldrin's model A-7L suit, serial number 077. Visitors to the museum naturally focus on these suits and the other lunar suits on display.

So that items in the collection can be prioritized for various reasons, including conservation, storage, and loans to other museums, the museum has established five categories of classification based on their historic significance.

1. The twelve suits and supporting gear worn on the moon.
2. Earlier research and development suits and one-of-a-kind suits and support equipment that represent the state of the art at that time.
3. All flown suits and support gear other than the lunar suits included in category 1.
4. All the backup and training suits and support gear that were not flown.
5. All other miscellaneous gear related to space suits that has unknown provenance. This includes items with little or no identification or that has illegible identification due to heavy training use.[5]

Time has not proven to be a friend of the fabrics and soft-goods materials made for space. As I pointed out in earlier chapters, the rubber convolutes in Apollo 14 suits were under attack almost from day one due to various factors, primarily copper contamination during manufacturing. Once that was understood, ILC removed all copper piping from their plant and added an antioxidant ingredient known as Agerite White to the rubber formulation. Suits and gloves from Apollo 15 and later missions contain rubber that are almost as flexible as the day they were made, and the National Air and Space Museum staff would like to keep it that way as long as possible.

Shortly after receiving the space suits and support items from NASA, the Smithsonian staff saw clear signs that many of the suit materials were breaking down—not just Apollo items, but also items from the Mercury and Gemini missions. The vinyl tubing used in the Apollo liquid cooling garments was starting to harden as the plasticizers started migrating out, which left a tacky feeling on the outside of the tubing and rust-colored stains on anything it contacted. Rubber items inside the suits started to become brittle and crack. Many of the other materials inside the Apollo suits, however, are relatively stable if they are not handled. The aluminized Mylar does not break down easily on its own nor does the Dacron and Kapton. If these material layers are not flexed and if they are kept at acceptable temperatures and humidity, they can expect a relatively long life. The challenge lies in the ability to preserve the vinyl and rubber in the suit parts.

Starting sometime around 1980, after museum staff realized that the suits needed to be stored in an environment that would improve their chances for a longer survival, they decided to place the suits and supporting items that were not on display in a large metal storage room maintained at approximately 13°C (55°F). Industry experts at that time determined these conditions to be the best for rubber, vinyl, and similar materials. However, even after a few years under these conditions, the materials were continuing to degrade.[6] It was understood that more needed to be done to preserve this valuable collection.

In March 2000, the museum initiated a special program called the Save America's Treasures Project, which was funded through a public-private partnership between the White House Millennium Council and the National Trust for Historic Preservation. The National Air and Space Museum's Space History Division secured $200,000 from that program specifically for the threatened artifacts of the Apollo program, and Hamilton Sundstrand, a United Technologies Company, matched the funds.[7]

At the time this program kicked off, curator Amanda Young oversaw the space suit collection. She had witnessed the degradation that was occurring in many of the suits, particularly the Apollo suits. She soon brought on Mrs. Lisa Young, a conservator, and together they focused on solutions to the many issues. They quickly enlisted a variety of experts from around the world who could shed light on why the different materials were decaying and what could be done to preserve them. Along with experts in the rubber business, there were metal experts to look at why the aluminum disconnects on some suits were corroding. I was contacted around that time and was honored to help in any way that I could as a representative of ILC Dover. Although I was not around during the Apollo years, I was able to round up many of the past experts from the Apollo years, including Paul Slovic, the chemist who worked on the rubber compounding for the Apollo suit program. He provided insight into the rubber-dipping process at ILC and the problems they faced at the time.

The result was a much better understanding of the conditions that were needed to preserve the space suits. It was found that the rubber would start to crystallize at temperatures below 10°C (50°F); this explained why the suits that were stored at 10°C or lower were hardening and were no longer flexible. The relative humidity was also critical because many of the materials absorb moisture that can result in mold and fungus. Thus, it was determined that a humidity around 35 percent with little fluctuation should be the target value. Other variables include:

Light: A level of 100 lux maximum is required because light can damage the exterior layers of space suits and supporting gear

Ultraviolet radiation: No more than 75 microwatts per lumen is recommended because UV radiation causes significant damage to exposed textiles.

Other factors such as pests and insects will cause damage. Food and drinks must be limited in all areas to reduce this risk.

Certain gasses such as ozone, carbon dioxide, sulfur dioxide, and nitrogen dioxide can accelerate the breakdown of polymer materials.

Dust and debris can contribute to the breakdown of materials over time.

The security of the objects is a high priority. Vandalism and theft are mitigated by proper display that limits accessibility. It is also important to consider display designs that adequately protect objects from natural disasters.[8]

During the earlier years, many of the Apollo suits and other support items such as gloves were loaned out for display at other museums around the world. Unfortunately, they were not maintained in ideal conditions. Suits were returned with plastic mannequins inside that were turning to dust, while others had old rags and newspapers or even metal pipes jammed inside to give them shape. Some had been exposed to high ultraviolet rays, fluctuating temperatures, humidity, and who knows what else. For those reasons, each suit required a complete assessment. Work began almost immediately to fix many of the problems after they were documented and understood. This was all carried out at the relatively primitive and cramped Paul E. Garber facility.

Fortunately, in December 2011, the entire space suit collection and other related items stored at the Paul E. Garber facility were moved thirty-seven miles west to the Steven S. Udvar-Hazy Center.[9] This museum boasts a state-of-the-art storage area and is supplemented by a full conservation laboratory. This ensures that the suits will be preserved under the most ideal conditions for as long as they remain there. They have all had the proper soft mannequins installed to give the proper shape and reduce the stress throughout the suits. These are made from a museum-quality polyethylene foam that is covered with polyester batting and white nylon hose.

The suits are laid face up on an acid-free paper honeycomb board that is approximately one inch thick. Inert foam pieces are inserted under areas of the suit such as the knees to provide the natural shape that is desired. The suits are covered with a lightweight, acid-free muslin cloth to keep dust and debris off the surfaces. They are stacked on a mobile rack system that can be expanded for access when desired. Individual suits can be removed from the racks while on the honeycomb boards and placed on gurneys for transport to other locations with minimal disturbance.[10]

Preservation versus Restoration

Webster's Third New International Dictionary defines restoration as "a bringing back to a former position or condition." Seeing a World War II aircraft properly restored to resemble its pristine appearance in the 1940s provides an enjoyable experience for museum visitors and historians. However, in my opinion, it is undesirable to see a space suit that has been modified from the condition it was when the museum received it, and I'm sure most would agree. Instead, conservation efforts are undertaken to make sure that scholars and the public see the actual materials that make up the individual suits complete with stains, repairs, and other blemishes that were present at the end of a mission, including moon dust in some cases.

One unfortunate example of necessary restoration involves Ed White's Gemini suit. It had been on exhibit for many years in the National Air and Space Museum. That building had a lot of glass that allows the UV rays from the sun to shine in day after day. Sadly, the UV exposure deteriorated the Nomex material on the Gemini suit to the point where a thorough evaluation concluded that it could no longer be displayed as it was. A new cover layer complete with mission and flag patches was made to go over the pressure garment underneath. Doing this for an Apollo suit that has moon dirt present on the cover layers would make the suit far less interesting. In the case of Ed White's suit, the original thermal micrometeoroid garment was at least saved and is in storage to preserve what is left of it.[11]

Fortunately, the Apollo suits were made of a fiberglass-based fabric that can hold up better than the Nomex material. Many of the Apollo suits have a cover layer of Teflon fabric that appears to have held up well over time and under the right conditions. While the fiberglass Beta cloth has other undesirable characteristics—the glass fibers can break if flexed excessively and the material does not hold up well to abrasion—these conditions are relatively easy to guard against using the proper handling and display techniques. The materials contained under the outer cover layer of the thermal micrometeoroid garment will likely never be seen again since there is no value in taking that garment off any of the suits in the National Air and Space Museum collections. More harm may be done to a suit by doing so than any value that might be gained. Several suits in the collection do not have the thermal micrometeoroid garment installed, so academics many years from now have these objects for evaluation. Some limited insight may be gained by the CT scans that National Air and Space Museum staff has undertaken on Armstrong's suit and others. They do provide a glimpse into the suit that the public will find of interest.

The Apollo 11 Space Suits

The Apollo 11 suits took a somewhat longer route to their final resting place at the National Air and Space Museum. NASA tested and inspected all three of these suits after the mission and NASA personnel in Houston inserted mannequins into them. They were then placed back into the Columbia command module in October 1969 to support a 48-state tour that lasted through 1970 and into early 1971.

Figure 11.1. This photo taken from inside the Apollo 11 command module shows Neil Armstrong's suit positioned in the seat approximately three months after the mission. The next suit is outside ready to be passed in. This exhibit took a 48-state tour that lasted into 1971. Courtesy Ulrich Lotzmann.

After the nationwide tour, NASA sent the command module and all of its contents to the Smithsonian. The Armstrong and Aldrin suits were removed from the capsule and placed in glass display cases at the Smithsonian Arts and Industries Building, where they remained until the National Air and Space Museum building opened in 1976. Once the new museum opened, both suits were displayed in a section on the second floor of the museum, while Mike Collin's suit remained in the command module located on the first floor. On my many visits through the 1980s and 1990s, I would make a beeline to that second-floor display, as would many others, just to see those amazing space suits. I recall the suits being in the dimly lit display behind a large glass window surrounded by a lunar-like setting. Even though at the time I was a novice regarding materials and suit preservation, I did realize that light could contribute to accelerated aging. Not until several years later, when I witnessed firsthand the efforts museum conservators were putting into the preservation efforts, did I realize everything that was involved in taking care of these fragile materials.

Since 2006, many visitors to the Washington, DC, museum have been disappointed because they could no longer see Neil Armstrong's suit on display. The removal of the Armstrong suit occurred during the Save America's Treasures project, when research revealed that the display conditions did not meet the new conservation standards. Additionally, conservators wanted to do a thorough assessment of Armstrong's suit to understand the nature of all degradation and slow it down. Buzz Aldrin's suit remained on display since it would be a huge disappointment to visitors if neither suit was available to see. Based on the results of the project, Armstrong's suit was placed in storage under very controlled conditions that conservators developed as part of their findings. This includes a temperature maintained from 17.2°C (63°F) to 18.3°C (65°F). Relative humidity is maintained at 35 percent. The display area on the second floor was never capable of controlling these conditions.

The fiftieth anniversary of the Apollo 11 mission on July 20, 2019, resulted in a much-improved display for Armstrong's suit when it was finally revealed in a section of the Smithsonian. Public contributions supported the effort. In July 20, 2015, the Smithsonian launched a kick-starter campaign called Reboot the Suit that sought to raise funds for conserving, digitizing, and displaying the Armstrong suit. The museum was able to raise almost $720,000, far exceeding the $500,000 goal of the campaign, proving just how interested people from around the world continue to be in this

Figure 11.2. Ten ILC Industries Apollo veterans gathered around Neil Armstrong's Apollo 11 suit. From left to right: John Scheible, materials engineer; Tom Pribanic, configuration management; Joann Thompson, seamstress; John McMullen, systems engineer; Frank Napolitano, hardware assembler; Sid Williams, draftsman; Homer (Sonny) Reihm, engineering manager and future company president; Bob Penney, engineer; Austin Pase, production control; Bill Ayrey, author; and Ken Shane, lead test suit subject. Courtesy Sid Williams.

treasure.[12] Eventually Armstrong's suit will be featured in a new gallery called Destination Moon that is scheduled to open in 2021.

Photogrammetry, chemical analysis, 3D scanning, and CT scanning have established a much better understanding of the current condition of the Armstrong suit. Other data has been gathered and documented that is focused on the existing damage to the suit. This will all help provide baseline records that can be used for future generations who want to measure the condition of the suit and assess its state of conservation in comparison to this baseline data.

Sometime in 2020, Buzz Aldrin's Apollo 11 suit will be removed from display and will undergo a complete assessment much like the one that was done for the Armstrong suit. It will then be placed into storage with the possibility of being rotated into the Destination Moon display after the Armstrong suit is rotated out and placed in storage. This rotation concept may help extend the life of both suits while offering the public a chance to see one of these historic treasures.

Reuniting Apollo Veterans with Neil Armstrong's Suit

On November 25, 2017, I had the honor and privilege of chauffeuring ten retired veterans of ILC Industries to the Smithsonian's Steven F. Udvar-Hazy Museum. This visit was arranged with the help of the National Air and Space Museum conservator Lisa Young and National Air and Space Museum curator Dr. Cathleen Lewis. Each of these ILC veterans played a key role in making the Apollo suit program such a success. Imagine their thrill at being able to stand beside Neil Armstrong's Apollo 11 suit for the better part of an hour. The best part was that I was able to witness their delight and hear their stories, as was the National Air and Space Museum staff. My only regret was that I could gather only these ten individuals. I wish I could have shared this day with many others who played a part in this great story.

ACKNOWLEDGMENTS

I wish to acknowledge the many ILC Industries employees who shared their Apollo memories with me over the past 40-plus years. Regardless of whether I've met you or not, if you worked on the Apollo suits for ILC Industries or you were a family member that lived through these years, then this book is dedicated to you.

I believe I first realized how wonderful these stories were when Tom Sylvester hired me to work in ILC's test laboratory. Tom started at ILC as an Apollo suit test subject, and the film footage of Tom wearing the Apollo space suit on a football field as he demonstrated the mobility of the new model A-7LB moon suit while passing and kicking a football is my most memorable image of him. This film helped sell that model to NASA. Although Tom passed away recently, I could always count on him and his excellent memory to answer any questions I had about the Apollo suits and the challenges of the program.

I became good friends with others such as Apollo draftsman Sid Williams, who can tell some funny stories about what it was like to work at ILC during the 1960s and early 1970s. Many of these stories cannot be divulged because of the nature of the pranks and the situations folks got themselves into, but that was business back in the day. The result of behavior that happened fifty years ago would result in immediate termination in today's business world. In hindsight, I understand how this camaraderie resulted in a tighter bond between co-workers. Between the tension of the work schedule and the pressure to get it right, Sid and many others used humor and practical jokes to relieve the daily stress while turning out exceptional work.

John McMullen and Homer (Sonny) Reihm were deeply immersed in the program and I am greatly indebted for the help and stories they provided. Over the years, I turned to Mr. James McBarron, the NASA suit engineering representative that ILC interfaced with on a daily basis, for his perspective on the Apollo suit program. He provided great insight and inspiration. Richard McGahey, my boss for many years, shared his stories about the quality program he worked on and the many challenges he faced because the Apollo space suit was constantly under development and its engineering design and manufacturing challenges were so complex.

Special thanks are offered to the following Apollo veterans for their time in helping me gather the information and understanding I needed: Ron Bessette, George Gleadow, Al Gross, Richard Martin, Jim Miller, Frank Napolitano, Larry Ornston, Tim Parker, Austin Pase, Bob Penney, Dixie Rinehart, John Schieble, Ken Shane, and Ray Winward. Thanks to Russ Dion and Professor Ulrich (Ulli) Lotzmann for encouraging me as I pushed through this project. Your support was very much needed and appreciated.

I'd like to thank Dr. Cathleen Lewis and Lisa Young of the Smithsonian National Air and Space Museum (NASM) for inviting me into their world of space suit collections and preservation over the years. I can't leave out Amanda Young, who invited me to become part of her team at NASM in the late 1990s, when she took on the first challenges associated with the preservation of the suits in the collection. It was a daunting project, and she had laser vision when it came to carrying out the job. Her efforts laid the foundation for the collection today.

There are many other veterans I'd like to acknowledge, but the list would go on for several pages. You know who you are, and I truly appreciate your help and input.

Finally, behind any project of this magnitude is often found the spouse who sacrifices valuable time together. Thanks go out to my wonderful wife, Cab, for her support. I couldn't have done it without you!

APPENDIX A

Technical Details of the Apollo Space Suits

The Apollo space suits were technical marvels for their time. No mission was ever cut short due to a suit problem and no astronauts sustained any injuries other than perhaps some bruised fingernails. Considering the potential for serious injury or certain death if failure occurred, this is a remarkable outcome. It is attributed to the work of the ILC engineers, seamstresses, and, other ILC personnel involved. The marriage of soft-goods engineering and the skills of the seamstresses and model makers made for a challenging process that ILC eventually mastered as much as something this complex could ever be mastered in such a short period of time.

I attempt to address as many of the technical aspects of the Apollo suit as possible. When a system for any of the suits was common to several or all the models, I have included it in one or another model but not multiple models unless the design is significantly different.

Fecal Containment Unit (FCU)

The fecal containment unit, or FCU, was provided for containment of solid waste when the astronauts were confined within the suits for periods of time when they could not readily take them off. Later in the Apollo missions, NASA changed the requirements so that the FCU was not worn unless there was a chance a crew member would have to be pressurized within their suits for 115 hours, as was possible only on the lunar missions or if a decompression event occurred on the spacecraft.

Prior to putting the FCU on, the astronaut was required to apply what was called a Silicote ointment to the buttocks and the entire perianal region. This salve provided a barrier on the skin to protect from irritation due to contact with any waste matter that could potentially be there over a period of days. The directions were to apply about one-third of the contents of the tube, then put on the FCU garment.

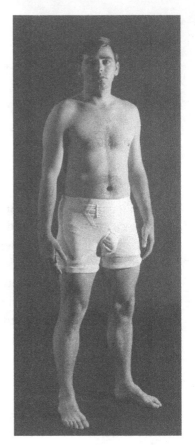

Right: Figure A.1. The fecal containment unit, a tight-fitting cotton brief that had absorbent layers. Courtesy NASA.

Below: Figure A.2. Various layers of the fecal containment unit. Courtesy NASA.

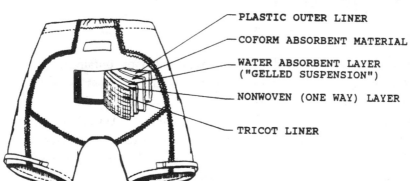

PLASTIC OUTER LINER

COFORM ABSORBENT MATERIAL

WATER ABSORBENT LAYER
("GELLED SUSPENSION")

NONWOVEN (ONE WAY) LAYER

TRICOT LINER

Buzz Aldrin specifically mentioned his FCU in his book *Return to Earth*. As he was suiting up for the launch, he noticed that his grandfather's Masonic ring that he had been wearing for over a year was missing. His plan was to carry it with him to the moon and now it was gone. Before panic set in, he realized that he had taken it off as he was in the bathroom applying the required ointment. One of the flight doctors in attendance was good enough to run down to the restroom to retrieve the ring for Buzz.[1]

The FCU garment was made of a two-way stretch fabric made of a blend of nylon, Spandex, and Olefin. The absorbency layer consisted of a plastic outer liner; a material known as Coform that absorbed water; a nonwoven, one-way layer; and a tricot liner. The garments were manufactured by Whirlpool Corporation in St. Joseph, Michigan.

Of course, the process that leads up to the development of anything like the fecal containment unit requires a lot of trials to figure out the best design and can sometimes end with lots of fun stories about how it was all accomplished (although testing this part would obviously not be considered fun). The manned testing for the FCU started on February 11, 1966. Model maker and frequent test subject Richard Ellis enjoyed telling me on more than one occasion about his association with the development of the FCU. Asked to evaluate the new FCU, Richard was told to put on all parts of the suit, including the new device. As he did, and before closing the zipper of the space suit for pressurization, an engineer reached into the opening in the back of the suit (through the opened zipper) and pulled out the waistband of the FCU garment. He then dumped 1,000 grams of cat food that the detailed test plan stated had a moisture content of 74 percent. This apparently came the closest to simulating poop in the suit. The engineer then closed the FCU waistband and zipped the suit closed. Richard had been briefed on the test plan, which included repetitive movements within the suit—bending, walking, sitting, lying down, and so forth. Richard answered a series of questions throughout the test about issues such as odor in the suit and any leaks that he could feel. Overall the evaluation was as favorable as could be other than some fit issues that could be resolved with little effort. The closing comments noted in the test plan stated, "The subject suffered no ill effects and other than his reluctance to sit down, was very cooperative."[2] Richard truly was a dedicated ILC employee.

Holding the Suit Together: Steel Cables and Restraint Systems

One of the most critical systems that was integrated in each Apollo suit from top to bottom was aircraft-grade steel cables that ran vertically from the boots to the helmet neck ring and horizontally from glove to glove. The combination of internal pressure and the physical loads the astronauts introduced during strenuous activity produced great strain throughout the suit that would literally tear it apart if the loads were not contained. The fabrics of the suit alone were no match for these stresses and strains, so the steel aircraft cables were designed to take these loads yet remain as invisible to the occupant as possible. The unfortunate side effect was that the astronauts would feel these cables pinching them at times, particularly when they were seated in the command module with the cables sandwiched between their backs and the couches.

ILC engineers worked continuously to redesign the restraint cable system throughout the Apollo program in response to various design weaknesses. This evolving challenge was brought about by cable failures during lab testing at both ILC and NASA.

The early suit designs and test trials indicated a number of failures in the Teflon guides the wires passed through in the crotch area of the suit. The guides were sewn into the outer restraint layer of the suit and controlled the positioning of the steel cables and kept them from abrading and cutting into the nylon fabric restraint layer. The area that sustained the greatest number of failures was in the crotch area because that is what was subjected to the higher loads due to strenuous activities such as walking, sitting, and bending.[3] In January 1967, astronaut Joseph Kerwin outlined the results of a four-hour test inside the NASA vacuum chamber in Houston, where he had participated in a test of the cooling ability of the new A-6L suit. He worked at various workloads inside the chamber on a treadmill, expending from 450 Btus up to as high as 2,000 Btus per hour. At the end of the four hours, when the suit was inspected, it was found that "the left thigh restraint cable [had] sawed its way through the Teflon tunnel and nylon restraint strap and pulled loose. Left leg mobility was lost almost completely."[4] This was remedied by replacing the Teflon guides with pulleys that would sustain the loads and roll over the cables as the legs were flexed back and forth. This eliminated any abrasion and friction that eventually wore through the Teflon guides.

By December 1971, just four months before the Apollo 16 mission, NASA engineers were requesting that ILC perform a number of tests on the cables and attachment swages using mechanical cycle testing equipment rather than manned tests to address the restraint cable failures in the crotch and knee. While the method was cheaper and delivered faster results, mechanical cycle testing was not necessarily a good representation of actual manned testing. The mechanical cycle rigs were failing cables, but manned tests appeared to indicate that the cables were fine. One of the solutions was to add a redundant cable over top of the knee convolute, which was already molded with two cables inside, thus providing four cables per leg: two on

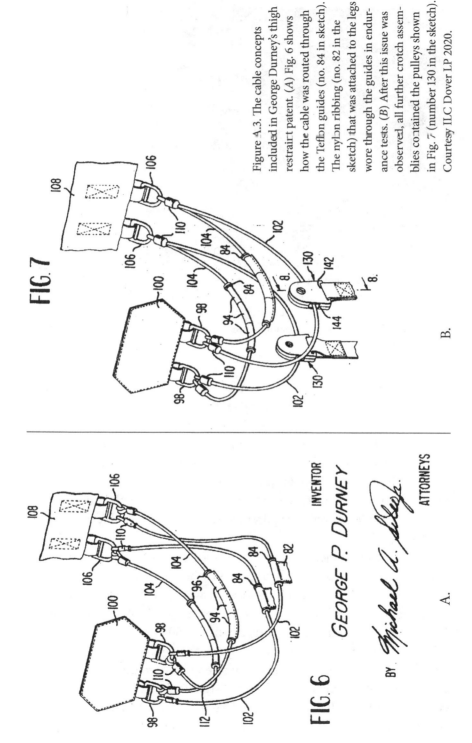

FIG. 7

FIG. 6

INVENTOR

GEORGE P. DURNEY

BY *Michael A. Selez.*

ATTORNEYS

A.

B.

Figure 4.3. The cable concepts included in George Durney's thigh restraint patent. (A) Fig. 6 shows how the cable was routed through the Teflon guides (no. 84 in sketch). The nylon ribbing (no. 82 in the sketch) that was attached to the legs wore through the guides in endurance tests. (B) After this issue was observed, all further crotch assemblies contained the pulleys shown in Fig. 7 (number 130 in the sketch). Courtesy ILC Dover LP 2020.

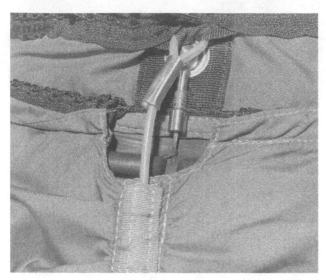

Figure A.4. A closeup photo of the redundant knee cable (the nylon-coated cable) that was added to back up the inner, primary knee cable that was molded into the rubber convolute. After the primary ⅟₁₆" diameter cable was replaced with a ³⁄₃₂" cable, the redundant cable idea was eliminated after testing showed that it was not necessary. Courtesy ILC Dover LP 2020.

the inside and two on the outside of each knee. If the primary steel cable molded into the convolute failed, the secondary cable would be attached to both the inside and outside of each knee assembly to back up the primary. Testing was carried out on this design and it was found to be satisfactory. In addition, the ⅟₁₆-inch diameter knee convolute restraint cable that was standard in these early model A-7LB suits was replaced with one measuring ³⁄₃₂ inches in diameter. After this larger-diameter cable proved to satisfy the design limit cycling requirements, the secondary cable concept was deleted.

Test data from sometime around 1972 shows the breaking strength of the various cables located throughout the suit when three samples were tested. The shoulder cable for example broke at an average 514 pounds with a minimum requirement of 480 pounds. The crotch cable broke at an average of 511 pounds with a minimum requirement of 480 pounds. The outer thigh cable broke at an average of 975 pounds with a minimum requirement of 920 pounds.[5] The minimum breaking strengths were calculated based on a worst case of the pressure loads of the suit combined with physical loads (which NASA and ILC engineers referred to as "man loads") that were added when the crew members stretched and strained the various parts of the suits during activities in the suit. It was likely that an additional safety factor was added to increase the margin of safety.

The other issue with the cable was that they were terminated at their ends by slipping a metal sleeve over the cable and bringing the cable end into the sleeve after wrapping it around a terminating point. This is called a swage fitting. A special tool was used to compress it closed, hopefully holding it in place. An Apollo engineer told me that for every section of wire cable to be used in a particular location of the suit, ILC would make ten units each and test nine of them by pulling them apart to failure. If the nine passed the minimum load test, then the one cable was acceptable for use in the suit. That is one of many examples of why space suits are expensive to make.

In the models A-6L and early A-7L Apollo suits, the steel restraint cable that wrapped horizontally around the upper torso and the shoulders could be disconnected from a shoulder cable disconnect bracket on the front chest. Disconnecting the cable ends would reduce the restriction the cable made, thus making it easier to get into and out of the suit. It also

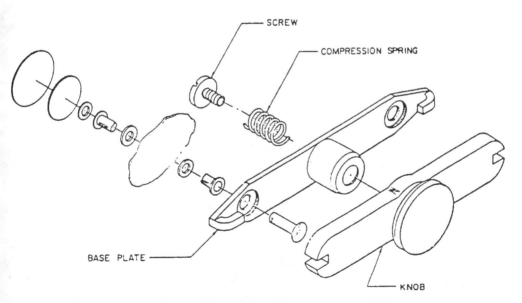

Figure A.5. Details of the model A-6L and early A-7L shoulder-cable disconnect assembly. The ends of the cables that wrapped around each side of the torso attached to the two base-plate tabs. They were released by pulling the knob hardware outward, which exposed the cable ends so they could be removed from the tabs. The cable ends always had to be attached any time the suit was pressurized. Source: Familiarization & Operations Manual, Model A7L, ILC document no. 8812700149B, June 6, 1969.

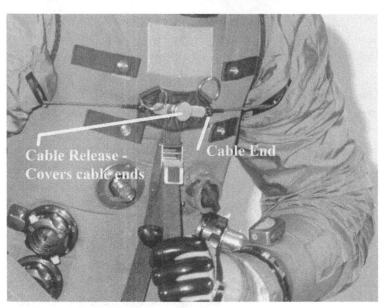

Figure A.6. The shoulder/torso restraint cable-release bracket and cable ends. This became fixed on all suits after Apollo 9. Note also the torso adjustment strap. Courtesy ILC Dover LP 2020.

made it more comfortable for crew members when they were working in the unpressurized suit. This feature was eliminated by Apollo 10, probably because the cables needed to be connected when the suit was pressurized because the mechanical loads that the torso was subjected to was significant at 3.75 pounds per square inch of pressure. There was always the chance that someone could forget to connect the cable, and it was better to eliminate the risk. Thus, the next generation of design had a fixed bracket that eliminated the chance of it being disconnected during a mission.

The adjustable torso webbing strap was considered part of the suit restraint system on the model A-7L suits. Because of the torso zipper and the absence of a rubber convolute in the lower torso/leg area, there was no easy way to flex at the waist, as when going from a standing position to sitting. The torso strap provided a mechanical means of bending the torso. Prior to Apollo 15 and the availability of the lunar rover, the primary purpose of this strap was to help crew members bend to a seated position when inside the command module.

Left: Figure A.7. Demonstration of the torso adjustment strap, which was used to force the upper torso to bend at the waist so the astronaut could take a seated position in the command module couches. It automatically held the suit in a bent position as the webbing was pulled tight. To get the suit back to the prone position, the astronaut simply pulled on the buckle at the top of the webbing to release the tension. Courtesy ILC Dover LP 2020.

Below: Figure A.8. Fit check of Apollo A-7L suit in the sitting position, as it would be on launch and reentry in the command module. The torso adjustment strap bent the torso at the waist to maintain this position. Courtesy ILC Dover LP 2020.

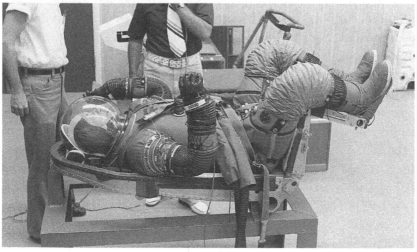

Zippers

The Apollo suits all used a two-zipper system, one overlapping the other. One was known as the pressure-sealing zipper, or closure, and the other was called the restraint zipper. The pressure-sealing zipper was located underneath the restraint zipper and was glued onto the pressure-bladder layer to form an integral, leak-free assembly. Together, the pressure bladder and the pressure-sealing zipper worked to retain the pressure of the gas inside the suit. The restraint zipper was sewn to the restraint layers of the suit and took the structural loads and prevented the pressure zipper from being subjected to any structural loads under pressure. That was because the patterns that made the pressure bladder were cut larger than the outer restraint layers, thus there was some slight excess material underneath so that no loading would be experienced.

Pressure-Sealing Zipper

Some items in the Apollo suits were considered to be more critical for the safety of the astronaut than others. If there was a list made up in order of concerns, the pressure-sealing zipper would likely be at the top. Although it proved to be very dependable throughout all of the missions, it presented challenges in the test laboratories at times and during post-mission testing, when the leakage rates were high due to lunar dust contamination and general use.

I have many folders containing memos expressing NASA's concerns about pressure zipper failures and leaks. ILC always managed to stay one step ahead, but it was close sometimes. No suits failed a mission due to a zipper problem, but you would not know it by looking at the correspondence at the time.

At the start of the Apollo suit program with the A-5L model suits, the purchase description called for a 38-inch Gemini G-4C pressure-sealing zipper manufactured by B. F. Goodrich, since it had been proven on the Gemini suits. This zipper was used until the debut of the model A-7L suit.

The model A-7L suits used a pressure-sealing zipper made by B.F. Goodrich (suit part number A6L-101105), but it performed well only when it was installed in a straight line. If it was bent to the left or the right of the center line, leakage could occur. But it could be flexed fore and aft, for example as it wrapped around the crotch area, and provided adequate sealing capability.

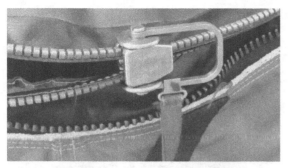

Figure A.9. ILC designed the slider that replaced the stock Talon Corporation unit they supplied with their pressure-sealing closure. the ILC slider had tighter machined tolerances that decreased leakage rates and provided a stronger seal between the two closure halves. The teeth of the outer restraint zipper are seen below the inner pressure closure. Courtesy ILC Dover LP 2020.

As the Omega suit was being developed in 1967 and 1968, a new alternative was found to the Goodrich zipper. One of the primary focus points of the new Omega suit was to take the zipper out of the waist and crotch area so the convolute area could be installed. This meant that the zipper would have to spiral around the upper torso section of the suit and would bend from side to side. After investigating various pressure-sealing zippers on the market, ILC engineers found the Talon OEB (for omni-environmental barrier) zipper. ILC obtained samples and installed one of the zippers in the torso of their experimental suit. It proved to work well at sealing the suit closed and keeping the leakage rates low.[6] As reported in an untitled ILC engineering memo from an unknown author, "When relaxed, both tapes lie on top of each other and form a very flexible band. Bladder patterns were developed that would allow this closure to assume this position and go down and around the torso in a relaxed position without bunching or attempting to change the length relationship of the tapes or tooth spacing. This technique removes all stress from the closure, permits easier operation, and extends closure life."

When ILC started working with the Talon closures, they found that they had a higher than desirable rate of leakage. After further investigating how the closures were designed, the ILC engineers realized that a slider fabricated with tighter tolerances should cut down on the leakage. Tests proved that theory to be true. As a result, ILC began machining their own slide fasteners and replaced the stock units provided by Talon. Ultimately the leak rates decreased. Figure A.10 shows the slide fastened on an A7LB model suit.

O. E. B. PRINCIPLE

Different from conventional zipper principles, the O.E.B. represents a new approach, a new design. The unique double surface seal provides a barrier from either side, and flexing, twisting, or bending does not impede its function. Pressure from either direction causes a hinging action of the inner locking elements, thereby further compressing the secondary sealing surface.

The fastener is made of two identical stringers, each consisting of rubber coated fabric tape, metal inner locking elements with corresponding outer clamps. The two stringers are engaged by the use of a slider. The inner elements hold the stringers together to compress the sealing surfaces, and the outer clamps provide a metal to metal surface for the operation of the slider. The fastener is sealed at the lower end with a molded bottom stop, and is available with either an open end or a closed molded end.

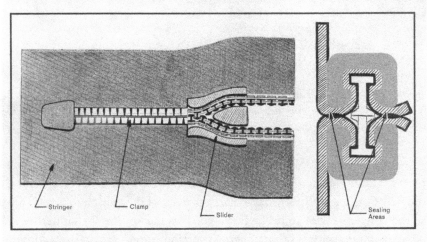

The O.E.B. fastener makes possible:
Complete protection of materials and equipment / easy storage or restorage / easy inspection / resulting in labor and material savings. The fastener provides a full seal, pressurized from either side, against air, dust, moisture, light, liquids, and gases.

O. E. B. TYPICAL APPLICATIONS

Sealed flexible covers and containers for engines, munitions, explosives and chemicals, electronic apparatus, gas masks, machine tools, rocket and missile hardware, and parachutes.
Protective clothing such as high altitude flying suits, underwater garments, survival suits, exposure suits and anti-contamination garments.
Air supported structures, radomes, tents, both personnel and emergency storage.
Ships equipment, air locks, hatch covers, etc.
Decontamination and sterilization bags.
Clean room to clean room transport containers.

Figure A.10. This 1960s-era Talon advertisement shows the characteristics of the OEB pressure-sealing zipper. Courtesy Talon Corporation.

Much correspondence from 1971 and 1972 addressed concerns related to failures of the OEB zippers during testing or training. The cause appeared in many instances to center on the improper operation of the zippers. By August 1972, training classes were being held at all NASA facilities and at ILC to train all personnel that had any contact with the suits about how to properly operate the zipper. Personnel were taught how to pull the closure in a straight line rather than at any angles; the latter method of closing the zipper was attributed to the failures in some cases. There was also a review of the inspection techniques to ensure that the zipper was in good condition.[7] Because of the elevated concerns with the OEB zipper, in the summer of 1972 ILC performed an evaluation on a B. F. Goodrich pressure-sealing zipper that could be used as a redundant restraint zipper. After many tests, ILC engineers decided that the zipper did not meet the operational and cycling characteristics required, so they scrapped that plan.[8]

By late 1970, this pressure closure was not a big sales item outside of ILC, so Talon informed ILC and NASA that they would soon be stopping production on the OEB zipper that everyone was depending on. Meetings with Talon management were held to make sure that a sufficient quantity of zippers was procured before production stopped. The parties reached an agreement by May 1971, and Talon settled on a contract to make 350 zipper assemblies and provide consultant services if the need arose. The cost would be $52,000 ($322,000 in 2019 dollars) but it would give ILC all of the closures they would need to build suits through the remainder of the missions. This came down to a cost of $149 ($922 in 2019 dollars) per unit, which was probably reasonable, all things considered.[9]

Restraint Zipper

The restraint zipper was also made by the Talon company and provided the strength necessary to withstand the mechanical pressure loads generated in the suit. The restraint zipper was tested by applying a tensile load across an area of about one inch to see where and how it would fail. The ultimate tensile load was typically in the range of 215 to 260 pounds per inch. It would fail when the metal teeth pulled off the zipper tape they were secured to. It was estimated that when the suit was pressurized at 3.75 pounds per square inch, the stress across the zipper was in the neighborhood of 26 to 28 pounds per inch. Momentary surge loads under dynamic conditions could produce loads of up to 60 pounds per inch. During normal operational use, the zipper provided a ten times safety margin and an adequate margin of safety during maximum dynamic loads. The zipper did not decrease in tensile strength even when it was cycled up to 250 times.[10]

Materials

Each of the materials used in the Apollo suit served a very specific purpose. Table A.1 outlines the significant materials as they were used in the Apollo 11 suits that Astronauts Neil Armstrong and Buzz Aldrin wore during their lunar mission.

Table A.1. Name, part number, function, and manufacturers of the materials used to make the thermal micrometeoroid garment

Material	Part number	Function	Manufacturers
Teflon fabric	T162-42	Thermal radiation, flame resistance, and outer protective layer	Stern & Stern Textiles Inc., Hornell, NY
Beta cloth (Teflon-coated yarn)	X4484	Flame resistance layer	Owens Corning Fiberglass Corp., Ashton, RI
Kapton-bonded Beta marquisette laminate	X993	Flame resistance and thermal insulation	G. T. Schjeldal, Northfield, MN
Perforated aluminized Mylar film	BN236-1	Thermal insulation	National Research Corp., Cambridge, MA
Nonwoven polyester cloth	E-1438	Thermal insulation spacer layer	Kendall, Inc., Fiber Products Division, Walpole, MA
Neoprene-coated nylon ripstop	CRP 2910 or Cat. 4096	Micrometeorite barrier and inner liner	Chemical Rubber, Beacon, NY, or Hodgeman Rubber Co., Framingham, MA

Thermal Micrometeoroid Garment

The thermal micrometeoroid garment (TMG) is typically one of the last items developed for a space suit. This garment requires somewhat less engineering design work, because it does not have the significant mechanical challenges of the pressurized torso, limb, and gloves of the suit. It is also developed based on the requirements of each mission, which sometimes were not established until later in the design process. As I write this, NASA has no firm commitments that require an advanced EVA suit, so most suit design work that NASA is currently involved with is focused on the mechanics of the pressure garment and items such as the gloves. The challenges of designing TMGs relate to the types of materials and numbers of layers that will protect against abrasion, solar and cosmic radiation, heat conduction, micrometeoroid impacts, and perhaps flames for short times. One of the major challenges of the Apollo suit program came about after the Apollo 1 fire when it was realized that the Nomex, as good as it was at protecting against high temperatures, was no match for the estimated 1,200 degrees Fahrenheit that was seen in the capsule. The solution was Owens Corning Beta cloth, but there were several manufacturing challenges related to the cutting and sewing of this fabric. More details on this are provided in Chapter 7.

In 1964 and 1965, when Hamilton Standard was in charge of the suit development, there was an effort to identify what the initial TMG would look like. The first concepts included the idea that it would consist of two separate pieces: an external thermal garment that would be covered by another separate garment called the meteoroid protective garment. The thought was that there would be a need for intravehicular thermal protection that would also serve as protection during extravehicular activities. During EVA work, the crew member would wear the second layer to guard against the possibility of damage from the impact of high-speed small micrometeoroids.

A NASA memo dated February 1965 clearly states that any previous reference to the two separate garments would be amended to reflect that the new garment would be one garment consisting of the micrometeoroid protection layer integrated over the thermal layers. This also alleviated the problems NASA was having with stowage space in what was then called the lunar excursion module. Overall weight dropped by about five pounds per garment, which was a lot considering that every pound saved had an impact on the cost to the program. This approach to TMGs continues to the present-day suits.[11]

The TMG was tested under various conditions to make certain that it would protect the crew in the most challenging heating and cooling extremes. This included both heat loading from the sun and heat loss in deep space. For extreme heat loading, the materials of the TMG were subjected to conditions that simulated the sun at a 70-degree angle on the moon while the astronaut faced a reflection off a crater at various aspect ratios. This was considered a worse-case condition. It was concluded that the seven layers of insulation were adequate to protect up to a maximum of 250 Btus per hour and a heat loss of -300 Btus per hour for the deep-space environment, where the suit would be losing internal heat.[12] After the Apollo 1 fire, work began at ILC using the new Beta cloth in March 1967. It had an Owens Corning part number of X4190D. This first style of material consisted of the woven Beta fibers that were coated with Teflon. In the next generation of Beta cloth, the individual yarns of Beta cloth were coated with Teflon and were then woven; the end result was rolls of fabric. It took ten months for ILC to product the first TMG, in January 1968. A lot of learning happened over those ten months.

The intravehicular cover layer was designed to provide a cover of minimal bulk and weight that would protect the command module pilot from heat and flames in the event of a fire on board. This layer was not designed to provide the protection needed during an EVA until the Apollo 15, 16, and 17 missions, when the command module pilot performed an EVA outside the capsule to retrieve film canisters. The cross-section consisted of the outer layer made up of Beta-cloth fabric. A second outside cover layer of Beta-cloth fabric was provided in areas of high abrasion. Later that was replaced by Teflon fabric that covered all of the Beta-cloth fabric. This was followed by an inner layer of Nomex fabric. On Apollo 9, the suit for the command module pilot also contained a layer of Kapton-bonded Beta marquisette between the Beta cloth and Nomex to reduce thermal conduction, since this mission included the first EVA from the capsule that exposed the command module pilot to potential thermal hazards with the hatch opened. On the Apollo 15–17 missions, the suit for the command module pilots had the added insulation because they each performed an EVA to retrieve film cannisters from the service module. (See Table A.2.)

The A-7L model extravehicular TMG had the same cover layers of Beta cloth, but below that were seven layers of aluminized Kapton film with a Beta marquisette fabric sandwiched between each Kapton layer (for a total of 6 layers) to reduce heat conduction. The A-7LB models used the Teflon cloth outermost layer followed by the Beta cloth and other layers below. Because the aluminized film was so fragile and could tear easily, grids of quarter-inch polyamide tape were applied to the film in 17 by 17 rows per every square yard of film to act as ripstop, which is just what the tape did if a tear started.

ILC and NASA quality personnel expressed a concern early in the production process of the TMG about the threads used to sew the garment together. Beta-cloth thread was the only thread that NASA had approved for use, since any other thread would not stand up to the potential heat of an on-board fire. Beta sewing thread was provided in sizes E-6 and E-12 and were coated in 10 percent nylon to protect the thread during the sewing cycle.[13] It was not out of the question that an operator could use a basic nylon thread by accident. Without a Beta-cloth thread securing the sections of the exposed surface of TMG together, they would literally separate when it was exposed to heat, thus likely exposing the crew member(s) to severe injury or death. Fortunately, a chemical known as "optical bleach," or MDAC, was used in the process of manufacturing the Beta threads. This chemical caused the threads to fluoresce under a blacklight. As a result, all TMGs were inspected using a blacklight to make sure the outermost threads used were actually Beta cloth.[14] Because the Beta-cloth threads had

Gridding: Application of ¼" Polymide Tape
onto aluminized Film. Purpose to install
a ripstop characteristic into the aluminized Film. Templets are used to tape off
areas of square yards. 17 rows x 17 rows
of ¼" tape per square yard. (Usage insulation for I/TMG, EV Boots, IV Spats &
EV Gloves)

Figure A.11. An ILC document outlining how the thermal micrometeoroid garment was made. It describes how the ¼" wide polyamide tape that was used to protect the very fragile insulation film was installed on each layer of film. Courtesy ILC Dover LP 2020.

such a low breaking strength, however, a nylon thread was used first to secure the section and the sections were folded under to conceal the nylon to any fire potential. Then the Beta threads were used to provide a relatively secure seam.

The first production cover layer for the TMG was completed on January 11, 1968 at the relatively new ILC facility in Frederica, Delaware.[15] That garment was installed on the model A-6L suit, serial number 009, and presented to NASA for first-article configuration inspection on May 15–16, 1968.[16]

Table A.2 provides a list of all the layers of materials in the TMG for all the various missions. The numbering of the layers starts from the outside and works its way to the inside, closest to the wearer. Changes to the TMG accounted for the most significant variations among the suits of various missions other than the complete model change from the A-7L to the A-7LB for Apollo 15 and subsequent missions.

Table A.2. Layers of the thermal micrometeoroid garment cover

Mission		Layup description	Part number	Features
Apollo 7	IVCL[a] (A7L-101100-12)	1, 2: Beta cloth (Teflon-coated yarn)	ST11G332-01	
		3: Nomex cloth, 5 oz.	ST11N330-01	
	ITMG[b] (A7L-200000-09)	1: Beta cloth	ST12G316-01	
		2, 4, 6, 8, 10, 12, 14: Gridded Kapton film (aluminized)	ST13H292-02	
		3, 5, 7, 9, 11, 13: Beta marquisette	ST11G273-01	
		15: Rubber-coated nylon ripstop	ST12N023-01	
Apollo 8	IVCL[a] (A7L-101100-12)	1, 2: Beta cloth	ST11G332-0	Same as Apollo 7
		3: Nomex cloth, 5 oz.	ST11N330-01	
	ITMG[b] (A7L-200000-10)	1: Beta cloth	1: ST12G316-01	
		2, 4, 6, 8, 10, 12, 14: Gridded Kapton film (aluminized)	2, 4, 6, 8, 10, 12, 14: ST13H292-02	
		3, 5, 7, 9, 11, 13: Beta marquisette	3, 5, 7, 9, 11, 13: ST11G273-01	
		15: Rubber-coated nylon ripstop	15: ST12N023-01	
Apollo 9	IVCL[a] (A7L-101100-11)	1: Beta cloth	1; ST11G332-01	New change in material in middle layer for thermal protection (designed to support first earth-orbit EVA)
		2: Kapton-bonded Beta marquisette	2: ST12H327-01	
		3: Nomex cloth, 5 oz.	3: 11N330-01	
	ITMG[b] (A7L-200000-12)	1: Beta cloth	ST12G316-01	Same as Apollo 7 and 8
		2, 4, 6, 8, 10, 12, 14: Gridded Kapton film (aluminized)	ST13H292-01	
		3, 5, 7, 9, 11, 13: Beta marquisette	ST11G273-01	
		15: Rubber-coated nylon ripstop	ST12N023-01	

Mission		Layup description	Part number	Features
Apollo 10	IVCL[a] (A7L-101100-12)	1, 2: Beta cloth 3: Nomex cloth, 5 oz.	ST11G332-01 ST11N330-01	Same as Apollo 7 and 8
	ITMG[b] (A7L-201100-13)	0: Teflon fabric patches	ST11T391-01	For abrasion resistance
		1: Beta cloth	ST11G332-01	Added fire protection
		2, 3: Kapton-bonded Beta marquisette	ST12H327-01	Added fire and thermal protection
		4, 6, 8, 10, 12: Perforated aluminized Mylar film	T17M042-02	Added fire and thermal protection
		5, 7, 9, 11: Nonwoven Dacron	ST11D024-01	Added new thermal spacer material
		13: Rubber-coated nylon ripstop	ST12N023-01	
Apollo 11	IVCL[a] (A7L-101100-19)	1, 2: Beta cloth 3: Nomex cloth, 5 oz.	ST11G332-01 ST11N330-01	Same as Apollo 7, 8, and 10
	ITMG[b] (A7L-201100-26)	0: Teflon fabric patches 1: Beta cloth	ST11T391-01 ST11G332-01	Same as Apollo 10
		2, 3: Kapton-bonded Beta marquisette	ST12H327-01	
		4, 6, 8, 10, 12: Perforated aluminized Mylar film	ST17M042-02	
		5, 7, 9, 11: Nonwoven Dacron	ST11D024-01	
		13: Rubber-coated nylon ripstop	ST12N023-01	
Apollo 12	IVCL[a] (A7L-101100-19)	1, 2: Beta cloth (Teflon-coated yarn)	ST11G332-01	Same as Apollo 7, 8, 10, and 11
		3: Nomex cloth, 5 oz.	ST11N330-01	
	ITMG[b] (A7L-201100-28)	0: Teflon fabric patches 1: Beta cloth	ST11T391-01 ST11G332-01	Same as Apollo 10–11
		2, 3: Kapton-bonded Beta marquisette	ST12H327-01	
		4, 6, 8, 10, 12: Perforated aluminized Mylar film	ST17M042-02	
		5, 7, 9, 11: Nonwoven Dacron	ST11D024-01	
		13: Rubber-coated nylon ripstop	nylon ST12N023-01	

(*continued*)

Mission		Layup description	Part number	Features
Apollo 13	IVCL[a] (A7L-101100-19)	1,2: Beta cloth	ST11G332-01	Same as Apollo 7, 8, 10, 11, 12
		3: Nomex cloth, 5 oz.	ST11N330-01	
	ITMG[b] (A7L-201100-28)	0: Teflon fabric patches	ST11T391-01	Same as Apollo 10–12
		1: Beta cloth	ST11G332-01	
		2, 3: Kapton-bonded Beta marquisette	ST12H327-01	
		4, 6, 8, 10, 12: Perforated aluminized Mylar film	ST17M042-02	
		5, 7, 9, 11: Nonwoven Dacron	ST11D024-01	
		13: Rubber-coated nylon ripstop	ST12N023-01	
Apollo 14	IVCL[a] (A7L-101100-19)	1, 2: Beta cloth	ST11G332-01	Same as Apollo 7, 8, 10, 11, 12, 13
		3: Nomex cloth, 5 oz.	ST11N330-01	
	ITMG[b] (A7L-201100-28)	0: Teflon fabric patches	ST11T391-01	Same as Apollo 10–13
		1: Beta cloth	ST11G332-01	
		2, 3: Kapton-bonded Beta marquisette	ST12H327-01	
		4, 6, 8, 10, 12: Perforated aluminized Mylar film	ST17M042-02	
		5, 7, 9, 11: Nonwoven Dacron	ST11D024-01	
		13: Rubber-coated nylon ripstop	ST12N023-01	
Apollo 15	IV CMP ITMG[c] (A7LB-101200-04 (Designed for EVA use)	1: Teflon fabric	ST11T391-01	Complete material layer for abrasion resistance.
		2: Beta cloth	ST11G332-01	
		3, 4: Kapton with Beta marquisette	ST12H327-01	Additional layer for fire/thermal protection
		5, 7, 9, 11, 13: Perforated aluminized Mylar film	ST17M042-02	Addition of film for thermal/radiation protection
		6, 8, 10, 12: Nonwoven Dacron	ST11D024-01	Addition of thermal spacers
		14: Rubber-coated nylon ripstop	ST12N023-01	Addition of the inner liner
	EV LITMG[d] (A7LB-201154-03)	1: Teflon fabric	ST11T391-01	Previously used as patches—now 1 complete layer.
		2: Beta cloth	ST11G332-01	

Mission		Layup description	Part number	Features
		3, 5, 7: Beta marquisette, laminated[e]	ST12G413-01	Different marquisette material with 1 additional spacer. Lamination = (Top): Beta cloth marquisette, sized (Bottom): Beta cloth ST11G245
		4, 6: Gridded Kapton	ST13H292-02	2 additional layers of film for fire & thermal protection.
		8, 10, 12, 14, 16: Perforated aluminized Mylar film	ST17M042-02	
		9, 11, 13, 15, 17: Nonwoven Dacron	ST11D024-01	1 Additional layer of thermal spacer
		18: Rubber-coated nylon ripstop	ST12N023-01	
Apollo 16	IV CMP ITMG[c] (Designed for EVA use)	1: Teflon fabric	ST11T391-01	Same as Apollo 15
		2: Beta cloth	ST11G332-01	
		3, 4: Kapton with Beta marquisette	ST12H327-01	
		5, 7, 9, 11, 13: Perforated aluminized Mylar film	ST17M042-02	
		6, 8, 10, 12: Nonwoven Dacron	ST11D024-01	
		14: Rubber-coated nylon ripstop	ST12N023-01	
	EV LITMG[d] (A7LB-201154-03)	1: Teflon fabric	ST11T391-01	Same as Apollo 15
		2: Beta cloth	ST11G332-01	
		3, 5, 7: Beta marquisette, laminated[e]	ST12G413-01	
		4, 6: Gridded Kapton	ST13H292-02	
		8, 10, 12, 14, 16: Perforated aluminized Mylar film	ST17M042-02	
		9, 11, 13, 15, 17: Nonwoven Dacron	ST11D024-01	
		18: Rubber-coated nylon ripstop	ST12N023-01	

(continued)

Mission		Layup description	Part number	Features
Apollo 17	IV CMP ITMG[c] (Designed for EVA use)	1: Teflon fabric	ST11T391-01	Same as Apollo 15 and 16
		2: Beta cloth	ST11G332-01	
		3,4: Kapton with Beta marquisette	ST12H327-01	
		5, 7, 9, 11, 13: Perforated aluminized Mylar film	ST17M042-02	
		6, 8, 10, 12: Nonwoven Dacron	ST11D024-01	
		14: Rubber-coated nylon ripstop	ST12N023-01	
	EV LITMG[d] (A7LB-201154-03)	1: Teflon fabric	ST11T391-01	Same as Apollo 15 and 16
		2: Beta cloth	ST11G332-01	
		3, 5, 7: Beta marquisette, laminated[e]	ST12G413-01	
		4, 6: Gridded Kapton	ST13H292-02	
		8, 10, 12, 14, 16: Perforated aluminized Mylar film	ST17M042-02	
		9, 11, 13, 15, 17: Nonwoven Dacron	ST11D024-01	
		18: Rubber-coated nylon ripstop	ST12N023-01	
Skylab	All suits TMG (A7LB-20119001	1: Teflon fabric	ST11T391-01	
		2: Beta cloth	ST11G332-01	
		3, 5, 7: Aluminized H-film	ST13H292-04	One extra layer of film and five fewer layers of Mylar film
		4, 6: Beta marquisette, laminated[e] Lamination = (Top): Beta cloth marquisette, sized (Bottom): Beta cloth ST11G245	ST12G413-01	One less spacer and five fewer layers of nonwoven Dacron
		8: Rubber-coated nylon ripstop	ST12N023-01	
ASTP	IVCL[a] (A7L-101279)	1: Teflon fabric	ST11T391-01	NA
		2: Beta cloth	ST11G332-01	
		3: PBI fabric[f]		

Source: Table compiled by author from various NASA and ILC Industries source documents.
Notes: a. IVCL = Intra-vehicular cover layer (command module pilot's suit)
b. ITMG = Integrated thermal micrometeoroid garment (commander and lunar module pilots' suits).
c. CMP ITMG = Intra-vehicular; integrated thermal micrometeoroid garment (command module pilot's suit).
d. EV LITMG = Extra-vehicular; lunar integrated thermal micrometeoroid garment.
e. Most of the Beta cloth is cut out of the panels after the suit is assembled. It remains only in the seam area.
f. PBI fabric was also known as polybenzimidazole (hence PBI). It had a very high melting point and retained its tough properties even at elevated temperatures. It has been used in clothing for firefighters for many years.

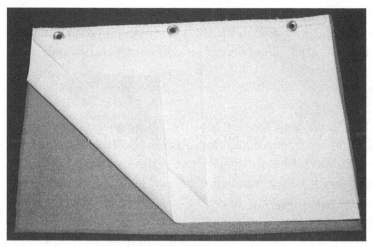

Figure A.12. The three layers of fabric used on the Apollo-Soyuz Test Program space suits. The inner layer of this thermal micrometeoroid garment consisted of a gold-colored fabric called PBI, a high-temperature material used by firefighters. Photo by Bill Ayrey.

Restraint Material

The restraint material consisted of a lightweight and low-elongation material that could stand up to both the internal pressures within the suit and the stress loads produced by the human working in the suit. It was important to design and cut the pattern pieces of the restraint material just a fraction smaller than the internal bladder pieces because once the restraint layer was assembled, it had to take structural loading when the suit was pressurized. Otherwise, the more fragile bladder material would be subject to loads that could possibly fail it. A failed bladder could result in the loss of the crew member.

By January 1967, ILC had settled on using a seven-ounce, dark-blue, Oxford-weave nylon fabric as a restraint material. For the model A-7LB, a nylon cloth that weighed 7.25 ounces per square yard was used. This heaver nylon had a tensile strength of 325 pounds minimum in the warp direction. A size E nylon thread was used to assemble the restraint pieces together.

At one point, NASA was interested in replacing the nylon cloth with a Nomex material because NASA engineers had expressed concern about the possibility that hot spots might develop due to heat leaks in the cover layers that would impact the strength of the nylon. The strength of the nylon fell off rapidly above 120°C (250°F). After further evaluation, ILC continued to use the nylon restraint material. That was likely because the TMG would provide the thermal protection needed and the nylon proved to be a superior fabric in terms of providing strength and its low elongation properties.

Bladder Material

The model A-7L pressure bladder consisted of neoprene-coated nylon cloth. This material was purchased from Reeves Inc. They manufactured it using what they called a Type W neoprene, which contained antioxidant 2246 added at two parts per hundred of rubber. It also contained carbon black and an inorganic pigment. Reeves felt that the usable shelf life would be ten years if it was protected from oxygen and light. It was reinforced in heavy wear areas by a layer of neoprene-coated nylon ripstop that was bonded over the bladder. On the model A-7LB suits, a nylon cloth scuff layer was bonded onto the entire inner surface of the neoprene bladder.

Liquid Cooling Garment Materials

The liquid cooling garment was made primarily of spandex, a DuPont material that was developed in 1959 by one of their scientists. The spandex ILC used could stretch as much as 600 percent and then fully recover its shape. Spandex was initially used to replace rubber in lady's undergarments but went on to be used in many other applications in the fashion industry.[17] The spandex garment was designed, cut, and sewn together in such a way that the final product looked much like a one-piece long-underwear garment with a zipper running vertically from the collar down to the lower chest area to allow the crew member to put it on and take it off. The liquid cooling garment also consisted of about 300 feet of eighth-inch-diameter vinyl tubing woven into the inside layers of the spandex that supplied the cooling water in and out of the garment. The very inside layer consisted of a chiffon material that provided comfort and prevented the vinyl tubes from getting caught up when the crew member was getting into and out of the suit, yet was very thin so that it did not interfere with the heat exchange process. The spandex had an open-weave structure that allowed the eighth-inch vinyl tubes to be woven to the inside layer in the regions where the most heat from the body could be extracted. Cold water from the primary life-support system was pumped through these tubes to remove body heat. This design worked more efficiently when the tubes were pressed against the skin so they could pick up the heat. The spandex was the best material for the job, since its elasticity kept the tubes close to the skin while also providing comfort to the wearer.

Chromel-R

Chromel-R was the trade name of a material manufactured by the Hoskins Manufacturing Company. It was made up primarily of 74 percent nickel and 20 percent chromium.[18] Essentially, it was fibers of steel woven together to form a cloth. It was quite heavy, weighing in at just over one pound per square yard. Of all the materials used in the Apollo suits, Chromel-R was the most expensive. The cost was as high as $2,500 per square yard in 1969 ($18,000 in 2019). As a precaution to guard against misuse, rolls of the material were stored in a vault so that no one would have access to it other than production control personnel, who could requisition out only what was needed for production.

Chromel-R first made its debut related to space suits in 1966, when the David Clark Company made a space suit cover layer for the entire lower torso for the Gemini 9 mission. The plan was that astronaut Gene Cernan was to perform an EVA to the back side of the Gemini service module and strap on an astronaut maneuvering unit to evaluate its capacity to assist future astronauts in performing EVAs. The AMU was essentially a jet pack that would propel the astronaut, but the propulsion system produced hot gases that could damage the lower torso of the space suit. The metallic Chromel-R material would stand up to this heat, so it was selected to serve as a protective cover layer. Unfortunately, Gene Cernan was never able to evaluate the astronaut maneuvering unit because the efforts required to perform this EVA were more than Cernan could tolerate. Before he could float to the back side of the Gemini spacecraft, he became overheated and his visor fogged up to the point where he could hardly get back to the hatchway and into the capsule. He was fortunate to survive the ordeal.

Early in the development of the Apollo model A-7L suit, it became obvious that abrasion against the Beta cloth resulted in excessive damage to the fibers. This was particularly true for the back of the torso, where the primary life-support system was rubbing against it. As a result, the engineers added a full cover layer of Chromel-R to stand up to this abuse. Later, they replaced the Chromel-R with layers of Teflon material, which resulted in a lighter-weight and less expensive suit.

In addition to providing abrasion resistance, Chromel-R did not puncture or tear easily and because of that, it was used to protect the main sections of the EVA gloves and the lunar boots. The gloves took a lot of abuse when crew members used tools or handled potentially sharp rocks. The boots were wrapped in Chromel-R to protect the pressure boots underneath against damage should they brush up against sharp rocks and boulders on the moon.

Convolutes

As with the AX-5L suit, the A-5L suit had what was known as an omnidirectional shoulder: aircraft wire secured to the upper front chest that was carried across the upper shoulder, where it was secured to the center of the upper back. The wire cable was routed through an aircraft-grade aluminum tube secured to the upper shoulder and precisely placed so that as the wearer moved their shoulder in all directions, the wire kept the shoulder convolute from moving outward and away from the torso, as it would be prone to do in an inflated condition. This meant that the wearer could move the shoulder frontward to backward as well as up and down while the cable kept it in position at all times. This was also known as a bi-stable restraint system because the cable held it more or less stable in the up and down positions. I have worn a pressurized Apollo suit and can attest that it took slight effort to raise the arm and shoulder from the lower position to above the shoulder level since the wire was taken to its ultimate loading at the midpoint. This also aided in keeping the arm in the overhead position since the wire helped hold the weight just over the center, as in Figure A.13.

The convolutes located throughout the suit were of the style that George Durney had been making all along and proved to be very reliable.

Vent System

The oxygen ventilation system used in the model A-7LB suit is pictured in Figure A.15. It was basically the same as that used in the other models; the only difference was how the ducts were routed inside the suit. This vent system was integrated into the torso, the arms. and the legs of the pressure garment assembly to take all the air, including moisture from sweat and exhalation, out of the suit and back into the primary life-support system for scrubbing and dehumidification. The oxygen hoses from the primary life-support system or the command or lunar module environmental control system were connected to one of the two blue inlet gas connectors made by Air-Lock. The setting of the diverter valve directed the gas so that 100

Figure A.13. The AX-5L suit with the omnidirectional-bi-stable shoulder design shown during flexing in the up position. Note the narrow shoulder width as the wire cable and turn-around tunnel holds the shoulder in closer. Courtesy ILC Dover LP 2020.

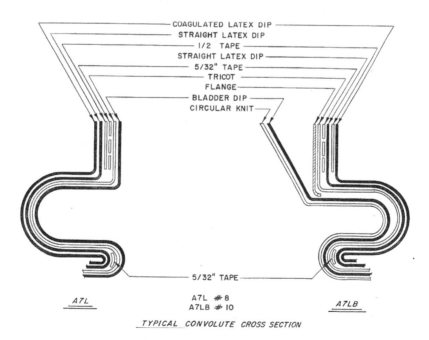

COAGULATED LATEX DIP
STRAIGHT LATEX DIP
1/2 TAPE
STRAIGHT LATEX DIP
5/32" TAPE
TRICOT
FLANGE
BLADDER DIP
CIRCULAR KNIT

5/32" TAPE

A7L

A7L #8
A7LB #10

A7LB

TYPICAL CONVOLUTE CROSS SECTION

Figure A.14. Cross-section of the rubber convolute. This rubber section, which was located in all the flexible parts of the suit, proved to be the key ingredient that made the ILC suit so mobile when pressurized. This sketch shows the difference between the A7L model suit and the A7LB models. Courtesy ILC Dover LP 2020.

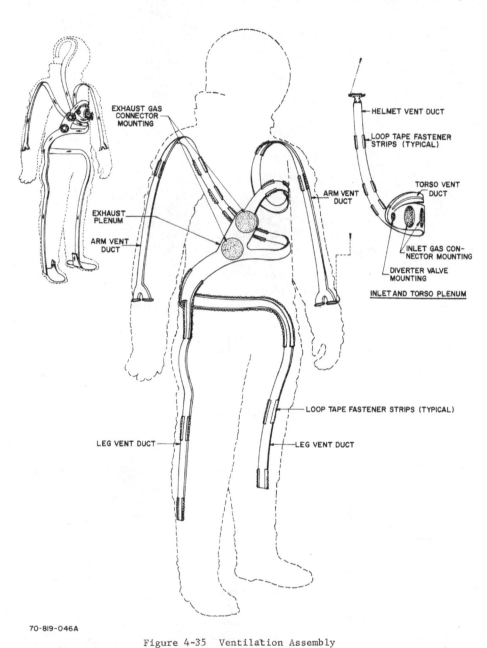

EXHAUST GAS
CONNECTOR
MOUNTING

HELMET VENT DUCT

LOOP TAPE FASTENER
STRIPS (TYPICAL)

ARM VENT
DUCT

TORSO VENT
DUCT

EXHAUST
PLENUM

ARM VENT
DUCT

INLET GAS CON-
NECTOR MOUNTING

DIVERTER VALVE
MOUNTING

INLET AND TORSO PLENUM

LOOP TAPE FASTENER STRIPS (TYPICAL)

LEG VENT DUCT

LEG VENT DUCT

70-819-046A

Figure 4-35   Ventilation Assembly

4-256

Figure A.15. The model A7LB gas ventilation system. Courtesy ILC Dover LP 2020.

percent of the inlet oxygen flowed to the outlet in the back of the helmet or 50 percent flowed to the helmet and 50 percent to the torso section. This was because the primary life-support system supplied a maximum of 6 cubic feet of incoming airflow to the helmet when in use on the lunar surface. This was enough because the liquid cooling garment cooled the body and gas flow contributed less significantly to cooling, as it had in the earlier space suits. When the astronauts were connected to the command module or the lunar module air system, it provided 12 cubic feet per minute, so the diverter valve was set so the suit maintained a 50–50 split—6 cubic feet to the helmet and 6 cubic feet to the torso. The crew members did not wear the liquid cooling garment in the vehicles, so the only cooling they had was from the airflow. This helped remove the body heat and humidity and was more than adequate since the crew was at a relatively low activity level and was not generating many Btus.

The vent system consisted of a number of small (⅜ inch) nylon spacer coils that were enclosed within a nylon mesh cloth and wrapped with a neoprene-coated cloth that provided an airtight duct. It was critical that

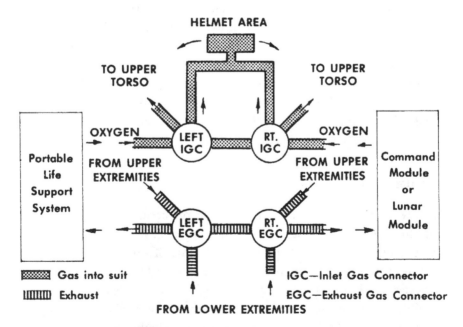

Figure A.16. This simplified diagram shows the flow of air into and out of the gas connectors and though the internal ventilation system for the extravehicular suits. Airflow into the suit was provided through the blue connectors and the exhaust gas was ventilated out of the suit through the red connectors, all located on the chest of the suit. Courtesy ILC Dover LP 2020.

Figure A.17. The layout of the model A-7LB gas connectors, the multiple water connector for the liquid cooling garment, and the electrical connectors. Also shown is the diverter valve used to direct the flow of oxygen inside the suit. Courtesy ILC Dover LP 2020.

all of the ducts remained fully opened and were not crimped closed to any degree; any crimping could result in pressure drop. Think of it as crimping a garden hose as water is flowing through it. The water pump has to work much harder to push the same amount of water all the way through. If there had been any crimps in the vent system, the backpack would not have been able to easily provide the correct flow of air for the astronaut. Throughout the Apollo program, ILC faced challenges as its engineers tried to ensure that the duct system would pass critical flow tests.

Pressure Relief Valve

The pressure relief valve was located under the access flap located on the right thigh on the A-7LB suits and on the left arm of some A-7L suits. It had an opening pressure of 5.0 and a reset pressure of 4.5 pounds per square inch by the time of the Apollo 15 mission. It had a rated flow of 12.2 pounds of oxygen per hour minimum at a pressure of 5.0 pounds per square inch absolute.[19] Copious documentation throughout the Apollo program provides evidence that NASA was concerned about whether these valves could repeatedly open and close within the specified range. Test results showed that they were not reliable in many cases. Valves were often replaced in many suits that were in use to support testing or in actual flight suits that were being tested just before their mission.

Urine Collection Device

The urine collection device was manufactured by Whirlpool Corporation and supplied by NASA as government property, but ILC was responsible for the interface of the unit with the crew member and the suit. Crew members wore a roll-on rubber prophylactic cuff. The end of this cuff emptied urine into a collection bag. The suit community furthered an ongoing joke about the sizes of these rubber cuffs an astronaut would select. In *Carrying the Fire,* Mike Collins says that the crew heroically referred to them as "extra-large, immense, and unbelievable." In reality, when the right size was not selected, leaks would inevitably occur and the urine would pool on the backs of the suit during training runs. Astronauts who did not choose wisely did not gain the affection of the suit technicians who were responsible for the care and maintenance of their suits.[20]

To remove the urine from the filled bag, a cap was removed on the outside of the suit (on the right thigh, under the pocket cover of the garment), and a vacuum hooked up to the connector removed the urine from the bag.

Figure A.18. The urine collection device bag worn over the liquid cooling garment. The tube hanging down on the wearer's right hip is the drain hose that connected to the pass-through of the suit and permitted the vacuum suction of urine from the bag if the crew members could not remove their suits for extended periods. Courtesy ILC Dover LP 2020.

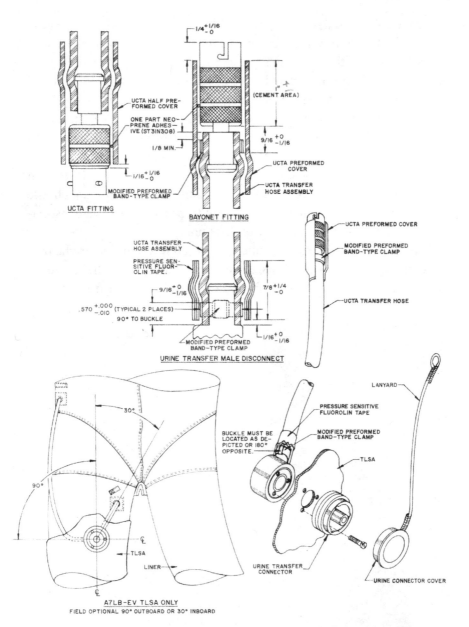

Figure 4-31   Urine Transfer Male Disconnect and Hose Assembly

Figure A.19. This sketch included in the maintenance manual shows the various details of the urine transfer system that would evacuate the urine from the collection bag. Courtesy ILC Dover LP 2020.

Medical Injection Patch

NASA doctors determined that there might be a rare instance when an astronaut would need to inject himself with medicine while in the suit. ILC engineers likely pointed out the obvious fact that sticking a needle into a pressurized suit was not a great idea. Upon NASA's insistence, however, ILC designed and installed a new medical injection patch in the suit: a very soft,

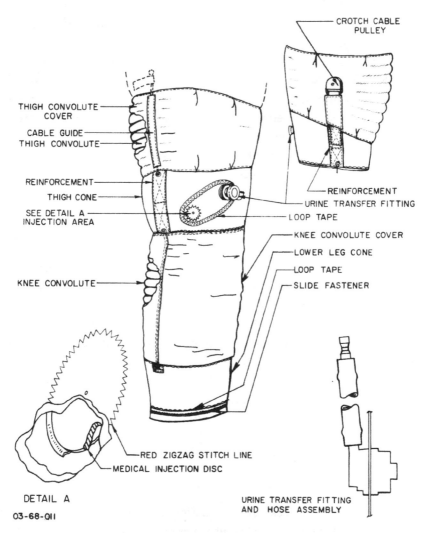

Figure A.20. Details of the leg including the medical injection patch and the urine collection pass-through. The red stitched line around the injection patch was provided to make it easy for crew members to find the location should they need to make an injection. Courtesy ILC Dover LP 2020.

round rubber patch two inches in diameter embedded in the leg that would seal shut when the needle was removed.

Test subject Richard Ellis had to demonstrate that the patch worked by injecting himself with an inert saline solution. Little more is included in any ILC documents from that period. An injection patch originally appeared in the David Clark Company's A-1C suit, so it is likely that NASA passed on what they knew about it to the engineers of ILC, who probably copied that idea and decided where to place it with the guidance of NASA doctors. It was located on the right thigh on the A-7L suit and on the left thigh on the model A-7LB suit.

Shane M. McFarland at the Johnson Space Center and Aaron S. Weaver at NASA's Glenn Research Center did research on the injection patch in 2013. They cut into the pressure bladder of an old A-7L space suit still in NASA's possession and removed the disk that the needle was intended to pass through. Their analysis found that it was made up primarily of a silicone rubber that was one-tenth of an inch thick. The disk was very soft so that it would self-seal as the needle was removed.[21]

In-Suit Drink Bag

Documentation does not provide many details related to the Apollo drink bags. They were likely a urethane material that could be heat-sealed together to form the optimal shapes to fit inside the upper torso of the suit and fasten with Velcro that was fixed to the inside of the neck ring. Fresh water could be added through a fill port, likely by a one-way sealed valve that allowed water to be added (perhaps through an injection needle of some kind) without any accidental discharge if pressure was applied when, for instance, the crew member might lean against it. The fill port was located beside the drink tube, which was referred to as the bite valve. The crew member would grasp the valve with their teeth and pull up to engage the flow.

Neck Dam

The neoprene rubber neck dam was designed to prevent any water from entering the suit should a crew member have to jump into the water during egress from the capsule after splashdown. Gus Grissom had experienced that problem during his egress from his Mercury capsule; water immediately started to fill his suit when it entered through the neck ring since he

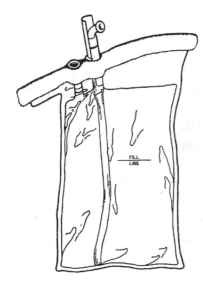

FILL
LINE

Left: Figure A.21. A sketch of the in-suit drink bag, designed by Walt Sawyer, one of the ILC engineers who worked out of the Houston Office. Courtesy ILC Dover LP 2020.

Below: Figure A.22. This sketch from an ILC field operations bulletin dated 4 October, 1968, shows the details of the neck dam used on the Apollo suits. Courtesy ILC Dover LP 2020.

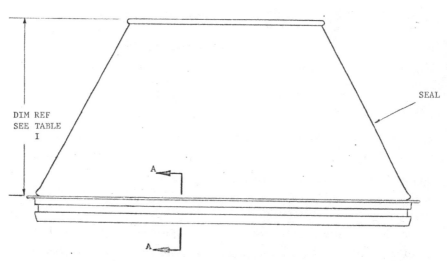

DIM REF
SEE TABLE
I

SEAL

A

A

TABLE I

SIZE DESIGNATION	DIM REF
A7L-1Z1036-02-1350*	5 3/4
A7L-1Z1036-02-1400*	5 1/2
A7L-1Z1036-02-1450*	5 1/4
A7L-1Z1036-02-1500*	5
A7L-1Z1036-02-1550*	4 3/4
A7L-1Z1036-02-1600*	4 1/2
A7L-1Z1036-02-1650*	4 1/4

NOEPRENE
SEAL

LABEL

ALUMINUM
ALLOY
RING

SECTION A-A

* Neck size

had his helmet off, as was expected. He almost drowned but was pulled up by the rescue helicopter at the last minute. To prevent a future recurrence, the neck dam was made to fasten to the neck-ring hardware of the suit with the small rubber opening, squeezing tight around the astronaut's neck.

Before entering the water from the capsule, a suited crewmember would remove the helmet assembly from the suit, pull the neoprene seal over the head, and position it comfortably around the neck. He would then engage the neck-dam ring with the helmet-attaching ring on the pressure-garment assembly.

In November 1968, an update was made to the drawings for the check-list pockets that added a snap inside the pocket to mate with a snap located on the end of a long cord attached to the neck dam. This was designed to help ensure that the crewmember would not lose the neck dam should he have to pull it out in an emergency situation. It was recommended that the crew have the neck dam in the check-list pocket for launch and then remove it and stow it on board the command module until preparations for reentry to Earth, at which time the crew members would put the neck dams back into the pockets. On later missions when the suits were not worn during reentry, the neck dams obviously were not considered for use.

Oxygen Purge System

The Apollo 9 and 11 through 17 missions all carried an oxygen purge system that used two high-pressure oxygen tanks mounted on the top of the primary life-support system when it was used on the lunar surface. These interconnected tanks were pressurized to 5,880 pounds per square inch absolute and could provide about thirty minutes of oxygen when they took the place of a malfunctioning primary life-support system. The command module pilots on Apollo 15 through 17 also used this system when they performed their EVA to retrieve the film canisters from the service module. They were the only parts of the primary life-support system that were used on the moon and brought back to Earth.

The oxygen purge system could be activated if a problem was encountered with the primary life-support system or if additional oxygen flow was required to maintain suit pressure due to excessive leakage during lunar activities. This was accomplished when the crew member pulled the oxygen purge system actuator located on the right side of the remote control unit attached to their chest; this opened the valve from the oxygen bottles. The flow rate into and out of the suit would be controlled by activating the

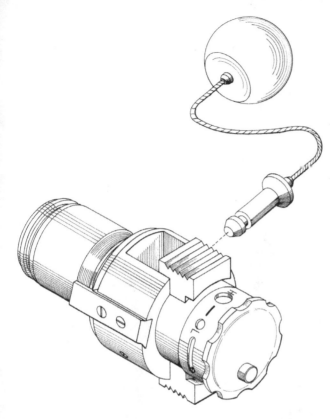

Left: Figure A.23. This sketch, made by ILC illustrators, shows the purge valve that was attached to the torso. The crew member would pull the red ball to permit the oxygen purge through the suit. Courtesy ILC Dover LP 2020.

Below: Figure A.24. Details of the oxygen purge system. High-pressure oxygen was stored in two tanks located in a separate back-pack just above the primary life support system. When a valve was activated from the remote control unit on the front of the chest, oxygen flowed into the suit and exited through the purge fitting located on one of the two outlet (red) hardware connectors on the suit torso. Courtesy NASA

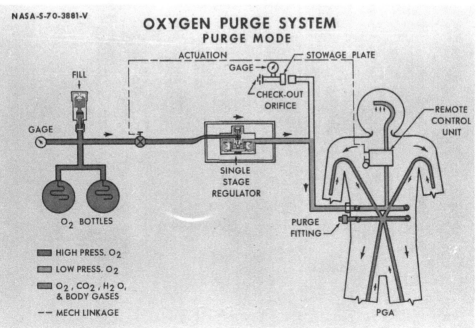

NASA-S-70-3881-V

OXYGEN PURGE SYSTEM
PURGE MODE

ACTUATION — STOWAGE PLATE
GAGE
CHECK-OUT ORIFICE
FILL
GAGE
SINGLE STAGE REGULATOR
REMOTE CONTROL UNIT
O_2 BOTTLES
PURGE FITTING

HIGH PRESS. O_2
LOW PRESS. O_2
O_2, CO_2, H_2O, & BODY GASES
— — MECH LINKAGE

PGA

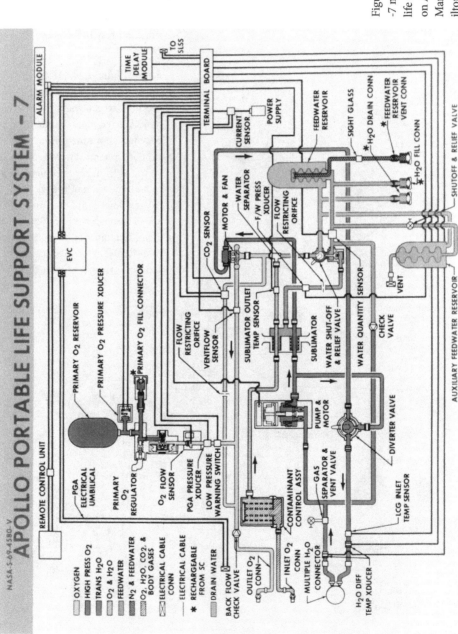

Figure A.25. Details of the -7 model of the primary life support system used on Apollo 16 and 17. Manufactured by Hamilton Standard. Courtesy NASA.

purge valve that was located on one of the two outlets of the suit (the crew member's right-side connector). To do this, the crew member would pull on a red ball that released a pin from the body of the valve.

They would then grasp two lock tabs and pull the valve outward. This would control the flow from the high-pressure oxygen tanks. A single-flow valve was provided for Apollo 9, 11, and 12 and a dual-flow valve was provided for Apollo 14 through 17.[22]

Portable Life-Support System

I am not an expert on the portable life-support systems, and literature exists to explain this separate system. Figure A.25 shows the schematic of the model used in the later lunar missions for those who are interested.

Electrical Harness Assembly: The Biobelt and the Communication Carrier Assembly

The electrical harness contained the main pass-through hardware that allowed electrical signals to be passed in and out of the pressurized suit. NASA supplied these components to ILC. The bio-instrumentation belt had pockets that held individual electronic modules that took the heart rate from the pads placed on the astronaut's torso and converted the input to an electronic signal to be carried to the primary life-support system for broadcast back to the doctors in Houston. This electronic harness also provided a connector for the communication carrier assembly so crew members could communicate with each other and NASA personnel in Houston. See Figure A.26.

Proposed Vomit Removal Apparatus

Even though a vomit removal device was never included in an Apollo space suit, it is worth mentioning since serious thought was given to including one based on what NASA learned early in the manned space program.

In 1961, the second human in space, Russian cosmonaut Gherman Titov, was the first to experience space adaptation syndrome, otherwise known as space sickness. Since then, countless others have succumbed to this illness. It is likely that most of the Apollo astronauts who suffered from it did not openly report it because they were afraid they would be removed from missions; some of the early Apollo astronauts who did suffer from it

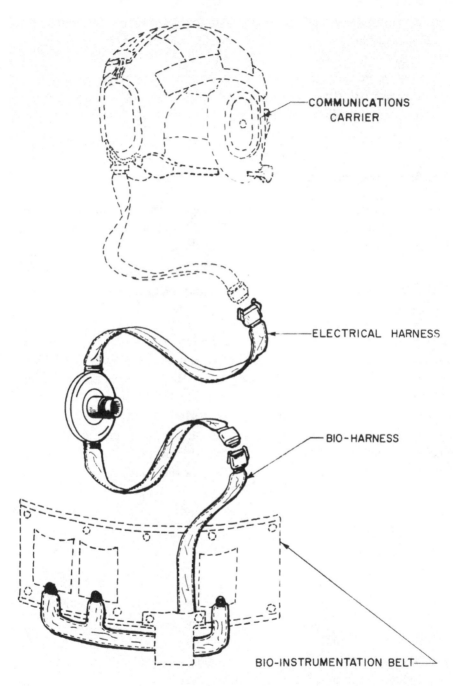

COMMUNICATIONS
CARRIER

ELECTRICAL HARNESS

BIO-HARNESS

BIO-INSTRUMENTATION BELT

Figure A.26. The electrical harness, the biobelt, and the components of the communication carrier assembly that carried the electronics into and out of the Apollo suits. Courtesy ILC Dover LP 2020.

never flew again once the NASA doctors became aware of their condition. Early into the U. S. manned space program, NASA physiologists found that the explanation was rather simple. The vestibular system in the inner ear contains a fluid called endolymph; this system helps the brain determine the position and movement of the head and the body. In weightlessness, however, that fluid is floating around similar to the way it might be moving if a person was in an automobile that took a sharp turn. When the eyes do not see the movement that should be taking place based on the feedback from the positioning of this fluid, the disconnect with the brain manifests itself as motion sickness. The effects can last for hours or days. Some never suffer from it, while others wish they had never left good old terra firma.

In 1969, concerns arose that a suited astronaut might have a severe case of this space adaptation syndrome while pressurized inside their space suit (a particular concern when planning for the longer-term EVA missions beyond Apollo 17). This motion sickness issue had the potential to create a host of problems that could have proven fatal if some systems within the suit were clogged with the resulting debris. Thus, you can bet that someone was out there looking for a solution. This time it was not the ILC people.

An organization named Bellcomm, Inc. (a subsidiary of Bell Labs that was a contractor for NASA) provided NASA with studies and analysis of various issues impacting the Apollo program. One report was issued in July of 1969 titled "Vomit Removal Apparatus for an EVA Pressure Suit: Design Concept Case 710."[23]

The proposed devices were well thought out but would have added a great deal to the complexity of the suit system. In the first system, vomit would be expelled to the vacuum of space by way of a rather large tube that was permanently mounted squarely in the front of the helmet. This no doubt would have served as a constant reminder to the astronaut about the possibility that they might toss their cookies while confined in the suit since it would have been squarely in their line of vision at all times. If vomiting turned out to be unavoidable, any TV cameras that provided a closeup of the suited crew member might have recorded a memorable sight not fit for broadcasting, not to mention the stains vomit would have left on the suit.

NASA showed no interest beyond this research phase. I read somewhere that humans do not control the natural vomit response easily and the idea that they could willingly employ this device was not reasonable. I'm not sure how the system was tested, if in fact it ever was, but surely a few test

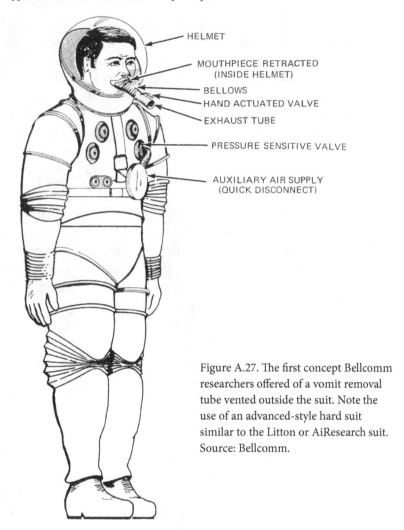

HELMET

MOUTHPIECE RETRACTED
(INSIDE HELMET)

BELLOWS

HAND ACTUATED VALVE

EXHAUST TUBE

PRESSURE SENSITIVE VALVE

AUXILIARY AIR SUPPLY
(QUICK DISCONNECT)

Figure A.27. The first concept Bellcomm researchers offered of a vomit removal tube vented outside the suit. Note the use of an advanced-style hard suit similar to the Litton or AiResearch suit. Source: Bellcomm.

VOMIT REMOVAL APPARATUS FOR EVA PRESSURE SUIT.

subjects at ILC including Richard Ellis would have been happy to know that ILC was not considered to provide such a device.

Apollo 9 lunar module pilot Rusty Schweickart was scheduled to do an EVA to demonstrate the new Apollo suits. While circling the earth, he was to exit the lunar module and enter the command module in order to demonstrate the emergency transfer in case the crew of future missions could not dock the two craft after rendezvous from the lunar surface.

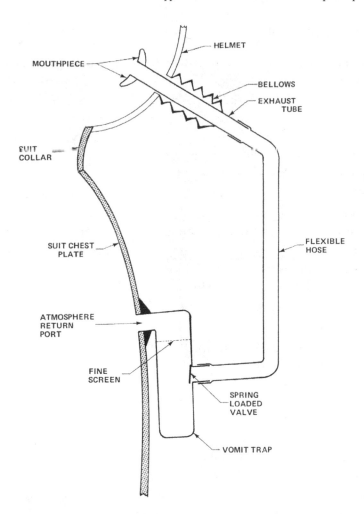

FIGURE 2 - A MODIFIED DESIGN USING RECIRCULATING SUIT ATMOSPHERE

Figure A.28. This was the second possible solution Bellcomm offered for dealing with vomit in the suit. This device would have contained it within the suit instead of expelling it to the vacuum of space. It might have also kept the suit cleaner for closeup TV and photo shots. Courtesy Bellcomm.

Unfortunately for Rusty, he came down with a severe bout of space sickness when the time arrived for him to suit up and perform the task. As a result, the EVA was canceled. He eventually did an abbreviated EVA outside the lunar module while crewmate Dave Scott did a stand-up EVA outside the command module. Schweickart was never placed in the lineup for future missions. Whether or not this fact was a result of his problem on Apollo 9 is unknown.

Miscellaneous Suit Anomalies

As I learned more about the Apollo suits over the years, I took special note of the various failures and anomalies that occurred. I found it particularly interesting that more problems did not occur, given the many systems in the suit that could have failed when subjected to the abuse they took while they were on the lunar surface. I recently had the opportunity to talk with Apollo 16 astronaut Charlie Duke and he asked me if I recalled seeing him and the other moon-walking astronauts falling down, pounding on tools, and generally working hard in the suits. He made it clear to me the astronauts gained a lot of confidence in the suits as they trained in them over the months leading up to their mission and neither he nor any of the other astronauts had a fear that the suit was going to fail. Charlie told me that his biggest fear was of rolling over onto his back as he fell during his lunar EVA because he worried that he could crack the fiberglass backpack and cause damage to components inside, but he said he never had a concern about the suit itself.

As newer materials and assembly techniques emerged into the early 1970s, NASA began to regard many of the features of the Apollo suit as obsolete or as providing an increased risk of suit failure, particularly as missions increased in duration. There was always a great deal of concern about the rubber convolutes even after the premature aging issues were somewhat resolved. Pressure-sealing zippers worked well, but there was always the concern about their dependability and about leakage that proved difficult to prevent. The wire restraint cables were another issue that NASA removed from the design of future suits. But for all of the shortcomings of the Apollo suits, they were still marvelous pieces of space gear.

Figure A.29 shows a malfunction report NASA put together sometime after Skylab that summarizes the issues experienced with the Apollo suits. What is immediately apparent are the higher numbers of problems with the early A5L and A6L model suits. However, these numbers decreased for most issues once the problems had been addressed and the deficiencies had been corrected. This is typical of any complex product that is invented and placed into use. Fortunately, just about all of the issues experienced with the suits were discovered in the labs during testing and a solution was found and implemented before flights.

Suit Component	Total	A5L/A6L	A7L	A7LB/S/L
Cables				
Crotch	19	17	1	1
Glove	4	1	1	2
Knee Convolute	7	3	1	3
Shoulder	10	10	0	0
Boot	2	2	0	0
Convolutes				
Shoulder	4	1	2	1
Knee	4	2	2	0
Thigh	2	1	1	0
Pressure Sealing Zipper Closure	24	13	2	9
Restraint Zipper	10	7	1	2
Vent System				
Torso	5	2	0	3
Boot	1	1	0	0
Boot				
Sole	17	9	8	0
Bladder	2	0	2	0
Restraint Fabric	4	3	0	1
Velcro Pad	1	1	0	0
Vent Attachment	1	1	0	0
Wrist Disconnect	14	10	4	0
Gas Connector	13	10	3	0
Pressure Relief Valve	30	13	6	11
Neck Disconnect	14	13	1	0
Helmet				
Shell	19	1	18	0
Ring Bonding	10	0	4	6
Feed Port	7	3	4	0
Vent Pad	3	2	1	0
Pressure Gauge	11	6	5	0
Glove				
Bladder	9	4	2	3
Gauntlets	2	2	0	0
Vent Attachment	6	6	0	0
Dust Seal	5	4	0	1
Capstan Adjustment	1	0	1	0
Slider Flap	5	N/A	N/A	5
Thermal-Meteoroid Layers*				
Abrasion Layer	5	3	1	1
Cover Layer	17	2	15	0
Hardware	5	1	4	0

*Includes Glove MR's.

APOLLO MALFUNCTION REPORT SUMMARY
Table 3-22

Figure A.29. An Apollo Malfunction Report Summary. Courtesy NASA.

APPENDIX B

Apollo Part Numbers and Serial Numbers

Many labels are found inside an Apollo space suit. Only a trained expert knows where to look for the various labels, such as the label for the pressure-garment assembly or the torso-limb suit assembly or the liner assembly. The Apollo suits are always referred to by the pressure-garment assembly serial number. Some look at the label for the torso-limb suit assembly and think that is the correct number, but it is not.

I am often asked questions about the suit serial numbers and I understand the confusion. They are not consecutive, as one might expect. The A-5L, A-6L, and A-7L suits each began at serial number (S/N) 001 and increased in order until the last suit of that model was made. The rationale was that the model number was the top-level identifier and the serial number identified the specific suit. When ILC started building the model A-7LB suits, they began with serial number 301 and ended with serial number 330, completely skipping the 200 series. An ILC memorandum dated September 3, 1970, suggests that the serial numbers for the new model A-7LB suits should start with 301, since the TMG that was being manufactured for the first suits were starting with the serial number 301. This would have clearly established this new model as a different variety based on the new series of numbers.[1] Thus, with the start of the A-7LB version, ILC chose to start the serial numbers at 301.

The next series of model A-7LB suits were the 401–404 suits, which were reserved for the command module pilot of Apollo 17 and his backup. These suits were made with the modifications needed to accommodate the spacewalk that was performed on this mission.

The next series of serial numbers was 601 through 637, used on the model A-7LB suits to support the Skylab missions. These suits were designed differently to accommodate these missions. The differences were mostly confined to the thermal micrometeoroid garment (TMG). Finally, regarding the 801 through 809 model A-7LB suits that supported the Apollo-Soyuz

Test Program, there is no clear reason why the 500 and 700 series of numbers were skipped. Knowing what I know about ILC configuration management, it makes sense to me that ILC chose not to assign 500 and 700 serial numbers in case NASA provided orders for more suits that would take the numbers up into these ranges. ILC management probably reasoned that it was better to leave room for growth since there was no penalty for doing so and the future was uncertain from where they stood. They may also have avoided these serial numbers because of some development suits ILC was trying out at the time. The numbers would then be available if those suits made it into production.

That brings me to another point I'd like to make. When a particular suit was modified for later missions, perhaps for training purposes, the serial numbers were not changed. The suit always retained the same serial number. A good example is Buzz Aldrin's serial number 036 pressure-garment assembly that ILC initially made for him when he was backup command module pilot on Apollo 8. ILC modified that suit for him to use for training on his Apollo 11 mission, including adding an extra set of connector hardware for the oxygen hoses, the pass-through for the liquid cooling garment, and other upgrades. ILC later modified that suit again as a training suit for astronaut William Pogue for his Skylab mission. Throughout these changes, the suit always had the same serial number, although the dash number following the part number changed to represent the configuration of the suit.

Table B.1 shows the various models of pressure-garment assemblies and their serial numbers and totals. I do not include anything ILC made before the 1965 contest that it won with the model AX-5L suit. This is based on the actual serial numbers of documented suits ILC made for flight, training, and backup and for other evaluation purposes.

ILC made other suits before 1968 that had the identification of "DMU" instead of the serial numbers that would typically be entered. These were called design mobility units, and ILC used them to evaluate new designs the engineers were trying out. They never made it beyond the ILC plant as far as I am aware. There may have only been a few of these and one DMU suit may have been used multiple times to evaluate different engineering updates that the production folks had incorporated. I addressed two examples of model A-6L DMU suits in chapter 5.

ILC was also required to retrofit many suits during the course of the contract. This meant that after ILC had manufactured a suit and sold it to NASA, NASA could return it to ILC with a contract change to change

Table B.1. Breakdown of the 227 suits ILC produced from model AX-5L to model A-7LB

Model	Serial Numbers	Description	Quantity
AX-5L		Suit for the 1965 competition	1
A-5L	001–019		19
A-6L	001–003	Qualification test suits	3
A-6L	004–025		22
A-7L	001–003	Qualification test suits	3
A-7L	004–096	Production suits	93
A-7LB		Design verification testing suits	4[a]
A-7LB		Qualification test suits	2
A-7LB	301–330		30
A-7LB	401–404		4
A-7LB	601–637		37
A-7LB	801–809		9

Note: a. DVT-001 was later used as a play suit for the development of the Skylab suits.

it from an intravehicular suit to an extravehicular suit or vice versa or to perform sizing adjustments to accommodate crew members for training purposes. ILC retrofitted a total of 111 suits during the Apollo contract. This gives you some insight into how much NASA was trying to cut costs by recycling the suits when possible.[2]

When sizing differences or configuration changes were made, ILC would revise the part number (P/N). For instance, one suit might be P/N A7L-100000-22 while another might be P/N A7L-100000-38. There was a total of fifty-two separate dash numbers in the ninety-six model A-7L suits ILC made. These differences between the suits had nothing to do with the serial number sequencing. Table B.6 provides more details.

Each end-item component was given its own serial number. That means that the pressure garment assembly, which is considered the top-level part of the space suit, had its own number. This is the number everyone refers to when talking about a specific suit. For instance, Neil Armstrong's pressure-garment assembly was serial number 056, part number A7L-100000-71. Armstrong's torso-limb suit assembly, which consisted of the entire pressure suit minus the TMG, also happened to be identified as S/N 056, part number A7L-100001. That was a lower assembly part number. Other completed parts that made up that suit each had their own serial numbers that were totally unrelated to the serial number of the pressure-garment assembly. The TMG for the Armstrong suit was serial number 063. The pressure boots were serial number 102.[3] The various parts were made at different times, and as each part was made, it was given its own sequential serial number. That is the reason the parts all had different numbers. Figure B.1 shows the major components used in the Apollo suit by part number at a given time. These part numbers changed from model to model as revisions and upgrades were made. Each of these part numbers included the part numbers and serial numbers of other sub-assemblies. Not included in this diagram are the helmet and gloves that also had their own separate parts with serial numbers or lot numbers.

Suit History and Usage

I was able to locate assorted information that showed suit usage throughout the Apollo program, including the final model A-7L and A-7LB suits. In the sections below, I outline these models and provide what history I know about each of the suit serial numbers. It is not complete, but it includes most of the suits ILC made.

Model A-5L Suits

Table B.2 is a snapshot of how many model A-5L suits were being used around April and May of 1966. The data was obtained from a NASA status report dated April 11–15, 1966. It is not all-inclusive and is somewhat vague concerning the details but it does show how NASA was putting these early model suits to use.

Model A-6L Suits

Table B.3 shows which model A-6L suits were in use in July 1967. At that time, NASA was asking ILC to furnish the new thermal micrometeorite garments that were being designed for the model A-7L suits so they could install them on these older version suits so that they would closely resemble the newer model A-7L suits. They were desperately needed for training the crews for upcoming missions.[4]

The A-6L suit serial numbers 001 was subjected to a burst-pressure test on July 18, 1967. A technical information release report generated by General Electric indicates that they were responsible for doing the testing and reporting the results to NASA. The suit was secured in a large oversized container and taken to its operational pressure of 3.7 pounds per square inch. It was then inspected to make sure it was in good condition. Once the suit had been inspected and given the go-ahead, the technician increased the pressure to 12.5 pounds per square inch, at which point the bladder

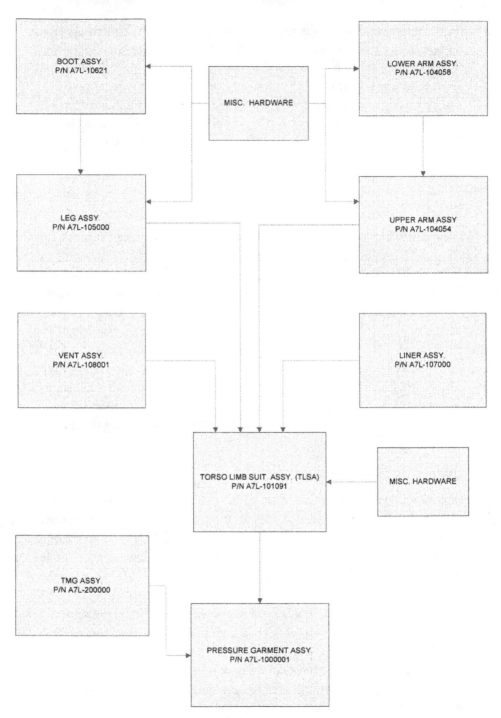

Figure B.1. This diagram gives a general idea of the upper-level parts that came together to form the pressure garment assembly. Courtesy Bill Ayrey.

Table B.2. Use of the model A-5L Apollo suits

Suit S/N	Ship date[a]	Used to support the following:
AX-5L-001	4/15	Testing at CDC[b]
A-5L-002	4/14	Crew member Gordon at Natick Labs
A-5L-003	4/18	Crew member Scott at Hamilton Standard for CO_2 purge testing on 4/16. Antenna testing on 4/15 & 4/16.
A-5L-004	4/11	Metabolic testing at NASA/JSC on 4/12.c Water floatation testing 4/15 to check life vest interface.
A-5L-005	4/26	Size "Peterson" to Grumman for interface testing
A-5L-006	4/27	Size "Jackson" to Grumman for interface testing
A-5L-007	4/22	Crew member Grissom to NASA/Manned Space Center
A-5L-008	4/20	Crew member McDivitt to NASA/Manned Space Center
A-5L-009	4/23	Crew member Armstrong to NASA/Manned Space Center
A-5L-010	5/17	Size "Newell" to Hamilton Standard
A-5L-011	5/1	Size med. regular to North American Aviation for interface testing
A-5L-012	5/6	Size med. long to North American Aviation for interface testing
A-5L-013	5/10	Size large regular to North American Aviation for interface testing

Source: "NASA Status Report: April 11–15, 1966," NASA document.
Notes: a. Proposed ship date from ILC Industries.
b. This is the contest-winning suit ILC supplied to NASA in March 1965. Suit shown as being returned to NASA/JSC to repair broken thigh cable then returned to CDC, Savannah, Georgia, on May 15, 1966. The CDC was the Communicable Disease Center of the U.S. Public Health Service, the predecessor of the Centers for Disease Control and Prevention.
c. JSC = Johnson Space Center.

layer of material failed between the neck ring and the shoulder convolute. Further investigation revealed that the bladder, which is supposed to be structurally supported by the outside nylon restraint layer, was sized too small. Because of that, the weaker neoprene bladder carried the pressure loads until it burst. After suit technicians repaired the bladder by making patches that permitted the loads to be taken by the restraint layer, a second test was done on July 20, 1967, after the repair had been made. This time, the pressure made it up to 17.5 pounds per square inch before leakage suddenly increased significantly enough to stop the test. Failure investigation indicated that the right ankle convolute restraint cable had failed. It was noted in the report that previous damage had occurred to this area from an earlier unrelated test The conclusion was that the ankle convolute restraint

Table B.3. Model A-6L Apollo suits in use in July 1967

Suit S/N	Used to support the following:
A-6L-004	Primary life support system qualification: Hamilton Standard
A-6L-005	Crew member John Bull: LTA-8[a]
A-6L-006	MSC support[b]
A-6L-007	Crew member Clifton Curtis (C. C.) Williams: training S/C 103[c]
A-6L-008	MSC support
A-6L-010	Crew member Wally Schirra training: S/C 103
A-6L-011	Crew member Dave Scott training: S/C 103
A-6L-012	Crew member Rusty Schweickart training: S/C 103
A-6L-013	Crew member Jim McDivitt training: S/C 103
A-6L-014	Crew member James Irwin: LTA-8
A-6L-015	Crew member Tom Stafford training: S/C 101
A-6L-016	Crew member Joe Kerwin: 2TV-1[d]
A-6L-018	Crew member Gene Cernan training: S/C 101
A-6L-019	Crew member Buzz Aldrin: 2TV-1
A-6L-020	Crew member John Young training: S/C 101
A-6L-021	Crew member Donn Eisele training: S/C 101
A-6L-022	Crew member Vance Brand: 2TV-1
A-6L-023	Crew member Walter Cunningham training: S/C 103
A-6L-024	Crew member Pete Conrad training: S/C 103
A-6L-025	Crew member Richard Gordon training: S/C 103

Source: "Suit usage—July 1967," NASA status report document.
Notes: a. LTA = lunar module test article; used to support training in Houston's thermal vacuum chamber.
b. MSC = Manned Spacecraft Center (the former name of the Johnson Space Center).
c. S/C = spacecraft. Each Apollo mission used a designated command module that was given a serial number during production. Apollo 7 used serial number 101, for example. S/C 103 was used on Apollo 8.
d. TV = thermal vacuum chamber. Used by the astronauts and other NASA personnel to test the suits and other equipment under vacuum conditions found in space and on the moon.

Table B.4. Projected deliveries of the Model A-7L suit on July 6, 1967

Suit S/N	Suit Delivery	Retrofit with new ITMG[a]	Use
A7L-001	7/20/1967	8/30/1967	Qualification
A7L-002	7/27/1967	9/6/1967	Qualification
A7L-003	8/3/1967	9/10/1967	Qualification
A7L-004	11/7/1967	1/23/1968	Apollo 7 prime suit
A7L-005	11/14/1967	1/29/1968	Apollo 7 prime suit
A7L-006	11/21/1967	2/2/1968	Apollo 7 prime suit

Source: "Supporting Data RECP 7E152: Incorporate Redesigned TMG on PGA," July 6, 1967, NASA document.
Note: a. Integrated thermal micrometeoroid garment. This garment used the new Beta-cloth cover layer. The first production ITMG was completed in January 1968.

cables in the pressure-garment assembly were likely the weakest point in the pressure-retaining portion of the suit. The report concluded that the results were very favorable, since the burst pressure was 75 percent greater than the contract end-item specification for the Apollo suit.[5]

A "Test Readiness Review" document generated by the NASA Crew Systems Division on November 20, 1967, provides the following information about three model A-6L suits:

> Suit serial number 001, which had been used for the pressure-burst test mentioned above, was repaired and later in the year subjected to design limit exposure testing at NASA's White Sands test facility and at Holloman Air Force Base.
>
> Suit A-6L-002 was used for an eight-foot altitude chamber test to verify its integrity. It was not subjected to thermal or solar simulation.
>
> Suit model A-6L, serial number 003 was subjected to design limit cycle testing at Natick Army Labs in Massachusetts. This consisted of cycling the suit both pressurized and unpressurized for a total of ten mission simulations, each of which included fifteen test profiles. It included donning and doffing the suit as anticipated on each mission.
>
> Suit A-6L-005 and A-6L-014 were used at Grumman Aircraft for altitude verification testing.[6]

Suit A-6L-002 is in the care of the Smithsonian and appears to have been sized for a NASA test subject. It is unclear what happened to the A-6L-003 suit, but it was likely destroyed in the testing that I outlined above. Suit A-6L-009 was modified at ILC to include the newly designed TMGs that were to be installed on the newer model A-7L suit. This suit also incorporated the new heel clips that would lock into the foot restraints of the couches to secure the astronauts' legs during launch and reentry. This suit was scheduled for command module interface testing at North American Aviation on August 17 and 18, 1967.[7] It was sized for Astronaut John Young. It is unknown where suit serial number 017 was at the time.

Model A-7L and A-7LB Suits

In July 1967, NASA outlined projected suit production and assignments for the model A-7L suits. It shows the projected delivery dates of pressure-garment assemblies A-7L-001 on July 20, 1967; suit A-7L-002 on July 27, 1967; and suit A-7L-003 on August 3, 1967. These first three models became the qualification suits that certified the design of the model A-7L suit. An

Table B.5. Record of fit checks for Apollo, Skylab, and Apollo-Soyuz Test Program astronauts

Mission (launch date) and position	Primary astronaut	Flight serial no.	Date of fit check at ILC	Backup serial no.	Fit check date at ILC
Apollo 7, 10/11/68					
Commander	Schirra	004	04/03/68	010	06/19/68
Command module pilot	Eisele	005	05/31/68	011	05/31/68
Lunar module pilot	Cunningham	006	04/29/68	012	04/29/68
Apollo 8, 12/21/68					
Commander	Borman	030	07/22/68	029	07/22/68
Command module pilot	Lovell	037	08/15/68	052	Unknown
Lunar module pilot	Anders	031	08/13/68	032	08/13/68
Apollo 9, 03/03/69					
Commander	McDivitt	020	08/08/68	014	08/08/68
Command module pilot	Scott	019	06/28/68	013	06/28/68
Lunar module pilot	Schweickart	015	12/01/68	021	07/12/68
Apollo 10, 05/18/69					
Commander	Stafford	047	12/11/68	042	10/17/68
Command module pilot	Young	043	10/25/68	048	11/08/68
Lunar module pilot	Cernan	044	10/26/68	049	12/19/68
Apollo 11, 07/16/69					
Commander	Armstrong	056	12/05/68	057	02/08/69
Command module pilot	Collins	033	02/10/69	034	Unknown
Lunar module pilot	Aldrin	077	03/24/69	076	02/26/69
Apollo 12, 11/14/69					
Commander	Conrad	065	03/03/69	068	04/21/69
Command module pilot	Gordon	066	03/03/69	059	Unknown
Lunar module pilot	Bean	067	02/20/69	070	05/20/69
Apollo 13, 04/11/70					
Commander	Lovell	078	Unknown	074	02/28/69
Command module pilot (prime)	Mattingly	059	8/18/1969	082	8/18/69
Command module pilot (backup)	Swigert	088	07/29/69	X	X
Lunar module pilot	Haise	061	08/20/69	060	12/05/68
Apollo 14, 01/31/71					
Commander	Shepard	090	01/28/70	084	06/16/69
Command module pilot	Roosa	085	07/07/69	091	12/29/69
Lunar module pilot	Mitchell	073	07/02/69	046	01/06/69
Apollo 15, 07/26/71					
Commander	Scott	315	02/26/71	319	04/14/72
Command module pilot	Worden	094	04/20/70	081	05/12/69
Lunar module pilot	Irwin	320	04/16/71	316	04/11/71

Training serial no.	Date of fit check at ILC	Backup astronaut	Primary serial no.	Date of fit check at ILC	Training serial no.	Date of fit check at ILC
		Stafford	007	Unknown		
		Young	008	11/17/69		
		Cernan	009	06/20/68		
024	Unknown	Armstrong	035	08/20/68		
055	Unknown	Aldrin	036	08/22/68		
025	Unknown	Haise	051	10/27/68		
		Conrad	016	06/27/68		
		Gordon	017	06/28/68		
		Bean	018	06/24/68		
		Cooper	045	11/07/68	071	04/24/69
		Eisele	011	05/31/68		
		Mitchell	046	01/06/69		
035	05/17/69	Lovell	074	02/28/69		
026	05/21/68	Anders	054	12/06/68		
036	05/19/69	Haise	060	12/05/68		
		Scott	063	07/16/69		
069	Unknown	Worden	081	05/12/69		
		Irwin	080	07/16/69		
075	Unknown	Young	086	08/06/69		

Bumped from mission

Flew mission

Training serial no.	Date of fit check at ILC	Backup astronaut	Primary serial no.	Date of fit check at ILC	Training serial no.	Date of fit check at ILC
		Duke	062	11/03/69	021	11/03/69
		Cernan	087	08/28/69		
028	Unknown	Evans	092	01/27/70		
041	Unknown	Engle	064	12/01/69	015	12/01/69
311	12/04/70	Gordon	317	03/05/71	313	12/02/70
058		Brand	096	06/15/70	055	Unknown
312	12/02/70	Schmitt	318	04/00/70	076	12/15/70

(continued)

Table B.5—*Continued*

Mission (launch date) and position	Primary astronaut	Flight serial no.	Date of fit check at ILC	Backup serial no.	Fit check date at ILC
Apollo 16, 04/16/72					
Commander	Young	322	07/16/71	326	09/03/71
Command module pilot	Mattingly	082	08/18/69	072	
Lunar module pilot	Duke	327	09/13/71	323	07/19/71
Apollo 17, 12/06/72					
Commander	Cernan	328	03/01/72	330	09/20/72
Command module pilot	Evans	404	04/26/72	402	03/07/72
Lunar module pilot	Schmitt	329	05/03/72	318	03/02/72
Skylab 2, 05/25/73					
Commander	Conrad	614	05/10/72	602	10/06/71
Scientist pilot	Kerwin	615	05/16/72	631	09/22/72
Pilot	Weitz	617	Unknown	616	05/22/72
Skylab 3, 07/28/73					
Commander	Bean	632	10/12/72	620	06/22/72
Scientist pilot	Garriott	634	Unknown	621	06/29/72
Pilot	Lousma	635	Unknown	622	07/10/72
Skylab 4, 11/16/73					
Commander	Carr	626	08/10/72	635	Unknown
Scientist pilot	Gibson	628	08/28/72	636	11/14/1972[a]
Pilot	Pogue	627	08/15/72	637	Unknown
Skylab R, Sep. 1973					
Commander	Brand	623	07/18/72	611	11/09/71
PLT	Lind	613	04/10/73		
Apollo-Soyuz Test Project, 07/15/75					
Commander	Stafford	801	07/30/73	802	Unknown
Command module pilot	Brand	806	09/21/73	807	10/18/73
DMP	Slayton	803	08/24/73	805	11/07/73

Miscellaneous fit check records			Notes/comments:
Kerwin	307	01/12/71 and 03/31/71	300-series: Apollo program evaluations for Skylab
Bean	308	02/17/71 and 09/8/71	
Garriott	633	10/06/72	
McCandless	619	06/15/72	
Schweickart	617	06/01/72	Fit check in flight Weitz used before mission
Musgrave	618	06/07/72	Evaluated same as McCandless on the same day
Jack Mays	601	8/27/1971	First fitting of a Skylab suit to NASA test subject

Training serial no.	Date of fit check at ILC	Backup astronaut	Primary serial no.	Date of fit check at ILC	Training serial no.	Date of fit check at ILC
303	10/16/70	Haise	324	07/22/71	305	06/01/71
059	8/18/69	Roosa	401	09/07/71	085	07/07/69
304	09/15/70	Mitchell	325	08/10/71	321	04/22/71
315	03/01/72	Young	326	09/03/71	303	10/16/70
064		Roosa	401	09/07/71	085	07/07/69
314	12/15/70	Duke	323	07/19/71	304	07/19/71
310[1]	10/28/70	Schweickart	605	10/19/71		
603	10/07/71	Musgrave	606	10/02/71		
604	10/13/71	McCandless	618	06/07/72	607	10/29/71
608	12/20/71	Brand				
609	11/10/71	Lenoir	624	08/16/72	612	12/01/71
610	11/17/71	Lind	625	Unknown	613	Unknown
		Brand	611	11/09/71		
309	10/29/70	Lenoir	624	08/16/72	612	12/01/71
		Lind	625	08/01/72	613	12/14/71

Not flown. Rescue mission originally planned to retrieve Skylab 3 crew.

		Bean	809	11/13/73		
403	Unknown	Evans	804	09/11/73		
		Lousma	808	11/13/73		

Notes: Note on Skylab suit program: Earliest contract note says suits must be made available for 07/71 altitude chamber testing. Altitude chamber testing took place on 01/17/73 with Skylab crew Conrad, Kerwin, and Weitz. They used suits 614, 615, and 616, respectively. Thus, it appears that 616 was marked as Weitz's flight suit but was changed out at some point to suit 617. Special thanks to Mr. Rolf Schoevaart for his assistance in reviewing and making additions to this chart.

a. The fit check of astronaut Gibson on 11/14/1972 was the final fitting ILC performed on an Apollo suit. A celebration was held at ILC to mark the occasion of this 190th fit check.

Table B.6. Model A-7L suit dash numbers and serial numbers

Dash no.	Serial nos.	Dash no.	Serial nos.
17	024, 025, 027	61	053
18	012	63	042, 044
20	003	67	011, 043
22	001	71	056
24	015, 016, 018	72	058
25	026, 028	80	049
28	006	83	077
29	005, 008	84	057, 075, 076
30	022	85	033
33	031, 035, 036, 051	86	074
34	040	87	083
38	002	89	055, 059
40	004, 029, 030	90	060
41	023	93	090
42	050	96	066
43	034	98	065, 067, 070
44	007, 009	99	088
45	041	104	062
47	047	108	032
50	038	109	010
51	073	110	020
53	054	111	052
54	021	112	048, 069
55	039	113	019
56	013, 037	114	064
57	014	115	085

Source: NASA and ILC documents.

ILC malfunction report dated December 4, 1968, referred to model A-7L suit S/N 002 as size "Young Special" and documents that a right boot sole failed during qualification testing (it cracked at the "Z" plate). The disposition was to perform a repair using a semi-flexible epoxy adhesive named Scotchcast 8).[8]

I compiled Table B.5 using fit-check records to determine where suits were resized and reused by various crew-members as NASA attempted to reduce costs by avoiding the need to procure new suits. When I was done putting the spreadsheet together, I found an interesting tale of how the suits were used. These fit-check records also provided the history of almost every suit ILC manufactured and who it was made for. The following model A-7L suits were not found in any of the records. However, in some cases I have the names of the crew members or others that would have appeared on the label or I located information about their use. They are as follows:

S/N 022: Fred Haise
S/N 023: Russell Schweickart
S/N 027: Unknown
S/N 038: Don Lind
S/N 039: Jack Mays (NASA test subject)[9]
S/N 050: Ron Evans
S/N 053: Used to qualify the new ILC arm bearing before the Apollo 11 mission[10]
S/N 069: Richard Gordon
S/N 079: Bill Anders
S/N 083: Harrison Schmitt
S/N 089: Unknown
S/N 093: Dave Scott

The Model A-7LB suits started with two suits made for design verification and certification testing. Serial numbers A-7L-302 and 306 were used for destructive testing. Because ILC used both of these suits for mission qualification testing of this suit model in the laboratory environment, they racked up significant hours of use. ILC performed a number of destructive tests after all manned qualification testing including cable tensile testing, seam tensile testing, and peel testing of the various fabric layers so that empirical data could be compiled in order to show the effects of long-term cycle usage on space suit materials.[11]

APPENDIX C

Apollo Contract Details

This appendix addresses specific details of the contracts NASA awarded ILC, including serial numbers, quantities made, costs, and contract challenges.

Costs

The price per space suit changed throughout the Apollo contract because of a host of issues. On the day in August 1968 that the Apollo suit was first presented to the public, ILC vice-president Nisson Finkelstein announced that the price was "about $217,000 per suit" ($1.6 million in 2019 dollars).[1] A NASA memo dated May 10, 1972 from Chief of the Crew Systems Division Robert Smylie broke down the price per suit. At that time, the cost for each suit, including all research and development, field support, spares, testing, retrofit costs, changes, maintenance, and repair, was calculated to be $300,580 ($1.8 M in 2019 dollars).[2] Cost of the suit alone was $150,000 ($928,000 in 2019 dollars).

The memo broke down the price for suits that were already in inventory but had been retrofitted "for dual usage as an additional suit." The inventory was 79 A-7Ls, 15 A-7LBs, and 17 modified command module pilot suits (total = 111) at a cost of $197,920 per suit ($1.2 million in 2019 dollars). This was for the base suit with no spares, field support, and so forth (as included in the first cost above) and included modifications only.

A suit consisted of the following elements:

Integrated torso-limb suit assembly
Helmet
Extravehicular activity gloves
Intravehicular activity gloves
Lunar boots
Lunar visor assembly
Liquid cooling garment
Purge valve
Urine collection transfer assembly
Fecal containment system

Contract Challenges

The Apollo suit program operated under the provisions of a cost-plus-award fee contract. This was not to ILC's advantage early in the program because of the many unknowns ILC and NASA encountered as they navigated the design and manufacture of such a new system.

Both ILC and NASA were learning that this process was nearly impossible to manage because of all the challenges associated with building space suits. There was no way to predict when a suit would be completed based on established schedules, primarily due to the need to accommodate the individual needs and desires of the astronauts, but also due to other changes to the suits necessitated by evaluations of how they interacted with the command and lunar modules.

A note found in a copy of the contract details from 1968 highlights the challenges ILC and NASA faced regarding how the fees were supposed to be awarded but how in reality they were not.

> The development of spacesuits that will serve the dual purpose of long duration flight plus extravehicular exploration of the moon, with its extremes in environment, poses an extremely complex undertaking. When these already existing complexities are greatly multiplied by having to develop and incorporate nonflammable materials, more ease of mobility features, and integration of the outer and inner pressure garments to eliminate excessive bulkiness, a task is created that requires outstanding management and technical capabilities to realize acceptable results. Within the first nine months, 36 configuration changes had been made and by December 1967, an additional 103 changes were directed. The magnitude of changes created substantial additional risks on the part of the contractor. He was working under a contract that no longer provided him with an opportunity to earn incentive fee. In spite of the incentive motivation being removed, the contractor continued his best efforts to accomplish the objectives of the contract and to commit all required facilities and resources to the contract effort. It is reasonable to assume that a contractor willing and capable of working under these circumstances, and still able to control his costs, would achieve a high average overall performance.[3]

As a result, NASA made changes to the contract so that ILC was in a better position to win incentive rewards, which they did as time progressed. Despite the absence of rewards during the initial years, NASA recognized

ILC for their ability to persevere despite the many obstacles. The part of the quote above that states "a task is created that requires outstanding management and technical capabilities to realize acceptable results" is a true compliment to Len Shepard, Nisson Finkelstein, and the workers who were doing their best to move the program along through a very dynamic process.

NASA paid the basic minimum fee per the delivery schedule, but ILC also earned an award fee based on their performance against a host of criteria that NASA management determined. An Incentive Evaluation Board was established at NASA to review ILC's progress in meeting the criteria at six-month intervals. The Incentive Evaluation Board included the chief of the Crew Systems Division and the appropriate administrative division chief. It is unknown who else many have been on this board. They received input about how well ILC was performing from the Incentive Evaluation Committee, which consisted of members who were involved in day-to-day interactions with ILC. It included the NASA resident engineer, the contracting officer, and so forth.

The award fee was based on a numerical rating that ranged from 60 (unsatisfactory) to 100 (excellent). A percentage of the possible maximum fee was established based on the numerical rating: at the rating of 60, NASA paid 0 percent of the possible incentive award and at 100, it paid 100 percent. On this sliding scale, a "good" rating of 85 would yield 62.5 percent of the possible award fee.

The categories that were evaluated included such things as program management, engineering, manufacturing, quality/reliability, and product support. Each category was weighted: program management had a weight factor of 30, engineering was weighted at 20, and product support was weighted at 10. The total was 100. As complex as this may have sounded, it worked out to everyone's advantage and ILC typically won decent award fees.[4]

Apollo Suit Requirements

Early in the Apollo program, NASA established requirements for all the systems to be used, including the spacecraft, all the miscellaneous support gear, and the suits. The following items highlight the most significant suit requirements (Note: It is very possible that any of the requirements that I list here were changed at some point in the process.)

The design had to include provisions to allow the astronauts to be continually suited and pressurized within the spacecraft if pressurization was lost onboard the command module during rendezvous activities with the lunar module. Provisions had to be made for the suit to accommodate the crew for up to 115 hours to return them to Earth.[5] This included the ability to remove urine and provide for food and water to be consumed, all of which had to happen while the crew was pressurized in their suits. A feed port was provided in the side of the pressure helmet that permitted the insertion of a straw-like device that supplied either water or food the consistency of toothpaste that could be squeezed into the mouth. A fecal containment bag was also supplied to be worn during suit-up and during certain phases of the mission.

A self-sealing injection patch had to be provided that would allow an astronaut to inject medications into their leg if needed yet seal the suit from leakage. This might include medicine for motion sickness. Again, this would be if the astronauts could not take off their suits due to the inability to pressurize the command module.

Suit Weights: In the Operations Handbook for the Apollo suits dated July 1968, the weight requirements for the model A-7L suits was listed as 41.87 pounds maximum for the extravehicular suits and 34.33 pounds for the intravehicular suits.[6] A NASA printout sheet from one year later that lists the actual weights of the Apollo 11 flight items shows that the extravehicular suit weighed 43.85 pounds, including the intravehicular gloves, the pressure helmet, and the communication carrier assembly. The Apollo 11 intravehicular suit weighing 33.43 pounds, including the helmet, the intravehicular gloves, and the communication carrier.[7] For the Apollo 13 mission, the weight requirements were listed as 55.29 pounds maximum for the extravehicular suits that included intravehicular gloves, the helmet, the communications carrier, extravehicular gloves, and lunar boots. The intravehicular suit was listed as 37.15 pounds maximum, which included intravehicular gloves, the helmet and the communications carrier. The liquid cooling garment had a dry weight of 5.0 pounds maximum.[8]

APPENDIX D

Astronaut Code Names

On May 10, 1968, NASA released a memo by Apollo Support Branch chief C. C. Lutz that provided a list of secret code names for the various Apollo astronauts. Alan Shepard, head of the Astronaut Office, drove the decision because he did not want anyone other than a few key people to know who had been selected for the earlier Apollo missions. He did not even want the astronauts he selected to know their assignment until the time was just right, in case changes had to be made for whatever reason. The Astronaut Office, however, needed to make sure that the customized suits were available when they were needed for missions, so they had to somehow ensure that ILC was in the process of making them even before the names were released. When ILC received the secret code names, they could start suit production. Len Shepard was the only one at ILC who was given the names, and it was up to him to make others under him aware on a need-to-know basis only. Once the Astronaut Office officially released the names for the missions, the code words were no longer used.

UNITED STATES GOVERNMENT

Memorandum

TO : Record

DATE: MAY 1 0 1968

FROM : EC9/Chief, Apollo Support Branch

SUBJECT: System for coding Apollo Crew members for future suit procurement

In order to establish a system wherein Apollo flight suits can be fabricated by ILC Industries without disclosing the identity of the selected crew members, the following system will be implemented.

    a. The contract presently specifies that suit sizes shall be as defined by NASA at the time of fabrication.

    b. A list of "code words" has been formulated by Crew Systems Division and assigned to each of the potential astronauts. The assignment of these code words will be known to an absolute minimum number of MSC and ILC Industries personnel. Initially these people shall be Mr. Alan B. Shepard (CB), Mr. C. C. Lutz (CSD), and Mr. Len F. Shepard (ILC Industries).

    c. When fabrication of the flight suits is required to begin, Mr. Alan B. Shepard will reveal the names of the selected individuals to the CSD representative. Fabrication will then proceed and all future actions concerning the suits will be conducted using the appropriate code words only. This procedure will remain in effect for each crew until official NASA release of the crew identities, at which time use of the code words will be discontinued.

The list of code words assigned to the potential crew members will be hand carried to the above mentioned personnel by the CSD representative in the near future.

C. C. Lutz

EC1:WLDraper:hd   5-9-68

Name	Code	Name	Code
ALDRIN	POLLUX		Earth
ANDERS	REGULUS	Musgrave	
ARMSTRONG	SIRIUS	MICHEL	ELNATH
BEAN	VEGA	MITCHELL	MENKAR
BORMAN	RIGEL	McCANDLESS	POLARIS
BRAND	SPICA	POUGE	ALGOL
BULL	ALTARES	ROOSA	CASTOR
CARR	THUBAN	SCHMITT	DENEB
CERNAN	SCHEAT	SCHWEICKART	HAMAL
COLLINS	MIZAR	SCHIRRA	MARKAB
CONRAD	PERSEI	SCOTT	MIRA
COOPER	RASALHAGUE	STAFFORD	SCOPII
CUNNINGHAM	ARCTURUS	SWIGERT	MIRFAK
DUKE	MEGREZ	WEITZ	ALKAID
EISELE	FOMALHAUT	WORRAN	ALCOR
ENGLE	ALTAIR	YOUNG	ALBIREO
EVANS	TAURI		
GARRIOTT	HYDRAE		
GORDAN	ORIONIS		
HAISE	PEGASI		
IRWIN	LYRAE		
KERWIN	DENEBOLA		
LIND	ALMACH		
LOVELL	BELLATRIX		
LOUSMA	BETELGEUSE		
MATTINGLY	CANOPUS		
McDIVITT	CAPELLA		
Lenoir	Alcyone		
Gibson	Merak		

Above: Figure D.1. NASA memo outlining the system for providing code names to astronauts in order to keep the crew assignments secret. Courtesy NASA.

Left: Figure D.2. List of astronauts and their assigned code names. Courtesy NASA.

APPENDIX E

A-7LB Suit Plug Loads and Fabric Stresses

Figure E.1 shows the loads and stresses the Apollo suit was subjected to when pressurized. Although the suits were pressurized to only 3.75 pounds per square inch in use, the report included loads at much higher pressures to make the point that the A-7LB suit could be used as an advanced suit if the need arose. Future suit development focused on operating pressures as high as 8.0 pounds per square inch.[1]

REPRESENTATIVE PLUG LOADS AND FABRIC STRESSES

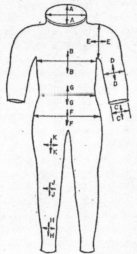

NOTES:

1. All measurements taken from A7LB-308 (Bean) at 3.75 PSIG.

2. Loads given are static plug loads (lbs)

3. Stresses given are fabric hoop stresses (lbs/in)

4. Longitudinal stresses are 1/2 of hoop stresses

LOCATION	CIRCUM.	SECT. AREA	LOADS & STRESSES				
			@3.75 PSIG	@6 PSIG	@8 PSIG	@16 PSIG	@20 PSIG
NECK RING A-A	31.0"	76.5 in²	18.5 lbs/in / 387.1bs	29.6 lbs/in / 459.0 lbs	39.6 lbs/in / 612.0 lbs	78.8 lbs/in / 1223.0 lbs	98.6 lbs/in / 1530.0 lbs
UNDER ARMS B-B	45.5"	164.8 in²	27.15 / 618.0	43.4 / 987.0	57.9 / 1319.0	115.8 / 2640.0	144.8 / 3290.0
WRIST OPENING C-C	12.17"	11.8 in²	7.58 / 44.2	12.12 / 70.9	16.16 / 94.5	32.32 / 188.2	40.40 / 236.0
ELBOW D-D	16.75"	22.25 in²	10.0 / 83.6	16.0 / 134	21.3 / 178.6	42.6 / 357.0	53.3 / 446.0
SHOULDER E-E	27.75"	61.5 in²	16.55 / 231	26.45 / 369	35.3 / 492	70.6 / 984.0	88.2 / 1230.0
BUTT F-F	46.50"	172.32 in²	27.75 / 646.0	44.4 / 1033.0	59.2 / 1378.0	118.3 / 2756.0	148.0 / 3444.0
WAIST G-G	43.0"	146.80 in²	35.63 / 552.0	41.1 / 880.0	54.7 / 1172.0	109.4 / 2344.0	136.7 / 2935.0
ANKLE H-H	18.0"	25.62 in²	10.74 / 95.0	17.2 / 153.50	22.93 / 205.0	45.86 / 410.0	57.3 / 512.0
KNEE J-J	22.5"	40.25 in²	13.4 / 151	21.5 / 241.0	28.62 / 322.0	57.24 / 445.0	71.6 / 805.0
THIGH K-K	28.0"	62.30 in²	18.7 / 233	29.73 / 373.0	39.61 / 497.0	71.22 / 994.0	98.85 / 1246.0

D-2

Figure E.1. This image shows the plug loads and fabric stresses in various locations on the Apollo suit. The plug load is defined as the area at the specified section of the suit multiplied by the internal pressure with the result being pounds per inch. The fabric stresses are the loads across the stitched seams at the specified locations expressed in pounds per inch.

TIMELINE OF SIGNIFICANT EVENTS

1937: Abram Spanel forms a new company in Dover, Delaware, named International Latex Corporation (ILC).

1947: Abram Spanel creates the Government Industrial Division at ILC; it begins work on MA-2 pressure helmets.

1952: ILC Engineer Len Shepard begins work on what he calls the first true space suit.

1956: George Durney is hired to work on pressure helmets and begins working with Len Shepard to design the early space suits.

1956–1961: Work continues on three development contracts to design a space suit for air force personnel. George Durney plays a critical role in developing pressure suits for these contracts.

1959: ILC Industries enters the prototype model SPD-117 space suit for the Mercury Program contest. Although their suit did not win, it was recognized as an advanced mobility space suit.

1962: ILC Industries enters the prototype model AX-1L suit for the Apollo contest and wins against seven other companies. They are forced to team as subcontractors with Hamilton Standard. From 1962 to 1965, they work on the prototype models A-1H to A-4H suits.

1963: ILC gets a small contract directly from NASA to build three prototype model A-2L space suits.

1964: NASA decides that all Apollo missions up to the first lunar landing will use David Clark Company suits. NASA refers to these as the Block I missions. It plans to use the Hamilton/ILC suits for Block II missions.

February 1965: Hamilton Standard drops ILC for nonperformance. NASA reopens the contest for the Apollo suit, assuming that the competition is just between Hamilton Standard and the David Clark Company.

Hamilton teams with B. F. Goodrich for the suit. ILC convinces NASA to allow them to enter a suit into the contest.

March 1965: ILC enters the contest and wins with their model AX-5L suit.

November 1965: ILC deliverers the first two model A-5L suits for testing. It builds a total of nineteen suits for training and other evaluations.

October 1966: ILC begins testing two model A-6L suits that are intended to be the first models to be worn on the moon.

January 1967: The Apollo 1 fire occurs, killing the three crew members and causing NASA to reevaluate the entire program. Changes to the Apollo suit design occur when NASA decides to drop the Block I and Block II designations and they contract with ILC to design the suit to be used on all future Apollo missions. This suit is designated the Model A-7L.

March 1967: ILC begins to work with a new high-temperature Beta-cloth cover layer.

November 1967: ILC engineers start looking into improvements to the model A-7L suit that will increase mobility and fix other problems. They call it the Omega suit.

January–August 1968: Qualification of the model A-7L pressure-garment assembly.[1]

September 12, 1968: ILC Personnel present their new and improved Omega III suit concept to NASA administration in Houston, Texas. NASA accepts the suit, which becomes the model A-7LB.

October 1968: Qualification for Mission C Prime (Apollo 7) suit

November–December 1968: Qualification for Mission D (Apollo 9) suit

February–June 1969: Qualification for Mission G (Apollo 11) suit

April 1969: Qualification for ILC-manufactured arm bearing for Mission G and up.

July 20, 1969: Wearing model A-7L suits, Neil Armstrong and Edwin (Buzz) Aldrin become the first humans to walk on the moon.

September 1970–June 1971: Model A-7LB suit qualification (Missions 15 through 17)

September 1971–February 1972: Model A-7LB suit qualification for design change for Apollo 16 (modification to the leg restraints of John Young's suit) and suit qualification for Skylab missions.

May 1973–November 1973: Skylab missions using model A-7LB modified suits

July 1975: Apollo-Soyuz Test Program missions using model A-7LB modified suits.

July 1975: ILC Industries moves their location to the Frederica facility and closes its doors at the Pear Street plant in Dover. The workforce is reduced to twenty-five people from the high of 900 in 1968.

1976: ILC wins the contract to build the soft-goods parts of the space shuttle suits as the subcontractor to Hamilton Standard. The workforce steadily increases and the product lines are significantly diversified over the next forty-five years.

GLOSSARY

Apollo-Soyuz Test Program: A single joint mission with the Russians that used excess Apollo rocket and capsule hardware to link US spacecraft with a Soyuz craft. The mission lasted nine days and symbolically provided an end date to the space race.

Astronaut life support assembly: This module, which the Skylab astronauts wore on the front waist section of the suit during their extravehicular activity, provided the oxygen, power, communications, cooling, and so forth that was previously contained in the backpacks of the crews on the lunar surface missions. Manufactured by AiResearch, it provided a more compact design and optimal placement.

Buddy secondary life support system: This emergency umbilical system could provide cooling water to the liquid cooling garments in two suits simultaneously when the primary life support system on one suit failed. The system was created in response to mission plans for crew members to venture increasing distances from the lunar module. Because of the heat energy generated inside the suit, any member who lost water cooling capability would face serious consequences.

Command module pilot: This crew member was in charge of the command module and stayed on board as the lunar module pilot and the mission commander flew the lunar module to the moon's surface to perform their mission there.

Communication carrier assembly (also referred to as the Snoopy cap): This headpiece contains the microphones and earcups that transmit all the communications between the astronauts and between them and the ground crew.

Constant wear garment: The cotton undergarment worn inside the suit during launch and other suited activities inside the command module and during all other unsuited activities on board throughout the missions.

Design mockup unit: The first version of a new model Apollo suit that was made to evaluate the strengths and weaknesses of the new version.

Environmental control unit: The air-conditioning system on the lunar module and the command module. The suits connected by way of the umbilical hoses that provided the airflow into and out of the suit.

Extravehicular mobility unit: The combination of the space suit and the primary life support system (also referred to as the backpack).

Extravehicular visor assembly: The helmet assembly that contains all the protective solar visors and shields. This fits over the pressure helmet (also referred to as the bubble helmet).

Fecal containment system: The underwear-like device that was designed to contain body waste and could be worn during extravehicular activity or in emergency situations where the crew might be suited for extended periods of time due to lack of pressure in the spacecraft.

In-suit drink bag: A bag containing drinking water that sits inside the suit in the front side just below the neck ring. A drink tube connected to the bag enables the astronaut to take a sip of water when needed during extravehicular activity.

Integrated thermal micrometeoroid garment: The outer layers of the space suit that are designed to be integrated onto the pressure garment assembly that is permanently installed throughout the mission. Cord is laced through loops stitched onto both the pressure garment and the thermal micrometeoroid garment, securing them together.

Intravehicular cover layer: The outer layers of the pressure garment designed for the command module pilot. The model A-7L suits used fewer layers than the extravehicular suits since this crewmember was not performing extravehicular activity and needed this garment only to protect from the potential of fire on board the capsule on the launchpad. The A-7LB suits used extravehicular cover layers for Apollo 15 through 17 since these suits performed extravehicular tasks.

Johnson Space Center, Houston, Texas: Originally named the NASA Manned Spacecraft Center; the name changed to the Johnson Space Center in 1973. This is the location where training and research for human space flight and flight control is done.

Kennedy Space Center: The launch facility for all of the manned missions located in Merritt Island, Florida. The name was changed from Cape Canaveral and was dedicated to President Kennedy in November 1963.

Liquid cooling garment: The spandex material garment that contains approximately 300 feet of 1/8" vinyl tubing and contacts the wearer's body. Cooled water running from the primary life support system through the garment removes excess body heat during an astronaut's extravehicular activity.

Lunar extravehicular visor assembly: The shell that covers the pressure helmet during extravehicular activity and contains all the visors.

Modular equipment transporter system (METS): Used on Apollo 14, this two-wheeled cart was pulled behind the crewmember during lunar surface extravehicular activity, enabling them to carry small tools, cameras and extra film cartridges, and sample containers. It also served as a portable work bench.

Manned Spacecraft Center, Houston, Texas: Opened in 1962 and renamed the Johnson Space Center in 1973. It still serves as the nerve center for human space missions.

National Advisory Committee for Aeronautics: Formed in March 1915 to study the problems and solution associated with flight. Absorbed by NASA in 1958.

National Air and Space Museum: Smithsonian museums that house and care for the nation's most significant aviation and space equipment. One museum is located on the Mall in downtown Washington, DC, and the other, the Steven F. Udvar-Hazy Center, is located in Chantilly, Virginia.

Oxygen purge system: This emergency system supplies a high flow of oxygen into the suit if a problem occurs with the primary life support system. Astronauts can pull a red ball to initiate the flow from a separate, high-pressure oxygen tank inside the primary life support system.

Portable life support system: The unit worn on the back of the suit that carries all the life support equipment the astronaut needs throughout extravehicular activity. It contains the breathing oxygen, liquid cooling garment water cooling system, batteries, and communications devices. It also scrubs out moisture and carbon dioxide gas that results from respiration. Also referred to as the primary life support system through-out the book.

Pressure-garment assembly: Typically refers to the entire space suit minus the primary life support system. Typically does not include the gloves, helmet, and other support gear.

Remote control unit: The controller attached to the front chest area of the suit with webbing straps and clips. The RCU displayed the oxygen level warnings and information about suit pressure conditions, controls for the radio, an oxygen purge valve mounted on the side, and a bracket to attach the Hasselblad cameras used to film many of the activities on the lunar surface during extravehicular activity.

Space suit assembly: All of the parts that make up the space suit, minus the primary life support system. Combined with the primary life support system, it is typically referred to as the extravehicular mobility unit (EMU).

Thermal micrometeoroid garment: The outer layers of the suit that protect against solar and cosmic radiation, micrometeoroid impacts, fire, and abrasion.

Torso-limb suit assembly: The Apollo pressure garment assembly minus the thermal micrometeoroid garment layers and support equipment such as the helmet and gloves.

Urine collection device: A bag within the space suit that collects the urine of male astronauts by way of a roll-on cuff.

Urine collection transfer assembly: The hardware that consists of the urine collection bag and the hardware that connects the bag to the outlet pass-through port on the right leg of the space suit so that urine could be removed from the bag and disposed of by venting outside the spacecraft.

NOTES

Introduction

1. "Successes of the 20th Century," in *Technology Triumphs, Morality Falters*, Pew Research Center, accessed July 3, 1999, www.people-press.org/1999/07/03/successes-of-the -20th-century.

2. D. F. Bruce, Contracting Officer, to Record, June 6, 1968, page 6 of enclosure 6, paragraph 3, NASA memorandum. The Apollo space suits could not function without the benefit of the oxygen, cooling capabilities, and communications that the primary life-support system, or backpack, as it is often referred to, provided. That device was a separate unit designed and manufactured by Hamilton Standard during the Apollo missions. Because my primary knowledge and interest lies in the dynamics of how the Apollo-era space suit evolved from the earliest designs to its manufacturing challenges and finally to its use on the moon, this work is focused on the ILC Industries space suits and the people who consistently met the challenges of making them. Aside from a few instances where I discuss the primary life-support system, I leave the details of these systems to those whose job it was to provide them.

Chapter 1. School of Hard Knocks

1. From a Playtex document titled "The Historical Development of Our Company," date and author unknown.

2. "Dover, Delaware, Becomes Home of New Basic Industry," circa 1937, typescript, Delaware Historical Society.

3. Westbrook Pegler, "More Details of Liferaft Deal," *Santa Fe New Mexican*, January 31, 1946, 4.

4. Ibid.

5. ILC Industries, "Mobile Anthropomorphic Protective Assembly for Space Exploration," technical proposal for RFP no. AF33(657)-62-5711Q, 62, ILC document.

6. *The Evening Sun*, December 3, 1962, Carroll County Section, 1.

7. Author's interview with Lawrence Ornston, October 21, 2006. Ornston was a vice-president of ILC throughout the Apollo program and a close friend of Shepard.

8. Author's interview with Edward Biter (stepbrother of George Durney), March 28, 2008.

9. Homer Reihm to Ms. Belena Chapp, University of Delaware, April 30, 1999, in author's possession.

Chapter 2. Developing the State-of-the-Art Space Suit

1. An excellent resource for further reading on this subject is Dennis R. Jenkins, *Dressing for Altitude* (Washington, DC: National Aeronautics and Space Administration, 2012).

2. Dennis R. Jenkins, *Dressing for Altitude: U.S. Aviation Pressure Suits, Wiley Post to Space Shuttle* (Washington, DC: National Aeronautics and Space Administration, 2012), 240–241.

3. Ibid., 243

4. NASA history interviews with Mel Case and Leonard Shepard, ILC Industries, April 4, 1972, Dover, Delaware, 24, NASA History Office, Washington, DC. Interviewer unknown.

5. International Latex Corporation, "Mobile Anthropomorphic Protective Assembly for Space Exploration," document IPD-21-62, 63. Addressed to Aeronautical Systems Division, Wright-Patterson Air Base, Ohio, April 9, 1962.

6. Author's interview with Homer Reihm, Apollo space suit engineering manager, ILC Industries, Dover, Delaware, July 23, 2018.

7. Interviews with Mel Case and Leonard Shepard, 24.

8. ILC used various synthetic rubber–based recipes in the construction of the suits, including neoprene, the trade name given to it by its inventor, DuPont, in 1930.

9. These failures extended even late into the development of the A-7L suit, although the later failures were attributed to the latex rubber formulation process, which resulted in premature degradation of the product, as detailed in chapter 10.

10. Interviews with Mel Case and Leonard Shepard, 25.

11. Each mission that NASA sets out on has its own unique set of challenges, and the thermal micrometeoroid garment is usually one of the last systems to be designed. That is because the design of a thermal micrometeoroid garment that will go to the moon is different from the design of the thermal micrometeoroid garment that will go to Mars, since Mars has a very slight gas environment. This gas contributes to some convective heating and cooling that helps carry heat to or from the suit surroundings, possibly requiring more insulation.

12. Abram Spanel to all ILC employees, July 21, 1969, ILC document.

13. Heather M. David, "AF Orders 'Quick Change' Space Suit," *Missiles and Rockets*, April 24, 1961, 18.

14. "Suiting Up for Space," chapter 3 of Loyd S. Swenson Jr., James M. Grimwood, and Charles C. Alexander, *This New Ocean: A History of Project Mercury*, NASA Special Publication-4201, http://history.nasa.gov/SP-4201/ch8-3.htm.

15. "Space Pressure Suit Design by Latex," *Aviation Week*, April 24, 1961, 33.

16. James M. Garwood, NASA history interview with Matthew I. Radnofsky, NASA Chief, Crew Systems, March 21, 1966.

17. President John F. Kennedy, "Address to Joint Session of Congress," May 25, 1961, https://www.jfklibrary.org/learn/about-jfk/historic-speeches/address-to-joint-session-of-congress-may-25-1961.

Chapter 3. The Turbulent Years, 1962–1965

1. James V. Correale and Walter W. Guy, NASA Fact Sheet #116, December 1962, 4. Correale headed NASA's Crew Equipment Branch and Walter W. Guy headed the Systems Analysis Section.

2. Len Shepard, "Proposal in Response to Request for Proposal No. MSC-62-26P," March 28, 1962, ILC document.

3. [John Allspaw], "Systems Engineering: A Great Definition," Kitchen Soap, July 18, [2010?], accessed January 15, 2020, www.kitchensoap.com/2011/07/18/systems-engineering-great-definition/.

4. "NASA 50th Anniversary Moment—Joe Kosmo," video, December 8, 2008. accessed January 15, 2020, https://www.nasa.gov/multimedia/podcasting/nasamoments_kosmo.html.

5. International Latex Corporation, "Project Apollo Space Suit Assemblies and Associated Life Support System," ILC document number IPD-17-62, March 28, 1962, 4-2.

6. Ivan D. Ertel, NASA interviews with Mel Case and Leonard Shepard, ILC Industries, April 4, 1972, 29, NASA History Office, Washington, DC.

7. Industrial Communication Branch to Public Affairs Office, October 12, 1962, NASA memorandum.

8. Human factors is the study and documentation of how the users (astronauts) interface with the suit in the environment(s) the suit is designed to perform in.

9. Donald E. Fink, "Hamilton Standard Starting Manned Tests of Advanced Apollo Suit Design Prototypes," *Aviation Week & Space Technology*, October 28, 1963, 49.

10. Interviews with Mel Case and Leonard Shepard, April 4, 1972.

11. M. I. Radnofsky, assistant chief, Apollo Support Office, to Record, n.d., "Apollo EMU," NASA memorandum.

12. W. O. Heinze, president, International Latex Corp., to W. E. Diefenderfer, president, Hamilton Standard, January 14, 1964.

13. R. D. Weatherbee to E. V. Marshall, "Summary of Recent Events Pertinent to Suit Procurement," February 7, 1964, 1, Hamilton Standard internal memorandum.

14. R. D. Weatherbee to E. V. Marshall, "Review of Space Suit Helmet Problems," February 7, 1964, 3, Hamilton Standard internal memorandum.

15. Ibid.

16. Richard S. Johnson, January 29, 1964, NASA memorandum.

17. Documentation of these space suits in the collection of the Smithsonian National Air and Space Museum also list the A-2L suits as the model SPD-766 on their labels, as per ILC's Specialty Products Division procedures. While suit number 002 still appears to be in the museum's collection, serial numbers 001 and 003 appear to have been in the collection at one time but were disposed of at some point for reasons unknown.

18. E. V. Marshall to R. E. Breeding et al., "Apollo Space Suit Task Force," March 31, 1964, 1, Hamilton Standard internal memo.

19. Grumman, located in Bethpage, New York, was developing the lunar module and the suit played a critical role as part of the systems needed to carry the moon landing out. The Gemini suit was developed by the David Clark Company.

20. M. I. Radnofsky, Assistant Chief, Apollo Support Office, to Record, December 21, 1964, 1, NASA memorandum.

21. Richard S. Johnson, Chief, Crew Systems Division, to Dr. Joseph F. Shea, "Block II Apollo Suit Program," January 25, 1965, 1, NASA memorandum.

22. Mathew Radnofsky, NASA Chief, Apollo Support Office "Apollo EMU," December 21, 1964, NASA memorandum to Record.

23. Author's interview with Homer Reihm, Apollo space suit engineering manager, ILC Industries, Dover, Delaware, July 23, 2018.

24. VIP stood for Vertical Internal Pivot, which was designed specifically for use in G-loading conditions such as in a spacecraft. It increased visual range and offered impact resistance.

25. Nisson A. Finkelstein, vice-president and general manager, ILC Industries, to Richard Johnson, Crew Systems Division, NASA, November 30, 1964.

26. Interview with Homer Reihm, July 23, 2018.

27. Heather M. David, "Apollo Suit Substantially Redesigned," *Missiles and Rockets*, April 26, 1965, 26–31.

28. J. C. Beggs to Distribution, "Minutes of Technical Review Meeting of the EMU Held January 25–26, 1965," January 29, 1965, 2–4, Hamilton Standard Memo.

29. D. Slayton, NASA, to Chief, Crew Systems Division, "Apollo Block II Training Suits," March 30, 1965, 2, NASA memorandum.

30. Matthew I. Radnofsky, untitled note, August 23, 1965, in author's possession.

31. M. I. Radnofsky, Assistant Chief, Apollo Support Office, NASA, to Record, "Apollo EMU," December 21, 1964, 1, NASA memorandum.

32. Dr. George E. Mueller, NASA, to Robert R. Gilruth, [Manned Space Center] director, "Procurement Plan for the Apollo Extravehicular Mobility Unit and EMU Ground Support Equipment Development and Fabrication," memorandum from September 17, 1965, 2.

33. My italics.

34. Robert Gilruth, Director, Manned Space Center, to Dr. George E. Mueller, NASA Headquarters, September 17, 1965, NASA memorandum EC911BG211.

Chapter 4. Second Chances: The Model AX-5L and A-5L Suits

1. Heather M. David, "Apollo Suit Substantially Redesigned," *Missiles and Rockets*, April 26, 1965, 26–31.

2. Ivan D. Ertel, NASA interviews with Mel Case and Leonard Shepard, ILC Industries, April 4, 1972, 29, NASA History Office, Washington, DC.

3. Ibid., 24.

4. Ibid., 29.

5. Author's interview with Homer Reihm, Apollo space suit engineering manager, ILC Industries, Dover, Delaware, July 23, 2018

6. Robert E. Smylie, Chief, Apollo Support Office, to Dr. R. Jones, et al., "Apollo Space Suit Evaluation Repair Policies," June 16, 1965, NASA memorandum.

7. Robert Jones, Crew Performance Section, to Chief, Apollo Support Office, "Response to PB-2-108," June 30, 1965, NASA memorandum.

8. Robert Gilruth, Director, Manned Space Center, to Chief, Procurement and Contracts Division, "Justification for Noncompetitive Procurement," September 20, 1965, 1, NASA memorandum.

9. Lillian Kozloski, *U.S. Space Gear: Outfitting the American Astronaut* (Washington, D.C. Smithsonian Institution, 1994), 81.

10. Robert Gilruth, Director, Manned Space Center, to Dr. George E. Mueller, NASA, September 17, 1965, NASA memorandum EC911BG211.

11. Richard Johnson to Chief, Advanced Technology Procurement, "Justification for Noncompetitive Procurement," July 30, 1965, NASA memorandum.

12. Ertel, interviews with Mel Case and Leonard Shepard, 29.

13. Gilruth, "Justification for Noncompetitive Procurement."

14. Kevin M. Rusnak, interview with James McBarron II, April 10, 2000, 9, Johnson Space Center Oral History Project.

15. Michael Collins, *Carrying the Fire* (New York :Ballantine Books, 1974), 134.

16. *Aviation Week & Space Technology*, June 3, 1968, 108.

17. Phone conversation with Dixie Rinehart, August 28, 2015.

18. Author's interview with Homer Reihm.

Chapter 5. The Model A-5L Space Suit Contract

1. Richard Johnson, NASA, to Mr. L. F. Shepard, ILC Industries. May 13, 1966.

2. Richard S. Johnson to Chief, Advanced Technology Procurement, "Justification for Noncompetitive Procurement," July 30, 1965, NASA memorandum.

3. L. F. Shepard. ILC AX-5L Purchase Description with cover letter dated August 24, 1965, ILC document.

4. The David Clark Company was contracted to provide suits up to the Apollo 10 mission.

5. Richard S. Johnson, "Close-Out and Final Payment under Contract NAS9-5332 with International Latex," September 26, 1967, 23, NASA memorandum.

6. Charles C. Lutz, Harley L. Stutesman, Maurice A. Carson, and James McBarron II, "Apollo Experience Report" February 1974, NASA document.

7. No author, "Problems: Crew Systems Division," January 6 to 13, 1966, NASA document.

8. No author, "Accomplishments: Crew Systems Division," May 5 to 12, 1966, NASA document.

9. Michael Collins, *Carrying the Fire* (New York: Ballantine Books, 1974), 113.

10. Chief, Crew Systems Division, to Apollo Support Office, NASA Trip Report, August 24, 1965, NASA document.

11. C. C. Lutz, Chief, Apollo Support Office, NASA, to Tony Riggan, General Research Procurement Branch, NASA, "DCASR [Defense Contract Administration Services Region] Philadelphia Price Analysis Report, case No. P-1576," May 19, 1966.

12. Minutes from an October 6, 1965, meeting, addressed to G. Burnett, NASA Representative at ILC, NASA memorandum.

13. C. C. Lutz, Chief, Apollo Support Office, NASA, to Tony Riggan, DCASR, [Defense Contract Administration Services Region] Philadelphia, "ILC Industries Contract Support," May 19, 1966, NASA memorandum.

14. No author, "Accomplishments: Crew Systems Division," February 24 to March 3, 1966, NASA document.

15. The cover layer of the TMG for the "Sylvester" A7L suit is now located at the Kansas Cosmosphere Museum in Hutchinson, Kansas. It is unknown where the pressure garment assembly is located. I recall seeing the complete suit on display at the Smithsonian National Air and Space Museum in Washington, DC, sometime around 1980.

Chapter 6. The Model A-6L Space Suit: Unveiling the First Moon Suit

1. Charles C. Lutz, Harley L. Stutesman, Maurice A. Carson, and James McBarron II, "Apollo Experience Report," February 1974, 24, NASA document.

2. Richard S. Johnson, Chief, Crew Systems Division, to L. F. Shepard, International Latex Corporation, "Inconsistency of ILC Contract and LTV Subcontract Delivery Requirements for EVVA," August 25, 1966, NASA memorandum.

3. N.a., "Feasibility Study of Utilizing Separate Intravehicular and Extravehicular Pressure Garment Assemblies," n.d., 7 pages, NASA report.

4. "Omega Suit Presentation," ILC technical outline, September 12, 1968, 4.

5. George Durney, Spacesuit Thigh Restraint Assembly, patent number 3,699,589, filed August 5, 1968.

6. C. C. Lutz to B. Cour-Palais, "Proposed Changes to the Apollo Pressure Garment Cross Section," January 17, 1967.

Chapter 7. The Model A-7L Space Suit, 1967–1971

1. John E. Flagg, president, David Clark Company, to H. T. Christman, NASA Small Business Specialist, March 11, 1966; C. C. Lutz, NASA, to H. T. Christman, June 9, 1966, NASA memorandum.

2. Jack Naimer, "Apollo Applications of Beta Fiber Glass," in "Proceedings of the NASA Conference on Materials for Improved Fire Safety," NASA document NASA-SP-5096, 161.

3. Ibid., 142.

4. Mel Case to Distribution, "Minutes of Meeting on Training Suit Coverall and Beta Cloth," March 6, 1967, ILC memorandum.

5. Training document written by Bob Wise, ILC Dover, ca. 1986, 2–4.

6. Ibid.

7. M. Kaplin to R. Tenaro, "Letter of Congratulations," January 11, 1968, ILC memorandum.

8. The TMG consisted of the torso, boots, gloves, and the primary life-support system cover layers.

9. George Durney, John Scheible, and Bob Wise, ILC Dover Training Manual, April 1993, page 4 of thermal micrometeoroid garment section.

10. J. Bateman and Al Gross to L. Adams, "Status of Suit Problems Noted during Lunar ITMG Testing at MSC," November 12, 1968, 4, ILC internal memo.

11. Air flowed into the suit through a vent located in the helmet just above and behind the top of the head. Air flowed over the face and into the torso area of the suit. The only way for air to exit the suit was through the vent system, which terminated at areas that ensured that air would reach the remote parts of the body such as the hands and feet.

12. Only the different suit models were certified, not individual suits. If design changes were made for a model that had previously been certified, testing would be performed around that change to verify that it would not fail on the mission.

13. Al Gross, ILC systems engineer, "Design Evaluation Test of a Modified ITMG," September 1968, ILC test report.

14. Author's interview with John McMullen, Apollo lead systems engineer, ILC Industries, July 18, 2017.

15. L. F. Shepard, "PGA Thread Verification," Project Directive: 1133, May 6, 1969, ILC memorandum.

16. Author's interview with Thomas Pribanic, ILC Industries configuration manager during Apollo, June 11, 2017.

17. H. D. Reihm to J. McBarron, April 4, 1972, ILC engineering memorandum.

18. One lot or batch of rubber would be made and used to produce many individual convolutes or other rubber parts located throughout the suit. Each part made would have The lot number of the batch each part was made of was recorded on the associated paperwork so it could be tracked if failure occurred. If that happened, all other items made from that batch would be located and held in quarantine, including entire suits if they had parts made from the defective batch.

19. Holly Hodges, NASA contracting officer, to ILC Industries, April 14, 1971.

20. Ray Winward, unpublished memoir, 2016.

21. Robert E. Smylie, Chief, Crew Systems Division, NASA, to L. F. Shepard, ILC Industries, April 21, 1970.

22. Author's interview with Ron Woods, July 9, 2018, Cocoa Beach, Florida.

23. Management Systems Study of the Apollo Suit Program, April 30, 1974, ILC Industries Dover, 50.

24. The acceptance data pack, or ADP, never left the suit it was associated with. It tracked every item that made up the suit by name, part number, and lot or serial number. It was continuously updated to note changes to the suit. The purpose was to show the current configuration of the suit and provide evidence of changes that had been made or of changes that needed to be made to bring the suit to the proper configuration. Unfortunately, these data packs were lost after they were delivered to the Smithsonian National Air and Space Museum.

25. No author, "Apollo/Skylab Suit Program Management System Study" (for NASA), April 30, 1974, 53, ILC Industries document.

26. Richard S. Johnson to Len Shepard, "Suit Fit Problems, PGA A6L S/N 012 and 013," February 29, 1968, NASA memorandum.

27. William Kincade, design certification review report, document CSD-A-643, May 7, 1968, 27.

28. T. K. Mattingly, "EMU Lunar Surface Qualification," March 12, 1969, page 1 of enclosure 2, NASA memorandum.

29. Associate Administrator for Manned Space Flight to George Mueller, "Lunar Exploration Program—Suit Procurement," October 16, 1969, NASA memorandum.

30. Contracting officer Charley D. Stamps, "Procurement Plan for the Development and Fabrication of Extravehicular Constant Volume Pressure Suit Assemblies," attachment to NASA contract approval form (MSC Form 352), April 18, 1969, 2–4.

31. L. F. Shepard to J. McBarron. "Structural Design Verification Testing (DVT) of the Arm Bearing/Upper Arm Cone Interface Configuration," July 14, 1969, ILC Industries memorandum.

32. Author unknown, "Design Description," ca. March 1969, 1. Document included in ILC file "Low Torque Arm Assembly Test Summary."

33. Data gathered from the table of operations used to record the work performed on the serial number 056 suit.

34. Rich Spolls and Bob Beauregard, spreadsheet listing differences between current configuration, configuration of CM 104 and CM 106, and quality configuration, April 4, 1969.

35. W. Draper, "CCBD 9E075" (details of costs and schedule), part of "Arm Bearing Retrofit Schedule," March 20, 1969, NASA document.

36. Normand J. Beauregard, NASA, to R. W. Currie, ILC Industries, "Contract Change Authorization No. 645, Revision A," April 28, 1969.

37. Robert W. Currie, ILC Industries, to N. J. Beauregard, NASA, "Contract NAS 9-6100, Schedule I, RE: CCA's 645 and 646," n.d.

38. Typed notes of Homer Reihm, ILC chief engineer on Apollo suit program.

39. L. F. Shepard to J. McBarron and T. C. Riggan, "Integrated Lunar Boots," April 29, 1970, ILC memorandum.

40. Author's interview with Richard Martin, March 20, 2019, Lewes, Delaware.

41. H. Reihm to M. Kaplan and R. Sartoris, "A-7L EV Glove Cross-Section," June 14, 1967, ILC internal memorandum.

42. T. K. Mattingly to All Astronauts, "Pressure Suit Gloves," July 22, 1968, 2, NASA memorandum.

43. Joseph P. Kerwin to Chief, Apollo Support Office, "Report on EMU Manned Qualification Test #4," 2, January 12, 1967, NASA memorandum.

44. D. E. Supkis, "Nonflammable Fluorel Compounds," Program Apollo Working Paper No. 1337, August 22, 1968, 1, NASA document.

45. L. F. Shepard to J. McBarron, "Engineering Evaluation and Resolution of IV Glove Fit Problem for Astronaut Scott," May 12, 1969, ILC Industries engineering memo.

46. C. K. Shepard and Cheryl Lednicky, "EVA Gloves: History, Status and Recommendations for Future NASA Research," NASA document JSC-23733, Final Report, April 1990, 20–21.

47. [First name unknown] Grimwood, NASA oral history interview with Mr. Forrest Poole, David Clark Company, location unknown. May 1, 1968, 2.

48. Harold R. Williams, "Water Pipes in Long Underwear May Keep Future Astronauts Cool," Associated Press news release, June 4, 1964.

49. "John Billingham," https://en.wikipedia.org/wiki/John_Billingham, accessed July 22, 2017.

50. A British thermal unit is the amount of heat required to raise the temperature of one pound of water by one degree Fahrenheit.

51. Eugene Cernan and Dan Davis, The Last Man on the Moon (New York: St. Martin's Press, 2007), 134.

52. Midwest Research Institute, "Liquid Cooled Garment," NASA Report No. CR-2509, January 1975, 8.

53. Ibid., 13.

54. No author, "Apollo EMU Garment Status Meeting," March 6, 1967, section II, page 1, NASA document.

55. Conversation with Ron Bessette, September 11, 2018.

56. Robert E. Smylie, Chief, Crew Systems Division, NASA, to Record, "Apollo Suit Costs," May 10, 1972, NASA memorandum.

57. Normand Beauregard to R. W. Currie, ILC Industries, NASA Contract Change Authorization No. 657, April 11, 1969.

58. The Apollo 7 through 10 suits used the link-net restraint on the arms of the EVA suits and had no bearings.

59. C. Lutz, H. Stutesman, M. Carson, and J. McBarron, "Apollo Experience Report," NASA Document NASA TN-D-8093, November 1975, 25.

60. Dwayne A. Day, "Spooky Apollo: Apollo 8 and the CIA," The Space Review, December 3, 2018, https://www.thespacereview.com/article/3617/1.

61. Lutz et al., "Apollo Experience Report," November 1975, 26.

62. C. Lutz, H. Stutesman, M. Carson, and J. McBarron, "Apollo Experience Report," NASA Document JSC-08597, February 1974, 4.

63. "Failure Summary for EMU Garments and Associated Equipment," MSC Crew Systems Mission D report, January 10, 1969.

64. Chief, Test Division, to Chief, Crew Systems Division, "Apollo 9 Mission Debriefing—Status of," April 1969, 3, NASA memorandum.

65. R. H. Steel and James McBarron, "Manned Spacecraft Center Certification Test Review Sheet," February 17, 1969, NASA document.

66. C. Lutz, H. Stutesman, M. Carson, J. McBarron, "Apollo Experience Report: Development of the Extravehicular Mobility Unit," February 1974, 4, Johnson Space Center document.

67. Mission Manager to Chief, Apollo Support Branch, "Apollo 10 EMU Post Flight Report," July 22, 1969, NASA memorandum.

68. Data obtained from copies of the various fit-check records.

69. "Qualified" means that the suit model had to be fully tested in the laboratory with subjects pressurized inside and performing rigorous tasks in order to prove the suit was capable of carrying out the missions without failure.

70. Robert E. "Ed" Smylie, NASA Chief, Crew Systems Division, interviewed by Carol Butler, Johnson Space Center Oral History Project, Bethesda, MD, April 17, 1999.

71. Kennedy Space Center Report following the Apollo 11 fit-crew function trials held February 26 to March 1, 1969; data provided by Dan Schaiewitz in correspondence with the author, October 23, 2015.

72. Ken Mattingly to Deke Slayton, "Connector Cover Problems," March 21, 1968, NASA internal correspondence.

73. Information provided by Dan Schaiewitz in correspondence with the author, October 23, 2015.

74. ILC Industries, "Proposal for Design and Development of Intravehicular Space Suit Assembly," May 26, 1970, A-3.

75. "Jim's Involvement in the Space Program," 1–2, personal notes provided by Jim Rutherford, QA supervisor, ILC Industries.

76. "Open Problem Pressure Garment Assembly Pressure Drop," January 3, 1969, NASA document. This meant that during training for the Apollo 9 mission, the suits failed the pressure drop when the flow of air through the suit was set at the nominal 6 cubic feet per minute. The problem occurred when the air flowing into and out of the suit met too much resistance because materials interfered with the flow stream or the orifices were too small for all of the air to pass through.

77. NASA Crew Systems Document number MSC-13-10A, January 3, 1969.

78. NASA Apollo 11 crew debrief conducted on July 31, 1969.

79. Preliminary NASA report on the Apollo 11 mission, n.d., 58.

80. Ibid., 52.

81. Ibid., 61.

82. Text accompanying Alan Bean's artwork at *Our Own Personal Spaceships*, The Greenwich Workshop: Fine Art Editions Made in America, accessed January 3, 2020, http://www.greenwichworkshop.com/details/default.asp?p=4539&a=5&t=4&detailtype= artist.

83. No author, "Apollo 12 EVA Technical Debrief," Section 10-47, "LM Ingress—2nd EVA," 10–56, n.d., NASA Document.

84. Mission Operations Branch, Flight Crew Support Division, "Apollo 12 Technical Crew Debriefing," December 1, 1969, 10-54–10-59.

85. Robert E. Smylie, Chief, Crew Systems Division, NASA, to L. F. Shepard, April 14, 1970.

86. NASA Test Preparation Sheet No. 11020403, January 27, 1970.

87. Eric Jones, "Commander's Stripes," *Apollo Lunar Surface Journal*, 2005, https://www.hq.nasa.gov/alsj/alsj-CDRStripes.html.

88. D. Rinehart, "PGA Glove Improvements Pressure Differential Comparison," ILC Test Report (Test No. 410-1), September 29, 1969.

89. The water connector from the primary life-support system consisted of only one connector that attached to the suit. It was referred to as the "multiple water connector" because the one metal fitting (made by Air-Lock) allowed water to flow into and out of the suit and the liquid cooling garment, thus eliminating two separate connectors. On both the suit side and the connector side, the water connector had a spring-loaded shutoff that sealed it closed, preventing water leakage upon disconnection.

Chapter 8. The Model A-7LB Space Suit: The Next Generation, 1971–1975

1. C. C. Lutz, Chief, Apollo Support Branch, to Branch Chiefs et al., "ILCI Presentation on Omega Suit," September 10, 1968, NASA memorandum.

2. Author's interview with Homer Reihm, September 7, 2017, ILC Dover.

3. C. C. Lutz, Chief, Apollo Support Branch, to Branch Chiefs, R. Johnson, et al., "ILCI Presentation on Omega Suit," September 10, 1968, NASA memorandum.

4. Author's interview with Homer Reihm, July 23, 2018, ILC Dover.

5. Obtained from the Omega III Operational Logbook provided by Richard Martin to author.

6. From Attachment I of "Engineering Change Proposal for a New EMU Qualification Unit," Engineering Change Proposal number 397-33, December 11, 1968, 10, ILC document.

7. Robert W. Currie, ILC Apollo Contracts Manager, to N. J. Beauregard, NASA Contracting Office, "Contract NAS 9-6100, Re: Engineering change proposal submission," ILC memo. December 11, 1968.

8. "Contract NAS 9-6100 Schedule I, Supplemental Agreement No. 201S, Exhibit A, Statement of Work for Design, Development and Qualification of an Apollo A9L Pressure Garment Assembly," February 14, 1969, NASA document.

9. Phone conversation with Jim McBarron, July 17, 2019.

10. George E. Mueller, Associate Administrator for Manned Space Flight, to "A/Administrator," "Lunar Exploration Program—Suit Procurement," October 16, 1969, NASA memorandum.

11. William E. Stoney, NASA Headquarters, Washington, DC, telex to Sam C. Phillips, Lt. General, USAF, Houston Manned Spacecraft Center, Attn: Dr. Robert Gilruth, May 1969.

12. Charley D. Stamps, "Procurement Plan for the Development and Fabrication of Extravehicular Constant Volume Pressure Suit Assemblies," April 18, 1969, 2–4, NASA document.

13. Robert R. Gilruth to NASA Headquarters, Attn: Associate Administrator for Manned Space Flight, "Procurement of Pressure Garment Assemblies (Space Suits)," June 11, 1970, 3, NASA memorandum.

14. R. H. Wood to R. H. Reihm, "A7LB CMP Separable Components Serialization," September 3, 1970, ILC Memorandum.

15. Robert R. Gilruth to NASA Headquarters, "Procurement of Pressure Garment Assemblies (Space Suits)," June 11, 1970, 3.

16. T. K. Mattingly, "EMU Lunar Surface Qualification," Enclosure 2, section II-C, March 12, 1969, NASA memorandum. In August 1971, President Richard Nixon proposed cancelling the remaining lunar missions beyond Apollo 15. Nixon's Office of Management and Budget Director, Casper Weinberger, persuaded him to continue with the two remaining missions.

17. Author's interview with Homer Reihm, July 23, 2018.

18. Robert Smylie, NASA Chief, Crew Systems Division, to Len Shepard, ILC, April 24, 1970, NASA telegram.

19. "Apollo Operations Handbook—Extravehicular Mobility Unit," March 1971, Revision V, 2–19, Section 2.3.1.1.1, NASA document MSC-01372-1.

20. John Young to Apollo 15 and Follow-On Persons, "The A7LB Pressure Suit Neck Convolutes," July 29, 1970, NASA internal memorandum.

21. L. F. Shepard, Apollo Program Manager, to J. McBarron, EC-9, NASA, "Engineering Evaluation and Resolution of IV Glove Fit Problem for Astronaut Scott," May 12, 1969, ILC engineering memorandum.

22. H. D. Reihm, Apollo Program Manager, to J. W. McBarron, EC-9, NASA, "Configuration of Apollo 15 CMP ITLSA's," January 22, 1971, 1–2, ILC engineering memorandum.

23. J. O'Kane, "Wrist Disconnect Dust Covers," August 8, 1972, ILC modified work request form.

24. Robert M. Rasmussan Jr., "Evaluation of Tape as an Effective Substitute for the Wrist Dust Covers," ILC-M-ER-00-345, November 7, 1972, ILC Industries test report.

Chapter 9. Skylab, the Apollo-Soyuz Test Program, and Other Development Suits

1. Matt Williams, "America's First Space Station: The NASA Skylab," Universe Today, May 29, 2015, accessed January 13, 2020, https://www.universetoday.com/57067/skylab/.

2. David S. F. Portree, "Assuming Everything Goes Perfectly Well: NASA's 26 January 1967 AAP Press Conference," Wired, July 19, 2014, https://www.wired.com/2014/07/assuming-everything-goes-perfectly-well-nasas-26-january-1967-apollo-applications-program-press-conference/.

3. Donald C. Elder, "The Human Touch: The History of the Skylab Program," accessed January 13, 2020, https://history.nasa.gov/SP-4219/Chapter9.html

4. Author's interview with Homer Reihm, July 23, 2018, ILC Dover.

5. Author's interview with Mr. John McMullen, former systems engineer, ILC Industries, February 7, 2017, ILC Dover.

6. Dale D. Myers, Associate Administrator for Manned Space Flight, NASA, to Dr. Robert R. Gilruth, director, Manned Spacecraft Center, August 31, 1970, page 3 of attachment.

7. B. A. Davis, Controller, to J. P. McDonald, ILC Industries, "Cost Proposal for Skylab Extravehicular Visor Assemblies," January 29, 1971, 1, Vought Missiles and Space Company document.

8. Asif A. Siddiqi, The Soviet Space Race with Apollo (Gainesville: University Press of Florida, 2000), 794.

9. Robert E. Smylie, Chief, Program Procurement Division, to Chief, Crew Systems Division, September 28, 1972, 2, NASA memorandum.

Chapter 10. End of a Historic Era

1. Video clip, Walter Cronkite, reporter, "USSR Intends to Buy United States Space Suits," CBS Evening News, March 15, 1973, Vanderbilt News Archive, accessed January 15, 2020, https://tvnews.vanderbilt.edu/broadcasts/227651.

2. Craig Covault, "Soviets Plan to Buy Apollo Suits for Comparative Study Purposes," Aviation Week & Space Technology, March 26, 1973, 17.

3. Michael J. Himowitz, "A Battered Space Suit for $100; Buyers in Orbit at Flea Market," Baltimore Evening Sun, March 20, 1975, C1.

4. Author's interview with Mr. John McMullen, former systems engineer, ILC Industries, February 7, 2017, ILC Dover.

5. Ibid.

6. An aerostat is a large, unmanned helium-filled balloon that is tethered to ground equipment and is used for surveillance or for monitoring electronics.

Conclusion: Preserving Our Treasures; The Smithsonian National Air and Space Museum

1. Only two pairs of lunar boots used by the Apollo 17 crew were brought back to Earth. I once asked Gene Cernan if it was in the plan to bring them back and he told me, "No, I made the decision to bring them back." Ten other pairs of lunar boots are scattered on the moon at each of the other five landing sites, including Armstrong's.

2. Media fact sheet: "National Air and Space Museum, a History of," Smithsonian, June 28, 2016, accessed January 13, 2020, https://newsdesk.si.edu/factsheets/history-national-air-and-space-museum.

3. Newsroom of the Smithsonian, "Visitor Stats," accessed January 13, 2020, https://newsdesk.si.edu/about/stats.

4. Lisa A. Young and Amanda J. Young, *The Preservation, Storage and Display of Spacesuits*, Collections Care Report no. 5 (Washington, DC: Smithsonian National Air and Space Museum, 2001), 3.

5. Ibid., 2–3.

6. Ibid., 4.

7. Douglas N. Lantry, "Dress for Egress: The Smithsonian National Air and Space Museum's Apollo Spacesuit Collection," *Journal of Design History* 14, no. 4 (2001): 343–359.

8. Young and Young, Collections Care Report, 21.

9. Amelia Brakeman Kyle, "That's One Small Step . . . ," Smithsonian National Air and Space Museum, December 12, 2001, accessed January 13, 2020, https://airandspace.si.edu/stories/editorial/%E2%80%99s-one-small-step.

10. Young and Young, Collections Care Report, 13.

11. Lillian D. Kozloski, *U.S. Space Gear: Outfitting the Astronaut* (Washington, DC: Smithsonian Institution Press, 1994), 172–173.

12. "Reboot the Suit: Neil Armstrong's Spacesuit and Kickstarter," Smithsonian National Air and Space Museum, July 20, 2015, accessed January 3, 2020, https://airandspace.si.edu/stories/editorial/armstrong-spacesuit-and-kickstarter.

Appendix A: Technical Details of the Apollo Space Suits

1. Edwin E. Aldrin with Wayne Warga, *Return to Earth* (New York: Random House, 1973).

2. ILC Test Plan for Test Number 186, January 11, 1966. No author listed

3. Charles C. Lutz, Harley L. Stutesman, Maurice A. Carson, and James McBarron II, "Apollo Experience Report," February 1974, NASA document.

4. Joseph P. Kerwin to Chief, Apollo Support Office, January 12, 1967, NASA memorandum.

5. Copy of test data labeled "Swage Fittings." No author or date other than a remark on data sheet that says "knee convolute A7LB New—Jan 7, 1972."

6. The design verification test procedure indicated that the maximum allowable leakage of the torso-limb suit assembly was 300 cubic centimeters per minute with the new pressure sealing zipper. This was while the suit was pressurized to 3.75 pounds per square inch.

7. R. H. Wood to Distribution, "Certification Classes for Actuating the A7LB OEB Zipper," August 24, 1971, ILC Industries memorandum.

8. P. Schneider, "Test Report on the Design Verification Testing of the B. F. Goodrich Pressure Sealing/Restraint Zipper," July 14, 1972, 10, ILC Industries report.

9. Chief, Crew Systems Division, to C. C. Lutz, "Talon (OEB) Zippers," May 19, 1971, NASA memorandum.

10. H. D. Reihm, Apollo Program Manager, to J. McBarron, NASA Technical Monitor, "A7LB Restraint Zipper Evaluation WRF DT-52A," June 15, 1972, 3–5, ILC Industries memorandum and test report.

11. Robert Smylie, Systems Engineering Division, to Chief, Apollo Support Office, February 13, 1965, NASA memorandum.

12. P. H. McKenney to W. W. Guy, "Feasibility of Reducing EMU Weight by Altering the Number of Layers of Insulation," September 17, 1969, 1–2, General Electric technical information release.

13. Edward S. Cobb Jr., "Preliminary Instructions for Sewing Beta Fiber Fabrics," n.d., Owens Corning document.

14. EC7 Chief to EC78 Head, Materials Development Section, "Composition of Finish on Glass Thread," August 11, 1967, NASA memorandum.

15. M. Kaplan to R. Tenaro, "Letter of Commendation," January 11, 1968, ILC memorandum.

16. Richard S. Johnson, Chief, Crew Systems Division, to Manager, Command and Service Modules Apollo Spacecraft Program, "Space Suit Requirements for Command Module Interface Verification," August 9, 1967, NASA memorandum.

17. "Spandex," What's That Stuff? 77, no. 7 (February 15, 1999), accessed January 13, 2020, https://pubs.acs.org/cen/whatstuff/stuff/7707scitek4.html.

18. F. J. Turnbo, "Measurement Report: Thermal Property Measurements of Manned Spacecraft Center Spacesuit Materials," NASA document 68-3346.11ja-31, April 1968.

19. J. D. Hollan to W. E. Ellis, "PGA Pressure Relief Valve Requirements," June 9, 1971, NASA memorandum.

20. Michael Collins, Carrying the Fire (New York: Ballantine Books, 1974), 200.

21. Shane M. McFarland and Aaron S. Weaver, "Breaking the Pressure Barrier: A History of the Spacesuit Injection Patch," paper presented at the International Conference on Environmental Systems, July 14–18, 2013, 6.

22. C. Lutz, H. Stutesman, M. Carson, and J. McBarron II, "Apollo Experience Report: Development of the Extravehicular Mobility Unit," NASA document TN D-8093, November 1975, 17.

23. M. A. Robinson, "Vomit Removal Apparatus for an EVA Pressure Suit: A Design Concept Case 710," July 29, 1969, Bellcomm, Inc. document.

Appendix B: Apollo Part Numbers and Serial Numbers

1. R. H. Wood to R. H. Reihm, "A7LB CMP Separable Components Serialization," September 3, 1970, ILC memorandum.

2. Robert E. Smylie, Chief, Crew Systems Division, "Apollo Space Suit Costs," May 10, 1972, NASA memorandum for Record.

3. Data obtained from the build paperwork for pressure-garment assembly S/N 056, manufactured for Neil Armstrong.

4. Owen E. Maynard to EC/Chief, Crew Systems Division, "Space Suit Requirements for the Command Module Interface Verification," August 18, 1967, NASA memorandum.

5. D. M. Robinson to J. Rayfield/R. H. Hester, "Burst Pressure Test of PGA A6L-001," Technical Information Release No. 556-D1.3.3-7347, July 24, 1967, General Electric document.

6. James McBarron, "Apollo Block II EMU Pressure Garment Assembly, S/N's A6L-005 and A6L-014, Test Readiness Review for Altitude Verification Test, for GAEC Internal Environmental Simulation Test and LTA-8," NASA document CSD-A-528, November 20, 1967.

7. Richard S. Johnson, Chief, Crew Systems Division to Manager, Command and Service Modules Apollo Spacecraft Program, "Space Suit Requirements for Command Module Interface Verification," August 9, 1967, NASA memorandum.

8. T. A. Vogel, ILC Industries Malfunction Report, ILC document number ILC-03689, December 4, 1968.

9. Suit S/N 039 was used for lunar-surface cycle testing on June 20, 1969. The objective was "to impose upon the S/N 039 PGA used during Mission G Design Limit Cycling Qualification tests, projected lunar surface lower torso leg & foot motions." "Quick Look Test Report," June 20, 1969, NASA document.

10. L. F. Shepard to Jim McBarron, EC-9, "Tests Conducted on New Shoulder Bearing Attachment Method," June 13, 1969, 2, ILC memorandum.

11. Phillip Schneider, "Test Report on Destructive Testing of A7LB EV TSLA's S/N 302 & 306," Report No. WRF DT-68A, July 13, 1972, 1, ILC document.

Appendix C: Apollo Contract Details

1. John Lannan, "Apollo Look Makes Debut," *Arizona Daily Star*, August 8, 1968.

2. Robert E. Smylie, Chief, Crew Systems Division, "Apollo Suit Costs," May 10, 1972, NASA memorandum for Record.

3. NASA Contract Record, June 6, 1968, page 19 of enclosure 9.

4. "Exhibit VI, Apollo Extravehicular Mobility Unit (EMU) Garment Assembly Program (GAP) Award Fee Evaluation Plan and Procedures," January 1, 1970, 1–13, NASA document.

5. C. Lutz, H. Stutesman, M. Carson, and J. McBarron II, "Apollo Experience Report: Development of the Extravehicular Mobility Unit," NASA technical note, Doc. no. NASA TN D-8093. November 1975, 7.

6. James L. Gibson, "Apollo Operations Handbook Extravehicular Mobility Unit, Volume I: System Description," NASA document no. CSD-A-789-(1), July 1968, 8.

7. "CSD/GFE Crew Equipment at Earth Launch," July 11, 1969, NASA document.

8. C. C. Lutz to M. Faget, A. Bond, et al., "EMU Weight Summary," July 3, 1969, NASA memorandum.

Appendix E: A-7LB Suit Plug Loads and Fabric Stresses

1. L. F. Shepard, H. D. Reihm, and G. P. Durney, "Advanced Suit Program, Project 01-502, Status Report," June 15, 1972, ILC Industries.

Timeline of Significant Events

1. "Qualified" means that the suit model had to be fully tested in the laboratory with subjects pressurized inside and performing rigorous tasks in order to prove the suit was capable of carrying out the missions without failure.

FURTHER READING

Abramov, Issak P., and A. Ingemar Skoog. *Russian Spacesuits.* Chichester, UK: Springer/ Praxis Publishing, 2003.

Ayrey, William. *ILC Space Suits and Related Products.* Technical Report, vol. 0000-712731. Frederica, DE: ILC Dover, 2007.

Cernan, Eugene, and Dan Davis. *The Last Man on the Moon.* New York: St. Martin's Press, 2007.

Christoffersen, Roy, John F. Lindsay, Sarah K. Noble, Mary Ann Meador, Joseph J. Kosmo, J. Anneliese Lawrence, Lynn Brostoff, Amanda Young, and Terry McCue. *Lunar Dust Effects on Spacesuit Systems: Insights from the Apollo Spacesuits.* NASA/TP no. 2009-214786. Houston, TX: National Aeronautics and Space Administration Johnson Space Center, April 2009.

Harris, Gary L. *The Origins and Technology of the Advanced Extravehicular Space Suit.* AAA History Series, vol. 24. San Diego: American Astronautical Society, 2001.

Jenkins, Dennis R. *Dressing for Altitude: U.S. Aviation Pressure Suits—Wiley Post to Space Shuttle.* Washington, DC: National Aeronautics and Space Administration, 2012.

Kozloski, Lillian D. *U.S. Space Gear: Outfitting the American Astronaut.* Washington, DC: Smithsonian Institution Press, 1994.

Lantry, Douglas N. "Dress for Egress: The Smithsonian National Air and Space Museum's Apollo Spacesuit Collection." *Journal of Design History* 14, no. 4 (2001): 343–359.

———. "From the Moon to the Museum: A Material History of Apollo Space Suits." PhD. diss., University of Delaware, 2010.

———. "Man in Machine: Apollo-Era Space Suits as Artifacts of Technology and Culture." *Winterthur Portfolio* 30, no. 4 (1995): 203–230.

Mallan, Lloyd. *Suiting up for Space: The Evolution of the Space Suit.* New York: John Day Company, 1971.

Monchaux, Nicholas de. *Spacesuit: Fashioning Apollo.* Cambridge, MA: MIT Press, 2011.

Thomas, Kenneth S. *The Journey to Moonwalking: The People that Enabled Footprints on the Moon.* N.p.: Curtis Press, 2019.

Thomas, Kenneth S., and Harold J. McMann. *U.S. Spacesuits.* Chichester, UK: Springer/ Praxis Publishing, 2006.

———. *U.S. Spacesuits.* 2nd ed. Chichester, UK: Springer/Praxis Publishing, 2011.

Young, Amanda J., and Mark Avino. *Spacesuits: The Smithsonian National Air and Space Museum Collection.* Brooklyn: Powerhouse Books, 2009.

Young, Lisa A., and Amanda J. Young. *The Preservation, Storage and Display of Spacesuits.* Collections Care Report no. 5. Washington, DC: Smithsonian National Air and Space Museum, 2001.

INDEX

Courtesy ILC-Dover-LP-2020

BILL AYREY was the manager of ILC Dover's Test Laboratory when he retired in 2019 after forty years of service. For much of his time there, he was responsible for testing the space suits the company made for NASA's space shuttles and the International Space Station. In his early years at ILC, he worked closely with veterans of the company who designed the Apollo space suits. He has also collected thousands of original documents related to the development and production of the Apollo suits. Over the past twenty years, he has assisted staff at the Smithsonian National Air and Space Museum in their quest to understand and preserve the Apollo suits in their collection.

Printed in the USA
CPSIA information can be obtained
at www.ICGtesting.com
JSHW021004051023
49385JS00003B/12

9 780813 080437